AROUND THE WORLD

THE ATLAS FOR TODAY

gestalten

AROUND THE WORLD

The world around us is a visual feast and we as humans are visual creatures. Long before we can speak or read, before we understand the concepts of words and numbers, our brains are already processing, storing, and contextualizing incredible visual input from the world around us.

When we were around five months old, we developed a sense of depth perception and a receptiveness to different colors. We learned how to start processing shapes, and to try to understand how they intersected with our needs. We learned the smile of another human being, the skin pattern of our favorite animal, and the color of a toy, a car, or an apple.

The older we became, the more we found ourselves surrounded by information. Today, we swim in floods of data and ideas, numbers and phrases that appear before our eyes and on our devices, on billboards and from the mouths of our friends, telling us what is happening to whom, who did what and where, why we should purchase this and not that, and how this thing changes everything and that thing changes nothing.

This is the age of information, and the avalanche of data shows no signs of slowing.

Fortunately, technology in the form of powerful computer software has also provided storytellers and visual interpreters with tools to help filter and contextualize massive amounts of data. Magazines, newspapers, books, and websites use sophisticated tools to go beyond the barest "this is what happened" and explain, graphically, how something happened and what went on, utilizing colors and sizes, and abstractions and representations to show individual elements in relation to each other and to ourselves. Individually, they help us understand a single issue. Together, they explain a slice of the world.

This book is filled with some of the most remarkable of these graphics, the most amazing and helpful that we've collected from media around the world (and where they weren't originally in English, we've translated the texts). Together, they create a snapshot of who we are right now, what we are doing with our work and leisure time, how we view ourselves and each other, and where things might be heading. The world is an amazing place, filled with complexity and contradiction. We never stop creating and learning.

Like any snapshot, it captures only what is within the frame of the camera. A book can contain worlds within its pages, but there will always be much activity that isn't included. Sometimes, it's because these stories aren't being told using visual graphics of this nature. Currently, the creation of such graphics is primarily taking place in those societies that have widespread access to the latest information technology, and have media that will reproduce and distribute them. These creations can be labor intensive — not all newspapers or magazines can afford to support the staff and time it takes to create them. And infographics — as they are often called today among the media — also tend to be created to explain what is in the news at the moment, aiming for ideas and trends that are circulating, as well as anniversaries and events considered worthy of commemoration.

That said, this book is still a pretty good representation of what the more fortunate sections of the world's population are currently thinking, watching, doing, and creating as we explore and expand our curiosity. The beauty of the tools used to create these graphics is that they are context neutral — they can be employed to describe sports or history, politics or religion — and they go far beyond what a journalist could explain, or a photograph could show.

These are the impossible images that place the world's skyscrapers next to each other, and supply a three-dimensional cutaway of what lies underneath Paris. They are a photo that could never be taken of the solar system and a visualization of the journeys of viruses far too small to be seen as they wreak havoc across the globe. Within each image are words of context and explanation, demonstrating the precision and care of each, created by those who occupy the new role of "visual journalist." A picture might be worth a thousand words, but these are graphics that might also contain them.

As a compilation, this book could be a time capsule for future generations, or the text sent out into space to explain Planet Earth to alien life. This is the world that we live in, beautiful and complicated, drawn out and explained. Take a good look.

CONTENTS

2 PREFACE

THE PLACE WE CALL HOME

8 Big Bang
10 The Universe
12 Planet Earth
14 Climate
16 Seasons
18 Seasons
20 Seasons
22 Seasons
24 Arctic Ice
26 Great Migrations
28 Continents
30 The End

LIVING TOGETHER

34 Greetings
36 Wedding Traditions
38 Happiness
40 Global Population
42 Migration
44 Megacities
46 Skyscrapers
48 Underworld
50 Metro Systems

52 Empires
54 The American Way
56 The White House
58 The Kanzleramt
60 The Left Wing
62 Nuclear Weapons
64 Casualties
66 Religion
68 Religion
70 Religion
72 Jerusalem
74 Picking the Pope
76 The Hajj
78 Holidays
80 Family Planning
82 It's a Woman's World

THE DAYS THE EARTH STOOD STILL

86 Titanic
88 Titanic
90 Pearl Harbor
92 D-Day
94 Berlin Airlift
96 Berlin Airlift
98 Mount Everest
100 Cuban Missile Crisis
102 JFK
104 JFK
106 Space Exploration
108 Space Flight
110 9/11
112 9/11
114 9/11
116 9/11
118 Fukushima
120 Killing Osama bin Laden

THE GOOD LIFE

124 Family Time
126 Family Spendings
128 Breakfast
130 Food
132 Tuna
134 Fruit
136 Organic Farming
138 Chocolate
140 Beer
142 Wine
144 Liquids
146 Travel
148 Flight Paths
150 Highways
152 Christmas
154 Fashion
156 Pets
158 Music
160 TV Shows
162 Twentieth-Century Painters
164 Twenty-first-Century Art Sales
166 Olympia
168 Football

FEAR AND LOATHING

172 Tarantino
174 Guns
176 Phobias
178 Plane Crashes

180	Pandemics	
182	Pandemics	
184	Smoking	
186	Drugs	
188	New Year's Resolutions	

MONEY MAKES THE WORLD GO ROUND

192	Pyramid of Wealth
194	B.R.I.C. States
196	Development Aid
198	Market Bubbles
200	Globalization
202	Drug Trafficking
204	Food
206	Food
208	Brands
210	Brands
212	Nuclear Power
214	Solar Power
216	Energy Market
218	Oil
220	Energy Flow

THE WORLD IS NOT ENOUGH

224	Waste of Energy
226	Resources
228	Resources
230	Carbon Emissions
232	Carbon Footprint
234	Natural Disasters
236	Plastic
238	Heat
240	Animal Biotopes
242	Water
244	Water
246	Water

OUR GREATEST IDEAS

250	Inventions
252	Nobel Prize
254	Light
256	Light
258	Submarine Cables
260	Internet
262	Digital Storage
264	CERN
266	EPILOG
267	INDEX
272	IMPRINT

MEASURING UNITS

We collected graphics from all over the world. According to their origin the different designers used different measures. To help you understand and compare the works shown in this book, we made a short list of some abbreviations in common use.

DISTANCE
mm	Millimeters	10 mm = 1 cm
cm	Centimeters	100 cm = 1 m
m	Meters	1,000 m = 1 km
km	Kilometers	1 km = 0.62 miles
in	Inch/inches	1 in = 2.54 cm
ft	Foot/feet	1 ft = 30.48 cm

AREA
m²	Square meters	1 m² = 10.76 square feet
km²	Square kilometers	2.59 km² = 1 sq mi
sq mi	Square miles	

VOLUME
ml	Milliliter	10 ml = 1 cl
cl	Centiliter	100 cl = 1 l
l	Liter	1,000 l = 1 m³
m³	Cubic meter	
km³	Cubic kilometer	

SPEED
km/h	Kilometers per hour	100 km/h = 62 mph
mph	Miles per hour	

TIME
Secs.	Seconds	60 secs. = 1 min
min.	Minutes	60 min. = 1 h
h	Hours	
yrs	Years	
BC	Before Christ	
AD	Anno domini	In the year of our Lord
a.m.	ante meridiem	
p.m.	post meridiem	

WEIGHT
μg	Microgram	1,000 μg = 1 mg
mg	Milligram	1,000 mg = 1 g
g	Gram	1,000 g = 1 kg
kg	Kilogram	1,000 kg = 1 t
t	Ton	

TEMPERATURE
°C	Degree celcius	0°C = 32°F
°F	Degree fahrenheit	

POWER
W	Watt	1,000 W = 1 kW
kW	Kilowatt	1,000 kW = 1 MW
MW	Megawatt	1,000 MW = 1 GW
GW	Gigawatt	
MWe	Megawatt electrical	Electrical output
MWt	Megawatt thermal	Thermal output
kWh	Kilowatt hour	
MWh	Megawatt hour	
GWh	Gigawatt hour	

DIGITAL DATA
MB	Megabyte	1,000 MB = 1 GB
GB	Gigabyte	
Mbit/s	Megabit per second	

AMOUNT
Mio.	Millions
Bil.	Billions
Tril.	Trillions

CURRENCY
$	U.S. dollar
€	Euro
£	Pound

ECONOMY
GDP	Gross domestic product
VAT	Value added tax
LDC	Least developed countries
PPP	Purchasing power parity

THE PLACE WE CALL HOME

When you feel upset about a soccer team losing, or a boyfriend ending a relationship, or just another parking ticket, remember this: you are a tiny speck on the cooled crust of a giant burning fireball that is constantly spinning as it flies around a vast inferno that has been burning for 23,000 times longer than the entire existence of humanity, is 28,080,000°F, and is the size of 1.3 million Planet Earths. Luckily, the development of the crust of our burning fireball created conditions for just the right mixture of oxygen, hydrogen, and other gases to be formed, combining with the gravitational pull of the inferno we call the Sun to protect us from deadly radiation, and to keep the gases trapped in at just the right levels to create conditions for life.

We have selected a series of images that remind us of these facts, and more. For instance, thanks to many interactions between the swirling mineral fires underneath the surface that create an invisible force we call magnetism, the gravitational pull on the oceans of the giant rock that orbits us that we have named the Moon, and too many other factors to count — only some of which modern science kind of understands — we have weather and earthquakes, polar ice (less than we'd like, and shrinking) and deserts (more than we'd like, and growing), rivers and mountains, and forests and oceans, all sustaining interdependent ecosystems of incredibly resilient creatures both tiny and huge.

Meanwhile, also spinning around that giant fireball are seven other rocks of sufficient size to be counted as significant by our own arbitrary measures. We aren't yet close to having the technological capacity to send a human to any of them, let alone to leave the orbit of this one fireball among approximately 100–400 billion in the galaxy, but there are at least a few small robots, shot into space and communicating back to us, that are flying around the vacuum of what we call our solar system, teaching us things via signals akin to the text messages of mobile phones. There is no way for any of us to truly understand what any of this means. How can we conceive of a distance through which it takes eight and a half minutes for the Sun's light to traverse, when we feel that two hours is too far to drive for dinner? We may as well try to explain the internal workings of an iPhone to a newborn kitten. The human brain is pretty limited — and it is, when you look at the bigger picture and everything that could possibly fit in it, really very small indeed.

Journey with us to the stars and into the oceans, and be amazed by the complexity and beauty of everything that there is.

The Story So Far

Something mysterious and unimaginably powerful happened 13.7 billion years ago. Whatever it was, it caused the laws of physics (as we know them) to separate and become defined forever. And the rest is history.

In the beginning there wasn't light. Whatever the Big Bang was, it was dark — it took 400,000 years for God to switch on the lights, or, if you prefer, for the necessary atoms to form so that light could start shining. Physicists can tell you all about what happened between now and a couple of billionths of a second after the Big Bang, when all the laws of physics were perfect and whole — the so-called singularity. After that, gravity broke off from the other laws of nature and did its own weird thing. And then it was up to all the quarks, protons, neutrons, electrons, atoms, and eventually molecules to sort themselves out, and collapse into stars. As the great Carl Sagan said, we are all star-stuff.

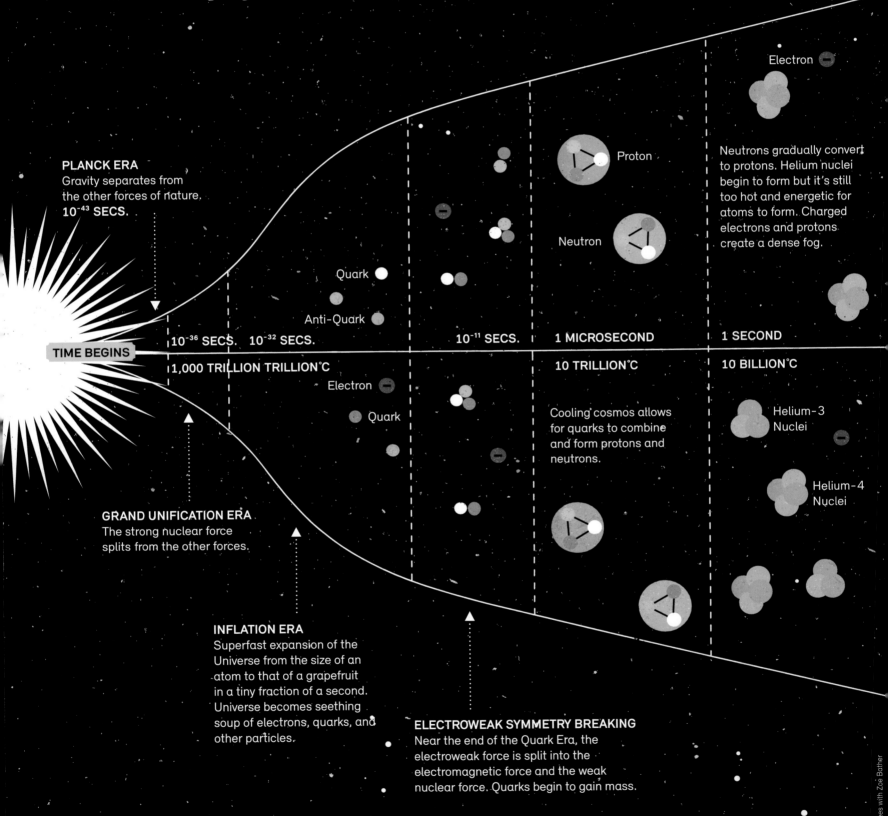

PLANCK ERA
Gravity separates from the other forces of nature.
10^{-43} SECS.

GRAND UNIFICATION ERA
The strong nuclear force splits from the other forces.

INFLATION ERA
Superfast expansion of the Universe from the size of an atom to that of a grapefruit in a tiny fraction of a second. Universe becomes seething soup of electrons, quarks, and other particles.

ELECTROWEAK SYMMETRY BREAKING
Near the end of the Quark Era, the electroweak force is split into the electromagnetic force and the weak nuclear force. Quarks begin to gain mass.

TIME BEGINS — 10^{-36} SECS. — 10^{-32} SECS. — 10^{-11} SECS. — 1 MICROSECOND — 1 SECOND

1,000 TRILLION TRILLION°C — 10 TRILLION°C — 10 BILLION°C

Cooling cosmos allows for quarks to combine and form protons and neutrons.

Neutrons gradually convert to protons. Helium nuclei begin to form but it's still too hot and energetic for atoms to form. Charged electrons and protons create a dense fog.

Design: Nathalie Lees with Zoë Bather

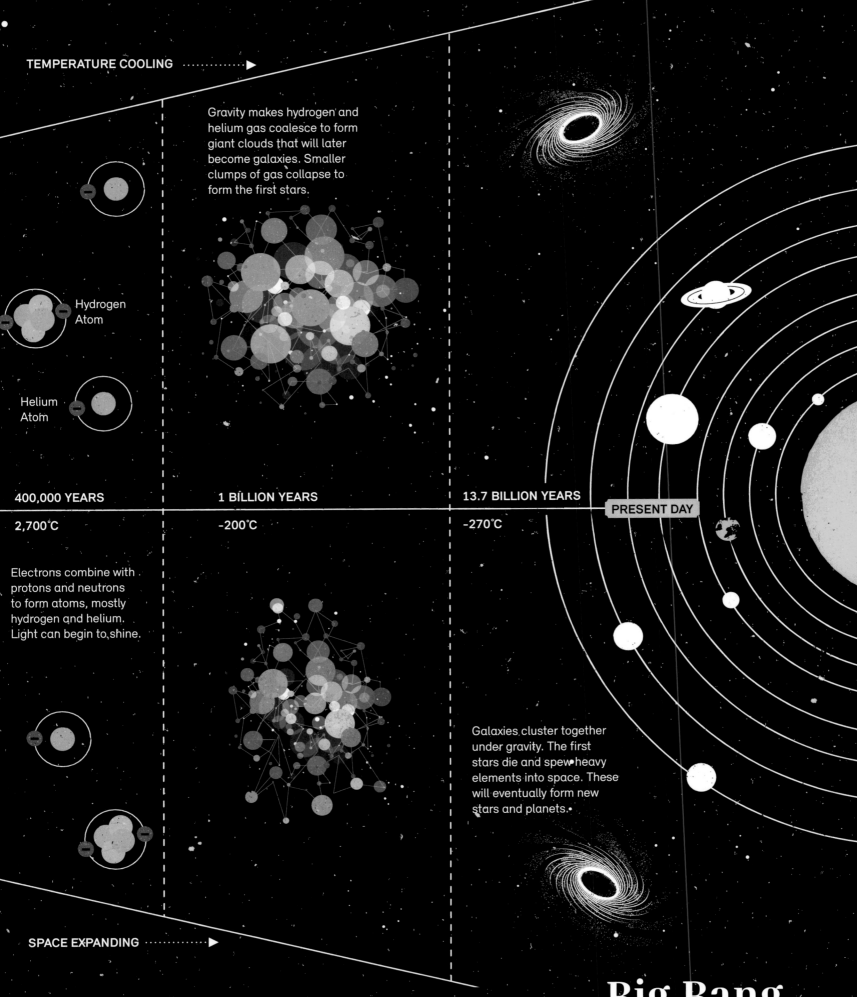

Free Throw This!

Brain-melting distances reduced to basketball size.

Place a basketball directly beneath the basket on a court. Then stick an ordinary quilting pin in the court exactly beneath the opposing basket. You have now created a scale model showing almost the exact distance between the Sun and the Earth, if the former were a basketball, and the colorful

IF THE SUN WAS THE SIZE OF A BASKETBALL ...
(SHOWN ACTUAL SIZE HERE)

THE EARTH TO THE SAME SCALE

← Orbit: 25.6 meters away from the basketball

Long distance: An Airbus A380 jet would take 205 days flying non-stop to circle the Sun, and another 5 million years to fly to the nearest star, Proxima Centauri.

RELATIVE SIZE OF THE OTHER PLANETS IN THE SOLAR SYSTEM

MERCURY   VENUS   EARTH   MARS   JUPITER   SATURN   URANUS   NEPTUNE

plastic pinhead was our planet — the third from the Sun. Of course, with a basketball-sized sun you won't be able to fit the whole solar system onto one court. You couldn't even do that if you had a football field. Or five football fields. In fact, if you wanted to mark the Sun's outermost planet Pluto (whether its planetary status is confirmed or not), you'll have to get another pin and stick it in the ground over a kilometer away. Time-scales along these lines are even more mind-boggling.

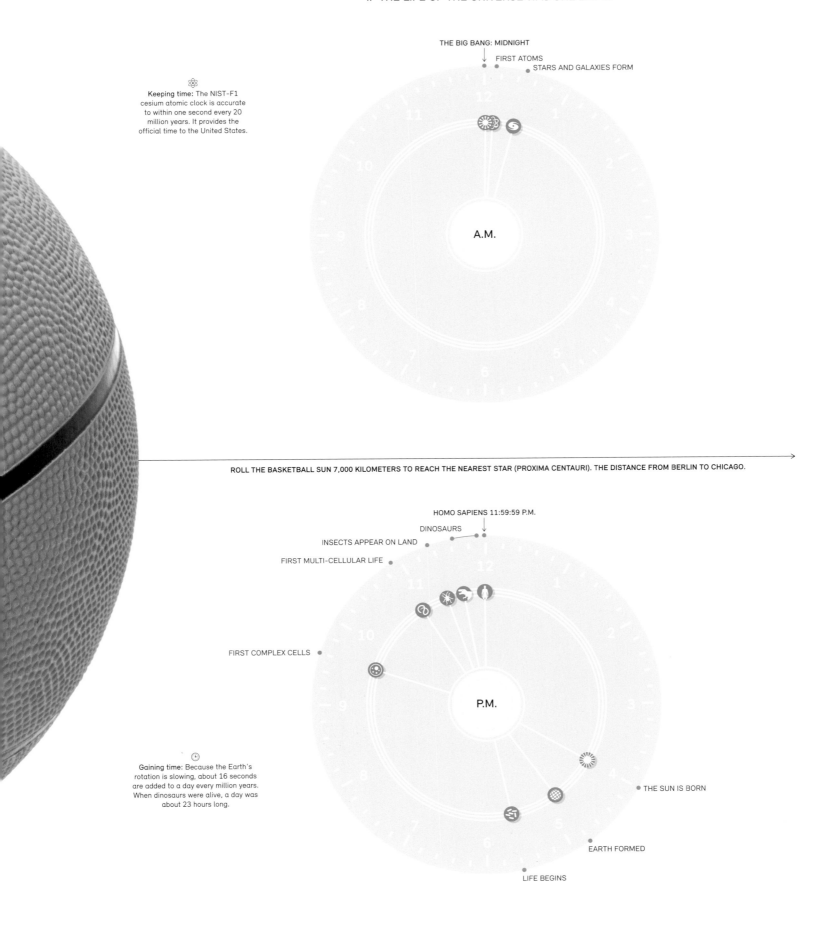

IF THE LIFE OF THE UNIVERSE WAS ONE DAY ...

THE BIG BANG: MIDNIGHT
FIRST ATOMS
STARS AND GALAXIES FORM

A.M.

Keeping time: The NIST-F1 cesium atomic clock is accurate to within one second every 20 million years. It provides the official time to the United States.

ROLL THE BASKETBALL SUN 7,000 KILOMETERS TO REACH THE NEAREST STAR (PROXIMA CENTAURI). THE DISTANCE FROM BERLIN TO CHICAGO.

HOMO SAPIENS 11:59:59 P.M.
DINOSAURS
INSECTS APPEAR ON LAND
FIRST MULTI-CELLULAR LIFE

FIRST COMPLEX CELLS

P.M.

THE SUN IS BORN
EARTH FORMED
LIFE BEGINS

Gaining time: Because the Earth's rotation is slowing, about 16 seconds are added to a day every million years. When dinosaurs were alive, a day was about 23 hours long.

The Universe
THE PLACE WE CALL HOME

A Dynamic Earth

All living things exist in a zone delicately balanced between two immensely powerful engines—the molten core of the Earth itself and the blazing sun. Each generates a host of forces, setting off the earthquakes, volcanoes, and extremes of weather that shape our lands and seas.

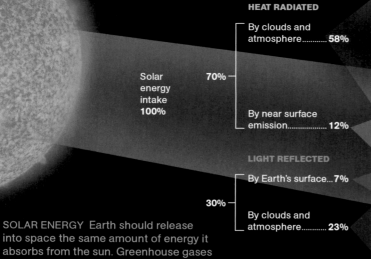

THE SUN A great ball of turbulent gases, our star supplies the heat and light needed to support life. Hydrogen atoms combine to make helium, producing an internal temperature of about 28,080,000°F.

HEAT RADIATED
- By clouds and atmosphere............ 58%
- By near surface emission................. 12%

LIGHT REFLECTED
- By Earth's surface... 7%
- By clouds and atmosphere............ 23%

Solar energy intake 100%
70%
30%

SOLAR ENERGY Earth should release into space the same amount of energy it absorbs from the sun. Greenhouse gases accumulating in the atmosphere upset that balance by blocking the energy release.

THE ATMOSPHERE A fragile layer of gas, the atmosphere protects our planet from the sun's radiation and other hazards. Stirred by the Earth's rotation and shifts in pressure and temperature, the atmosphere changes constantly, causing what we experience as weather.

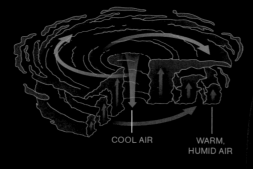

COOL AIR   WARM, HUMID AIR

HURRICANES Fueled by warm ocean water, hurricanes spin at more than 74 miles an hour. Heavy rains and an elevated sea surface add to the danger.

THE OCEAN Covering more than two-thirds of the planet, the ocean interacts with the atmosphere to make Earth habitable. Its currents distribute heat absorbed in the tropics to northern lands that would otherwise be icy year-round.

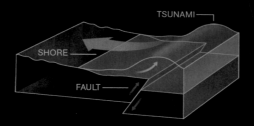

TSUNAMI
SHORE
FAULT

TSUNAMIS Large disturbances of the seafloor—especially earthquakes—create towering waves that can devastate the coastal communities they strike.

HURRICANE On the side of the Earth facing the moon, gravity pulls the ocean outward, creating high tides. Inertia on the opposite side of the planet has the same effect.

ATMOSPHERIC CELL

NORTH AMERICA

HURRICANE

OCEAN CURRENTS Evaporation, precipitation, cooling, heating, wind, up-welling, and other forces move water at the surface and stir deep circulation.

ATMOSPHERIC CELLS In constant loops, air warmed by the sun in the tropics flows to the poles as cold polar air travels to the tropics.

SOUTH AMERICA

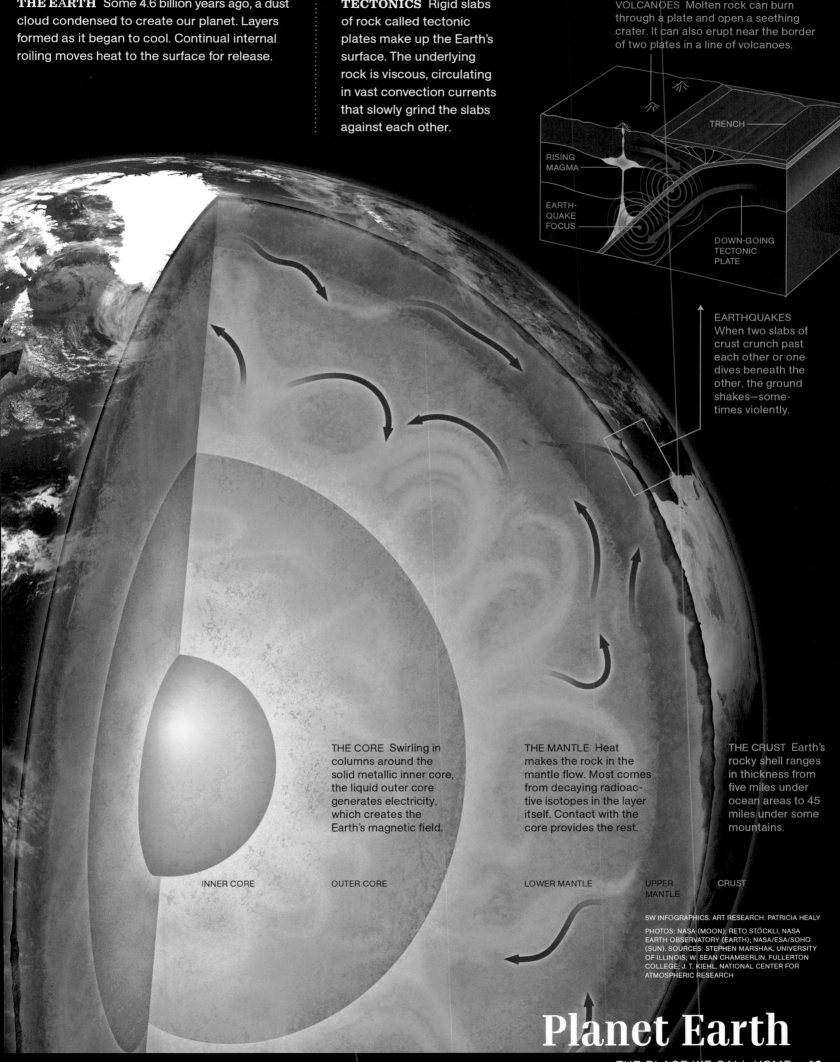

THE EARTH Some 4.6 billion years ago, a dust cloud condensed to create our planet. Layers formed as it began to cool. Continual internal roiling moves heat to the surface for release.

TECTONICS Rigid slabs of rock called tectonic plates make up the Earth's surface. The underlying rock is viscous, circulating in vast convection currents that slowly grind the slabs against each other.

VOLCANOES Molten rock can burn through a plate and open a seething crater. It can also erupt near the border of two plates in a line of volcanoes.

RISING MAGMA
EARTHQUAKE FOCUS
TRENCH
DOWN-GOING TECTONIC PLATE

EARTHQUAKES When two slabs of crust crunch past each other or one dives beneath the other, the ground shakes—sometimes violently.

THE CORE Swirling in columns around the solid metallic inner core, the liquid outer core generates electricity, which creates the Earth's magnetic field.

THE MANTLE Heat makes the rock in the mantle flow. Most comes from decaying radioactive isotopes in the layer itself. Contact with the core provides the rest.

THE CRUST Earth's rocky shell ranges in thickness from five miles under ocean areas to 45 miles under some mountains.

INNER CORE   OUTER CORE   LOWER MANTLE   UPPER MANTLE   CRUST

5W INFOGRAPHICS. ART RESEARCH: PATRICIA HEALY

PHOTOS: NASA (MOON); RETO STÖCKLI, NASA EARTH OBSERVATORY (EARTH); NASA/ESA/SOHO (SUN). SOURCES: STEPHEN MARSHAK, UNIVERSITY OF ILLINOIS; W. SEAN CHAMBERLIN, FULLERTON COLLEGE; J. T. KIEHL, NATIONAL CENTER FOR ATMOSPHERIC RESEARCH

Planet Earth

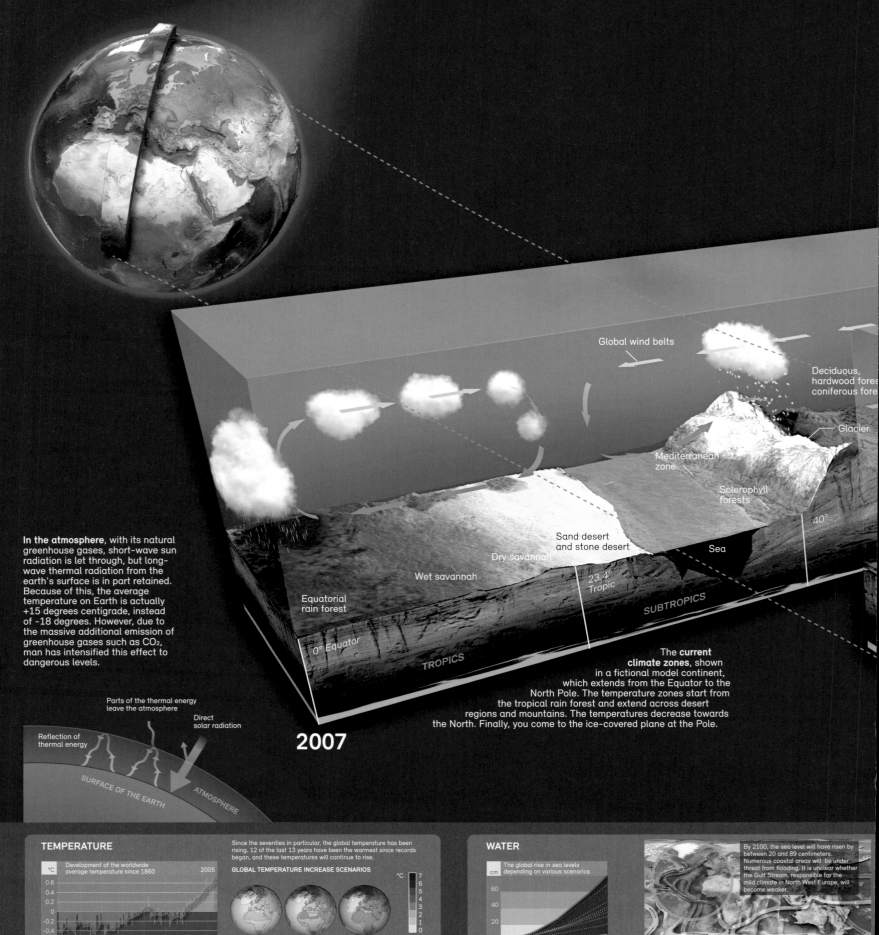

Yes, climate change is natural, but that doesn't mean we don't have an influence.

Climate changes naturally over time, and so we shouldn't be too worried by a little extra heat, you might think. That's certainly true — according to some clever climate scientists, we are technically still in one of the planet's ever-recurring ice ages (because for now at least we still have permanent ice at both poles), and we are due to enter a tropical phase anyway. But whatever the main reason for the change of climate is, it does not change the fact that releasing millions of tons of carbon dioxide into the atmosphere does, as it happens, change how the world breathes. This graphic traces the life cycle of our atmosphere.

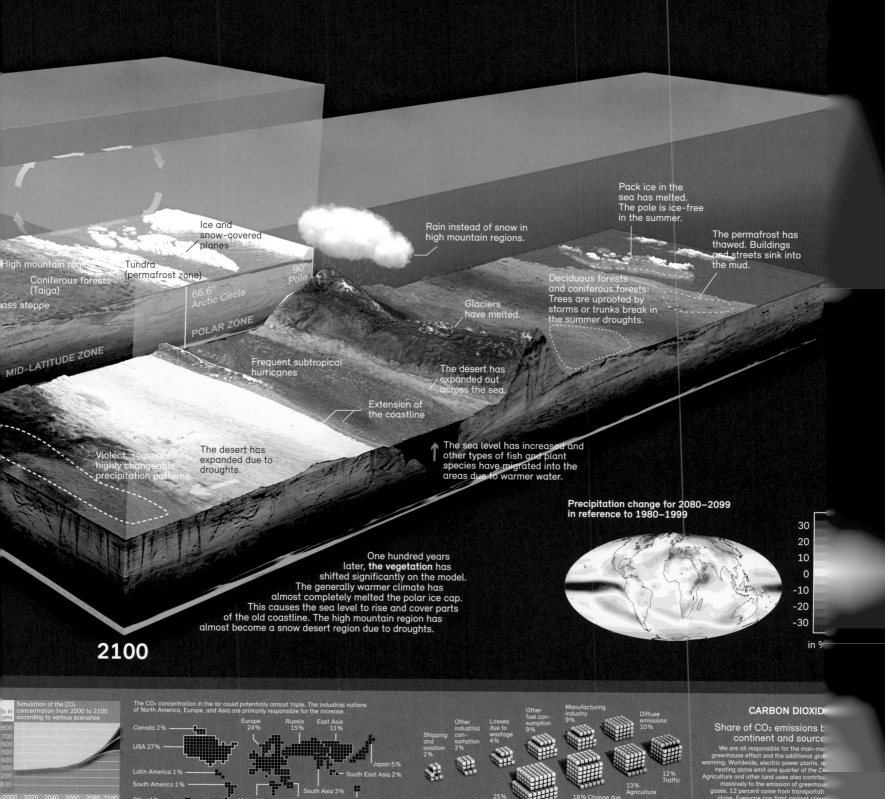

Spring

The world wakes up.

The four seasons seem to make such a harmonious cycle — offering the plants and animals on the planet a full complement of weather conditions to enjoy all year round — an image of perfection. But if it wasn't for an imperfection — the Earth is tilted 23.5 degrees off the perpendicular in relation to the Sun — we wouldn't have any seasons at all. It seems odd to think that if it hadn't been for what some scientists believe was a collision with a Mars-sized object during the solar system's formation, the Earth would have one smooth uni-climate all year round. It's hard to imagine what life would be like now if that had been the case. Apart from the consequences for nature, what of all the wisdom that the seasons teach us? The cycle of life, accepting the inevitability of death and so on? And apart from all that, we would have none of the joys of spring, displayed on this page. No awakening, no sappy songs about birds and bunnies, and sheep having babies, and no seasonal blossoming crocuses and so forth.

Gradual bloom

All plants compete for sunlight in order to create organic substances using photosynthesis. In order for small plants such as **spring whitlow-grass** to also get enough light, they must forge ahead and blossom before they are overshadowed by the larger species. Of course, the high season is during the sun-rich months. The autumn crocus is the last to unfold its petals.

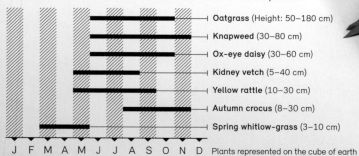

- Oatgrass (Height: 50–180 cm)
- Knapweed (30–80 cm)
- Ox-eye daisy (30–60 cm)
- Kidney vetch (5–40 cm)
- Yellow rattle (10–30 cm)
- Autumn crocus (8–30 cm)
- Spring whitlow-grass (3–10 cm)

J F M A M J J A S O N D Plants represented on the cube of earth

A nursery between green stalks

The 15 cm **Meadow Pipit**, which lives on insects and spiders, builds a nest of stems and grasses on the ground. There, the female hatches one or two broods of up to six eggs each per year. The young are able to fly after about two weeks.

4–6 eggs   13–14 days incubation period   13–14 days rearing period

Meadow knapweed

The **Meadow Pipit** hunts insects.

Ox-eye daisy

Autumn crocus

A **Ground b**
captures
an earthwo

The larvae of
a **June beetle**
feed on roots
and spend
more than
two years
in the ground.

◄ 50 cm ►

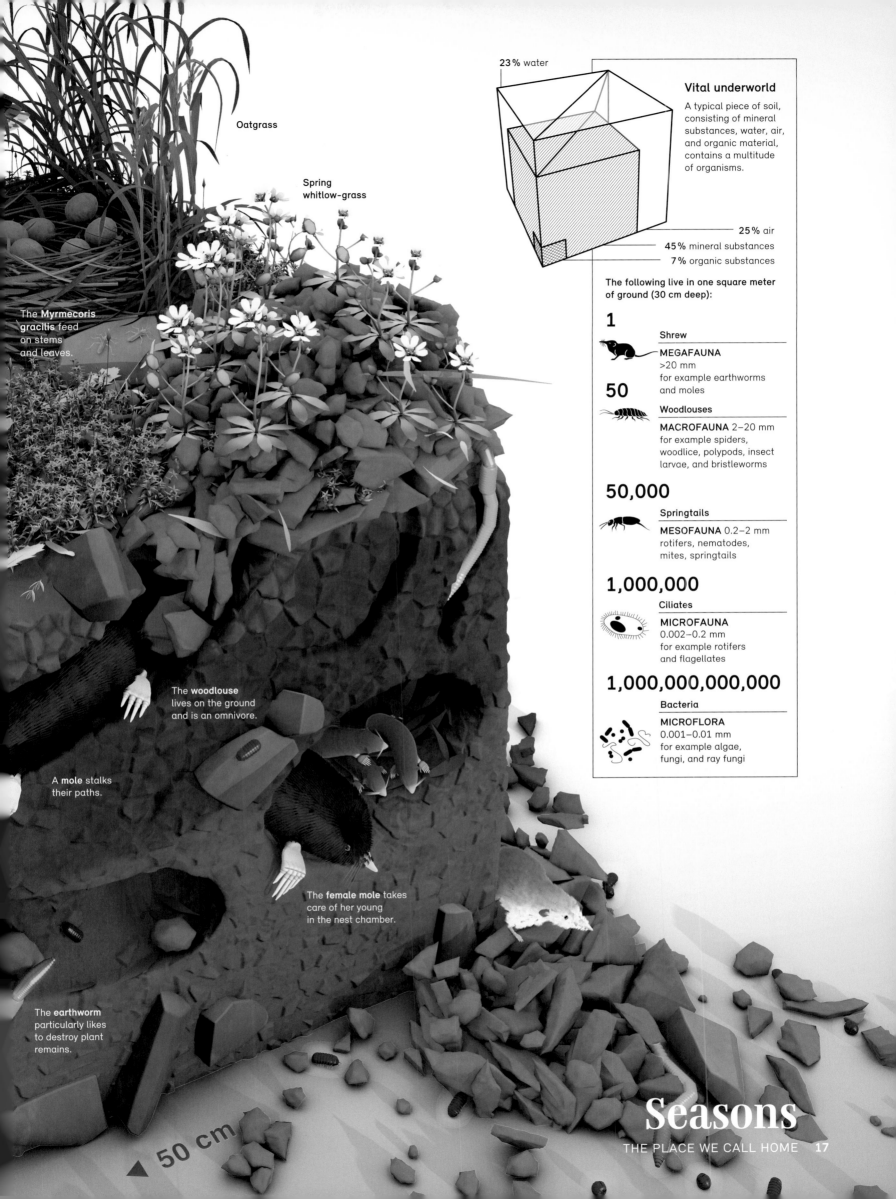

A **crab spider** sits in a daisy flower and waits for pollinators.

The praying mantis has trapped a blood cicada with its front legs and is sucking the insect dry.

Knapweed

Foam Cicadas eat the green parts of the knapweed.

A shrew has captured a **Ground beetle** and is eating it.

The larvae of the **foam cicada** produce a foam which contains protein and use this to build their nest.

Summer

Life is warm.

The summer is obviously a busy season, with all those insects, frogs, birds, and other animals that are seldom seen in the other seasons. Of course, the plants, now at their full height, are getting in on the action too, and the swarming insects play a vital part in the plants' life cycle. By July the crops begin to shed their seed, and soon it's time to reap what you have sown (if you want to get Biblical about it).

Masters of disguise

Crab spiders can change their body color to match the color of the flower on which they lie in wait. Approaching insects do not see the danger — until the hunter has caught its victim with its front legs and injected it with poison.

Deceitful safety

A female **Fruit fly** deposits its eggs into the head of a flower — in this way, the larvae grow in safety with the best of nourishment.

1. The Fruit fly pierces the wall of the flower and places an egg inside.
2. An injected message encourages the plant to grow abnormally here, resulting in the formation of a "gall."
3. The larva develops and eats from the gall.
4. By stinging into the plant, a braconid wasp locates the larva and lays its own egg inside. The larva is eaten from the inside out.

A night of love and lights

On warm summer nights, the female common **glowworm** lights up in the grass and signals that she is ready to mate. The males, who also glow, then circle in the air and allow themselves to fall. The creatures die a few days after mating. As an insect, it lives exclusively from fat reserves, which it collected as a larva.

Seasons

Fall

The first chill.

The days grow chillier, the kids go back to school, and the trees prepare to shed their leaves. And as they do so they decide to explode into a spectacular array of gold and brown, as if to match the early sunsets. You start to wonder what hornets ate before plumcake was invented. Mushrooms pop out of the forest soil and the first animals are about to leave or start to build a home for the even colder season ahead.

Oatgrass

A **Wasp spider** lies in wait for prey in its web.

Autumn crocus

Burying beetles have smelled prey.

Burying beetle larva

Cadaver of a **mole**

Tapered sapling

A **blindworm** searches for food in the Meadow Pipit's abandoned nest.

Kidney vetch

The caterpillar of the **Dwarf Blue** eats the seeds of the **Kidney vetch**.

Myrmecoris gracilis now turn brown in color.

Mown **oatgrass**

Rotating oatgrass

The "seed pods" of the plant are so light that the wind can disperse them far and wide. With the help of a winding appendage, the "grain," which expands in moisture and rotates while doing so, then buries the seeds in the ground where they then germinate.

Cadaver burial

The smell of decomposition from a dead mole attracts numerous Burying beetles. First of all, the males fight for the corpse, then the victorious male and his female bury the dead animal, so that it sinks into the ground to a depth of up to ten centimeters. This is the food supply for the beetle's larvae.

Dancing to a different rhythm

The autumn crocus, a plant which grows to a height of up to 30 centimeters and is poisonous to many animals and to humans too, develops to a different rhythm than most plants in the biotope.

It blooms in **autumn** up until October and is fertilized.

The blossom disappears in **winter**. The tuber and ovaries survive in the ground.

In **spring**, long green leaves grow.

The seed capsules grow high in the **summer** and release their contents.

Give and take

Plant root

Bacteria

Clover and certain ground bacteria live in symbiosis, a partnership of mutual utility: So-called nodule bacteria in the soil are attracted by substances in the roots of the plant. They penetrate the tissue and form a nodule-like thickening. There, the single-cell organisms make the nitrogen in the air available to the plant, which it cannot extract on its own. Later, the usable remains of the bacteria are also absorbed by the host plant.

Seasons

THE PLACE WE CALL HOME

Possible mutilation

The body of an earthworm consists of numerous segments with small bristles — a fully grown specimen has up to 160 segments. The worm can use these to move forwards and backwards. If a predator grabs its back end, the worm can detach this section, leave it to its attacker, and escape. The lost part later regenerates. However, his ability to regenerate has its limits: the end section cannot survive if the front end is detached. It is not possible to make two worms out of one.

Head
Segment
Reproductive organs

Regeneration limits

A **Myrmecoris gracilis** seeks protection in the oatgrass.

A **shrew**, which is also active in cold weather, has captured a spider.

Woodlice bury themselves in litter or in the ground.

Winter

The big time-out.

Winter is no fun, unless of course you're a kid and enjoy having snowball fights. Our fellow creatures, though, are less fortunate, which is why most of them — if they don't have the luxury of wings — just curl up in a ball, take a couple of sleeping pills (metaphorically), and wait it out until nature wakes up again next spring.

A drop of the ocean under its belly

Woodlice are small crabs which originally lived in the ocean. They populate the damp layer of organic matter on top of the soil and eat the decomposing plant tissue there. After mating, females form a fluid-filled pouch between their cursorial legs, called the "marsupium," in which they lay their eggs. 40 to 50 days later the young hatch.

Marsupium

Mosses form spear-shaped spore capsules.

Earthworms withdraw to deeper layers within the soil.

The **autumn crocus** spends the winter as an underground bulb.

The larva of a **June beetle**.

Seasons

THE PLACE WE CALL HOME 23

Twilight of the Arctic Ice

The empire of ice at the top of the world is shrinking. The Arctic Ocean's summer ice pack covers little more than half its former reach, as a sweeping satellite image from September 2008 documents. Atop Greenland's formidable ice sheet, melting has also quickened. Sea ice, naturally expanding and contracting with the seasons, has covered this ocean year-round for most of the past three million years. But the Arctic is uniquely sensitive to climate change (right). Ten years ago global-warming models predicted the Arctic Ocean could be ice free in summer by 2100. Then the date dropped to 2050, and now to 2030—or sooner. As climate scientist Mark Serreze puts it, "Reality is exceeding expectations."

Ice Sustains Ice
The brilliant white of ice and snow reflects more than 80 percent of incoming sunlight. This reflective quality is called albedo. The high albedo of an ice-covered Arctic helps keep its temperatures low and preserves its ice.

A Balance of Warmth
Some of the solar energy reflected by ice or reradiated as heat returns to space. Some is absorbed into the atmosphere by greenhouse gases like carbon dioxide and water vapor, whose heat-trapping qualities make life on Earth possible.

◀ More ice, more re

Northern Hemisphere
Land 39.4%
Ocean 60.6%
Sea ice* 6.5%
5.8 million sq mi

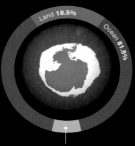
Southern Hemisphere
Land 18.5%
Ocean 81.5%
Sea ice* 7.1%
6.9 million sq mi

*Average maximum extent Winter 1978-2002

North and South
The Northern Hemisphere has experienced a greater temperature rise than the Southern, in part because it has more land, which warms faster than open ocean. Yet troubling signs of warming in Antarctica—where the vast continental ice sheet holds 85 percent of Earth's freshwater ice—make clear that the bottom of the world is also vulnerable.

Minimum extent September 2008
Minimum extent September 1980
Minimum extent September 2007

Relief vertically exaggerated

Arctic Retreat
Measured at the end of summer, the sea-ice minimum in September 1980 spanned an area slightly smaller than the contiguous United States. The September 2008 minimum was just over half that size. Regional weather patterns contributed to the even greater decline in 2007.

← Higher reflection

Year | Summer sea-ice extent

 1980 | 3.01 million sq mi

1985 | 2.66 million sq mi

 1990 | 2.39 million sq mi

 1995 | 2.36 million sq mi

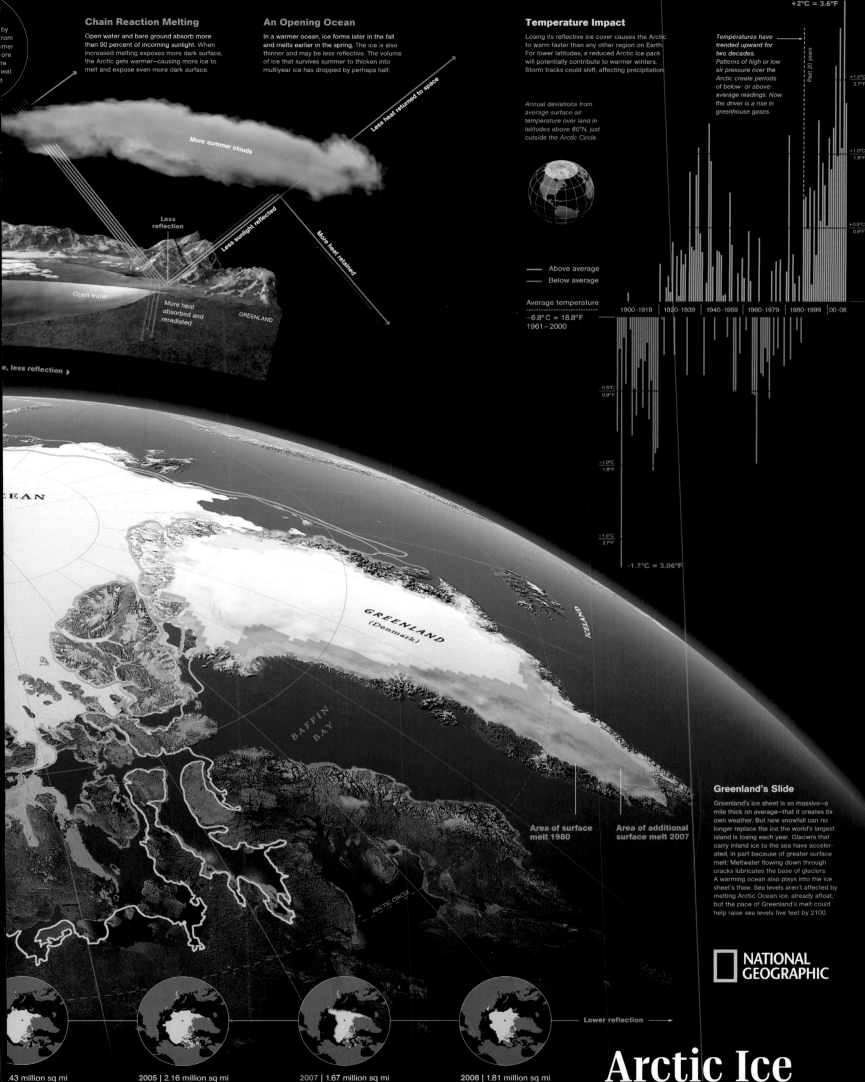

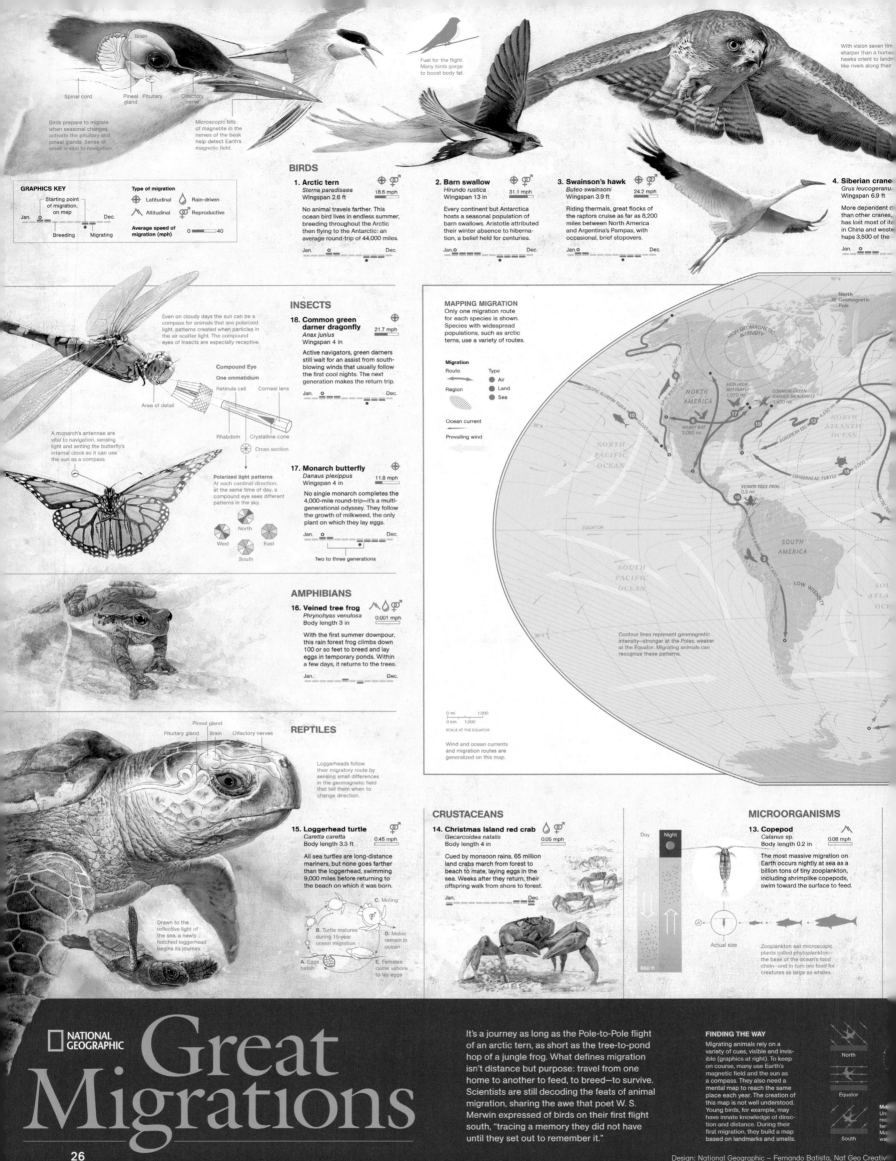

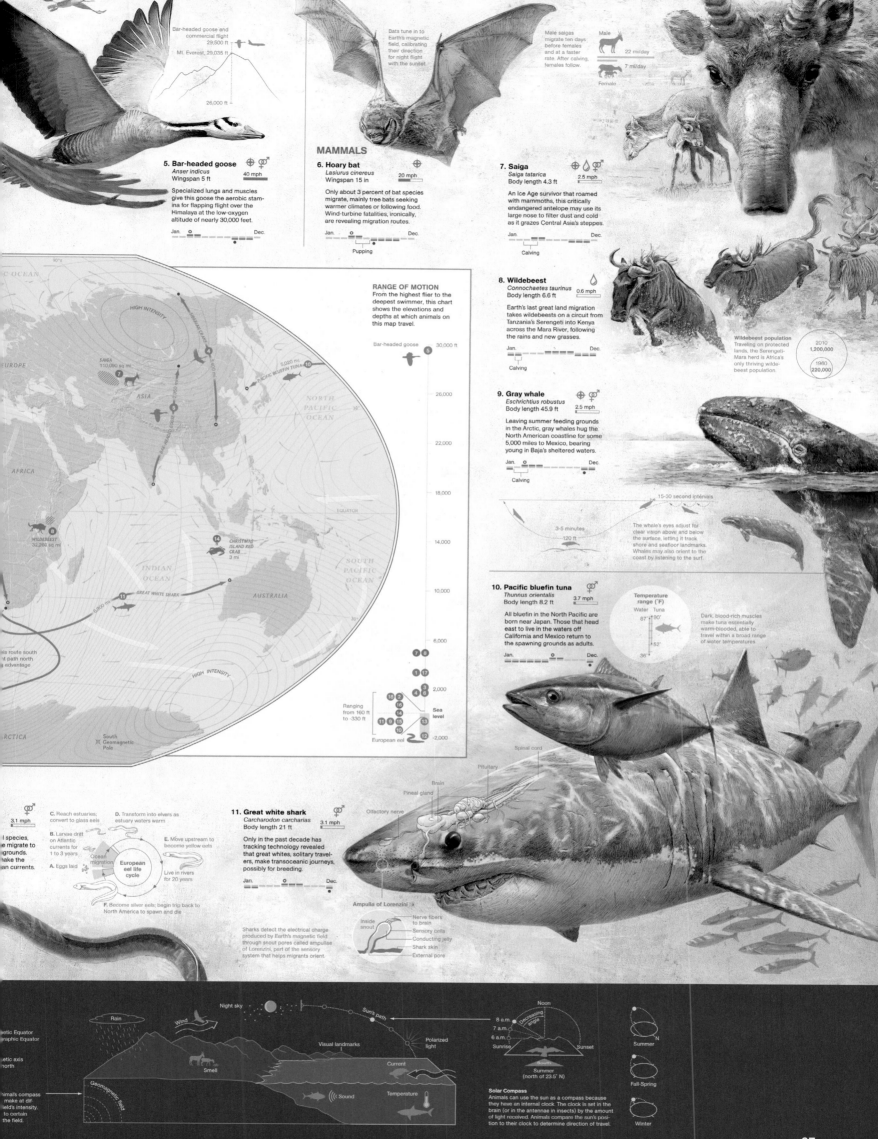

Continental Shift

The world map, rejigged as a play of scales and dimensions.

Most people can draw a rough map of the world — start with the upside-down pear of Africa, use the outline of the bite that's been taken out of it to trace the nose-shaped tip of South America, then add the skinny isthmus of Central America … etc etc. But even if we know more or less where everything is, let's face it: our planet's geographical facts are a mess. What if you really want to see the physical components of the world in relation to each other? Here, the world's biggest natural features have been plucked from their unruly geographical circumstances and laid out according to scale. It's the purity of size in infographic form — rivers, mountains, lakes, and islands expressed in two dimensions alone. And there's another dimension here too: human masses, measured both in absolute numbers and density. Major collections of people are presented here in direct comparison — the biggest cities (measured by population), and the biggest airports (measured by passenger count).

HOW TO READ IT

Is it possible to create a new map of the world visually comparing continents and their sizes; rivers, mountains, lakes, and island dimensions; city population, density, and even the number of passengers that transit every year at the most important airports?

Each continent is represented as follows:

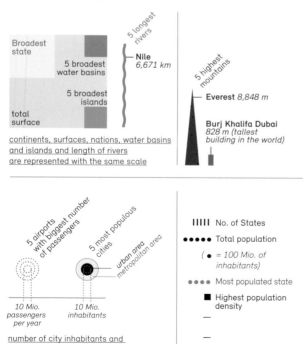

continents, surfaces, nations, water basins and islands and length of rivers are represented with the same scale

number of city inhabitants and airport passengers are represented with the same scale

	No. of States
•••••	Total population (• = 100 Mio. of inhabitants)
••••	Most populated state
■	Highest population density
—	
—	
—	No. of jet-lag crossed

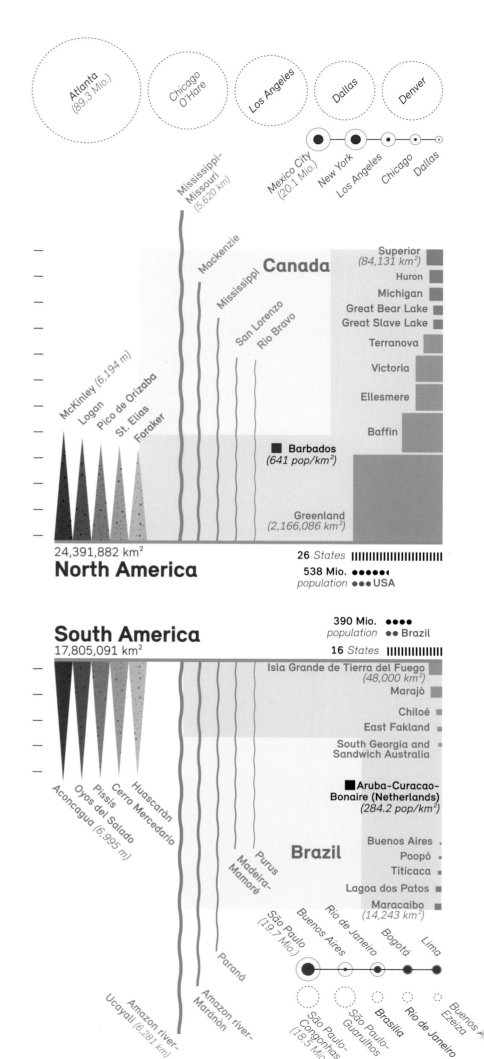

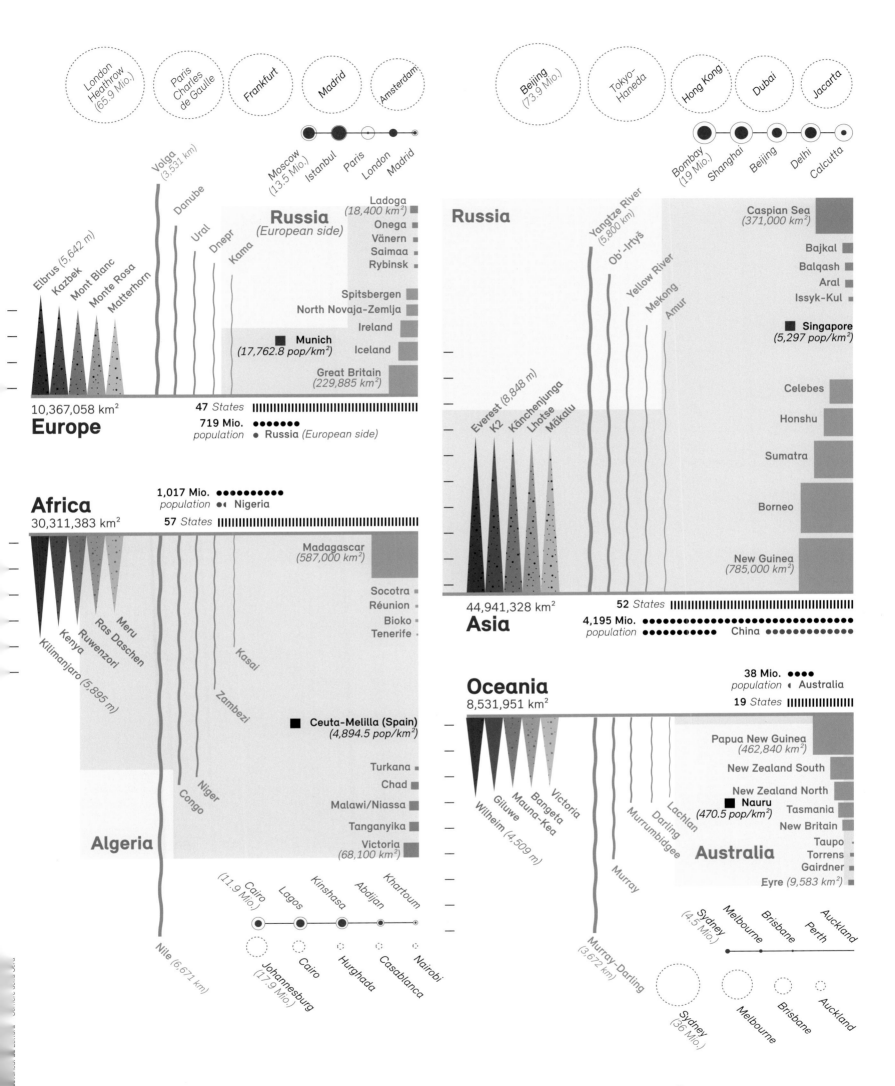

Live Very Long and Prosper!

There might be an end, but it is not near.

One day, about 10 billion years from now, the Sun will be smaller than the Earth. It'll be a dim white ball floating in space surrounded by a vast cloud of gas and dead, charred rocks. One of those rocks will be a black cinder that we once called Earth. We couldn't care less because according to *Star Trek* mankind should have explored and prospered the universe 10 billion years ago anyway.

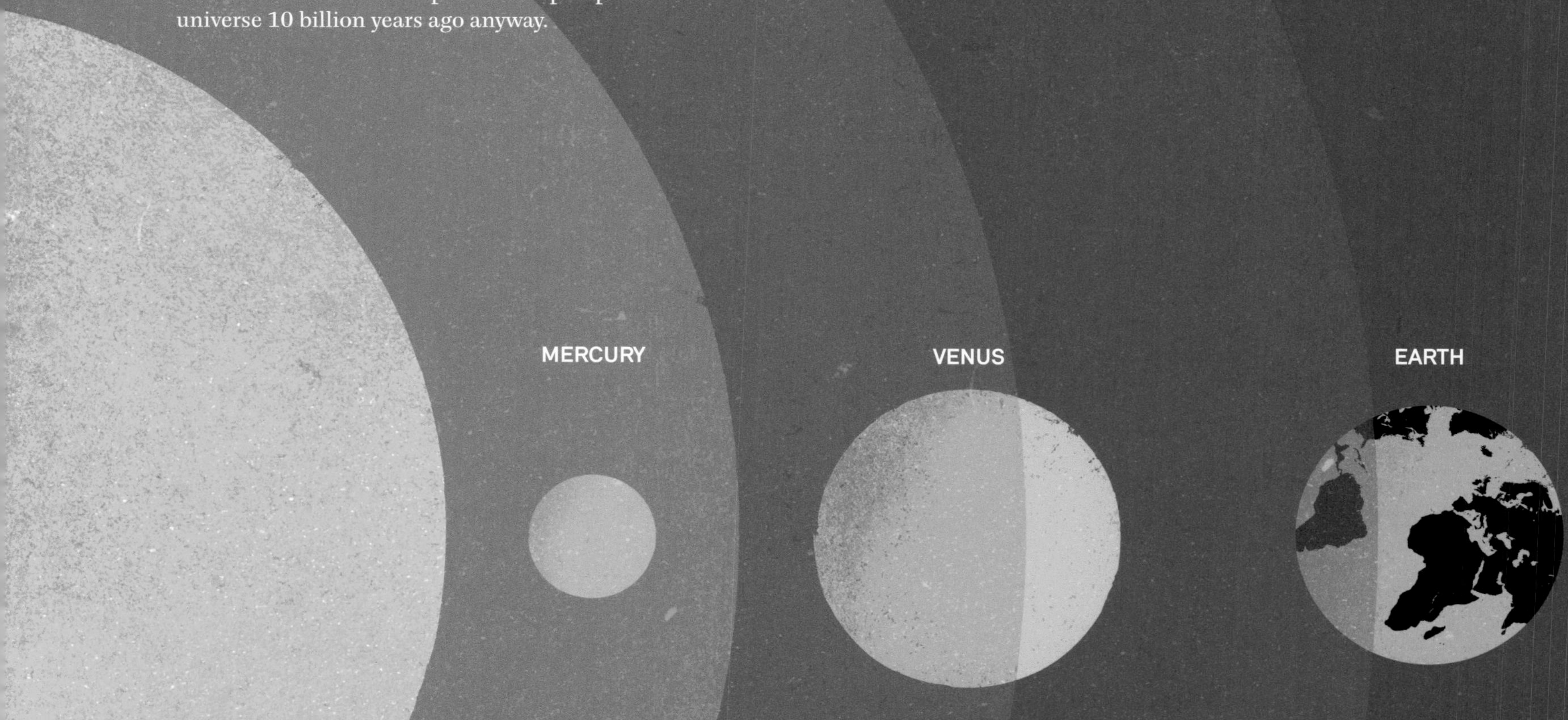

MERCURY

VENUS

EARTH

STAGE 1
MAIN SEQUENCE STAR

STAGE 2
RED GIANT

The Sun becomes a red giant as its stores of hydrogen run dry. Its core collapses and the extra heat this generates will cause its outer layers to expand.

Outer layers of the Sun expand, cool, and become redder. Its radius will increase around 250 times, taking it past Mercury, Venus, and towards Earth.

Earth will become increasingly hot until oceans boil away, the atmosphere escapes into space, and it is left as a charred, lifeless cinder.

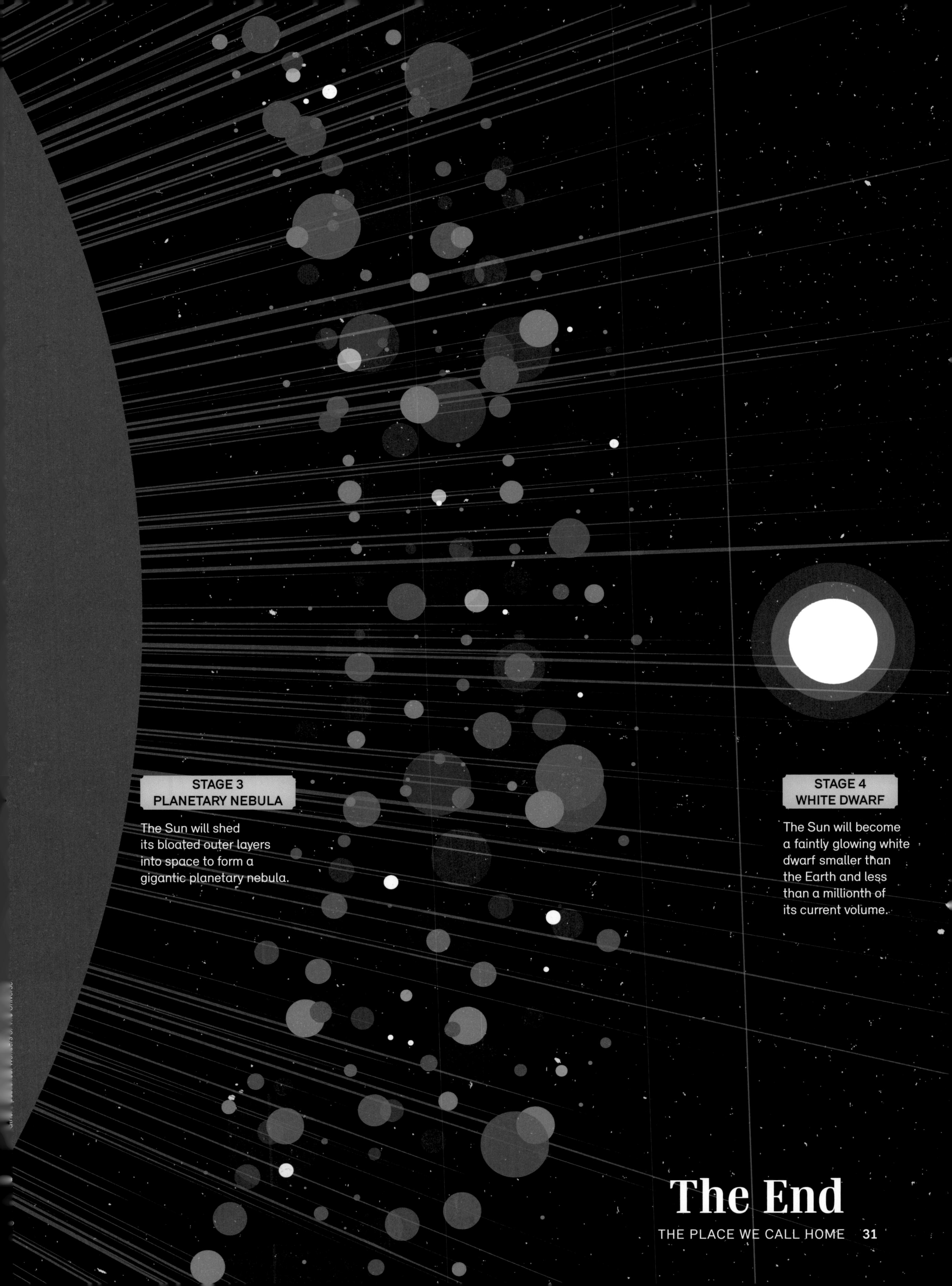

STAGE 3
PLANETARY NEBULA

The Sun will shed its bloated outer layers into space to form a gigantic planetary nebula.

STAGE 4
WHITE DWARF

The Sun will become a faintly glowing white dwarf smaller than the Earth and less than a millionth of its current volume.

The End

THE PLACE WE CALL HOME

LIVING TOGETHER

Strange coincidences happen, but none are more random, nor have more impact over every aspect of your life, than the facts of your birth. These accidents have already either empowered you or placed you at a significant disadvantage.

Every single person is incubated in a womb, and emerges alone. They are then swept up into a unit made up of an adult or adults who accept responsibility for this baby's development. These are adults collectively labeled according to geography (what nation they live in), belief (religion), resources and history (class), ability (age, language, perceived mental and physical capacity), and other arbitrary, invented divisions (race, culture) that make up human societies. We are now more than 7 billion individuals, grouped by many different labels. That doesn't mean that every newborn's fate is predetermined. So many lives are lived in pursuit of a measure of the power — economic, political, religious — that is granted at birth to others, and some will succeed in grasping it, either for themselves or for their children. The world's population and migration trends shown in this chapter suggest that the centers of power are shifting, as they always have throughout recorded history (try explaining to a serf under the armies of Genghis Khan that Mongolia is now a military nonentity).

Of course, some power centers hold out longer than others, adapting somewhat to the modern world while using the rhetoric of tradition to help justify their continuity. Whatever your religious beliefs, there is no doubting that the influence of Christianity, the endurance of Judaism, and the growth of Islam are major forces in the world today. This chapter will illuminate some of their commonalities and differences through core rituals and practices. It also reveals the rituals and practices of some of the world's political systems. The Temple in Jerusalem and the White House in Washington, DC might on the surface appear quite different, but at their hearts there is a similar faith in exceptionalism that justifies their power in the minds of those who wield it. Popes, priests, prime ministers, and presidents have more in common than you might think. And as for skyscrapers, aren't these temples by any other name, reaching for the heavens in the names of bravado and commerce?

The power systems highlighted in this chapter are predominantly those that dominate the urban, westernized world, a function of the graphics we have managed to track down. Existing forms of global power, as we see here, may get all the media attention, but it doesn't mean that Western society is somehow a natural or ideal state of being. In order for the power equation to balance, it requires that a few people have it, while the others do not. The more we can understand how these systems are formed, the better equipped we are to decide what works and what needs to change.

The more we create ways to divide ourselves in this incredible experiment called "humanity," the more we understand that we're exactly the same.

Greetings

LIVING TOGETHER

| GREAT BRITAIN | ITALY | NORTH AMERICA | ANCIENT TRIBES |

Mendelssohn's Wedding March

Mendelssohn's Wedding March was first performed at a wedding on June 2, 1847, during the wedding of Dorothy Carew and Tom Daniel in Great Britain. The march became popular after being performed at the wedding of Princess Victoria and Crown Prince Frederick of Prussia on January 25, 1858.

Wedding rings

In Ancient Rome, the groom sent an iron ring to the bride as a symbol of unbreakable bonds. Egyptians were the first to use rings in the wedding ceremony. The rings became widely used in the seventeenth century.

Bridal bouquet

The tradition of the bridal bouquet toss was born in North America, after a bride passed her bouquet to a friend who was soon happily married herself. Later, to avoid offending anyone, it became customary to throw the bouquet with the bride's back turned on the unmarried female guests.

The first dance

The wedding dance tradition dates back to the ancient tribes. At that time, the whole wedding ceremony consisted of a ritual dance performed by a young man with his chosen partner.

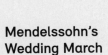

Garter

The garter toss tradition was born in fourteenth-century England. People believed that a piece of the bride's clothing would bring happiness and luck. To avoid damaging the bride's dress, the groom tossed a garter to the guests.

Veil

In Ancient Rome, the bride's veil was red and symbolized the bride's belonging to the groom, and in Ancient Greece a yellow veil covered the bride's entire body as a symbol of total obedience. The white veil is a symbol of pureness and innocence.

| INTERNATIONAL TRADITION |

Breaking champagne glasses

Breaking dishes is an international tradition. In Russian villages, clay pots were broken for good luck. In England, the groom tossed a dish with the remaining wedding bread on the road. In Germany, dishes were broken by relatives, and the newly married couple cleaned up the pieces as a symbol of overcoming their first troubles together.

Put a Ring on It

In the name of love — a truly universal bond.

Marriage — or at least some form of binding commitment — is common to almost every human culture, though its exact form has differed throughout time. The individual components still vary wildly across societies and tradition. The idea that marriage should not primarily be undertaken for social and political means was once a subversive one. For instance, the idea that love is — you know — an important element is a pretty new idea. Indeed, almost every culture in the world, apart from the ones that developed in the last couple of centuries, would say that marriage is a social and political contract — and nothing else. Not that married couples didn't occasionally fall in love in the old days — it's just that it was a happy accident if they did. The great Roman general Pompey's infatuation with his young wife Julia raised many eyebrows in the late Republic — "Check out this guy," they used to say with a mixture of amazement and ridicule, "he actually loves his wife." Here are a variety of ways humankind has celebrated getting hitched.

Unusual wedding traditions

| SOUTH KOREA | FIJI | KENYA | SWEDEN |

A goose as a gift for the mother-in-law

When entering the bride's house, the groom presented a wild goose to the future mother-in-law. The goose was a symbol of fertility and commitment.

A whale tooth for the bride's father

The groom presented a whale's tooth to the bride's father, a symbol of wealth and high social status.

Exchanging roles

In Kenya, the young husband had to wear female clothes for the first month of the marriage, thus "experiencing" the hard life of a woman.

A coin in the bridal shoe

The bride put a silver coin in her left shoe and a gold one in the right shoe. The coins were given to her by her parents.

Wedding Traditions

LIVING TOGETHER

Happy Together

Does the whole world really smile with you?

What makes us happy? What does that well-worn phrase "quality of life" mean in practice? What does it take to live a good life? People can be defined by how they answer these epicurean questions. Once a year, the intrepid pollsters at Gallup conduct hour-long interviews with around 1,000 people in over 130 countries to find out how happy we all are. The interviews are then rated on a simple 1–10 scale, and the results correlated with a variety of social and economic factors. Statistics like unemployment and gross domestic product per capita are taken into account, but the interviews themselves

reveal one thing — that the most important factors for personal happiness are income, life expectancy, and satisfaction with the government. There is also evidence that diversity in political representation breeds greater satisfaction — people are happiest when there are a higher proportion of women in parliament. So what makes us happy? It all comes down to democracy, freedom, and money apparently. It's that complicated — and that simple.

Happiness
LIVING TOGETHER

Crowded House

This is likely to be the century when the world population reaches its peak.

The world population doubled, from 3.5 billion to 7 billion, in the 36 years between 1976 and 2012. However we did it — improved health care, technological advances, an accumulation of material wealth — the fact is that homo sapiens are now, after some 200,000 years of existence, finally overrunning the planet. We have spread everywhere, become the top (more or less) of every food chain. Some academics predict that the world's population will reach 15 billion, while others say it will probably flatline around a mere 10 billion. Better living conditions and education will have only a mild effect on the growth rate, especially in the now poorer countries. And sure, in the long run there is plenty of evidence to suggest that less clever life-forms — the rat, the ant, the common cold virus — will outlive us and take over. But, for time being, we're very successful at reproduction and at sticking around.

How to read this infographic

These maps show the world's population, by country. In both maps each square represents one million people and each country's population figure is rounded to the nearest million. Countries with a population under 500,000 are excluded. By comparing the maps to a regular map of the world (above), the size of the population in different countries and the change since 1950 can be seen.

☐ Each square equals one million people

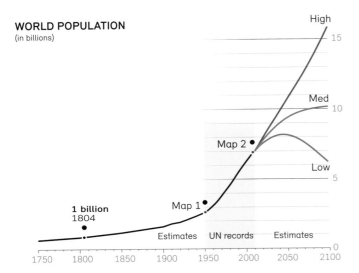

WORLD POPULATION (in billions)

GROWING UP
The most populous countries, according to projections by the UN. (based on medium variant projection)

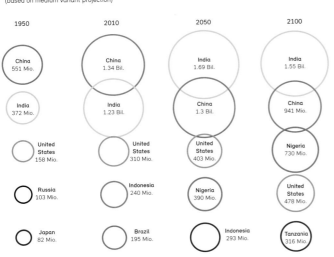

1950
Year UN population records begin

Population
2.53 billion

2011
6.97 billion

Britain — Third most populated nation in Europe, after Germany and France.

Canada — Lowest population density in the Americas.

United States — Number of people up twofold since 1950.

Brazil — Population increased by three times since 1950.

78 Mio. Increase in the world's population between July 2010 and July 2011.

A CROWD OF SEVEN BILLION

The world may seem crowded, but the seven billion of us don't take up as much space as you'd think. According to the Jacobs crowd-counting formula, devised by Herbert Jacobs of the University of California, Berkeley, we could fit into an area the size of Hong Kong.

Resulting crowd size

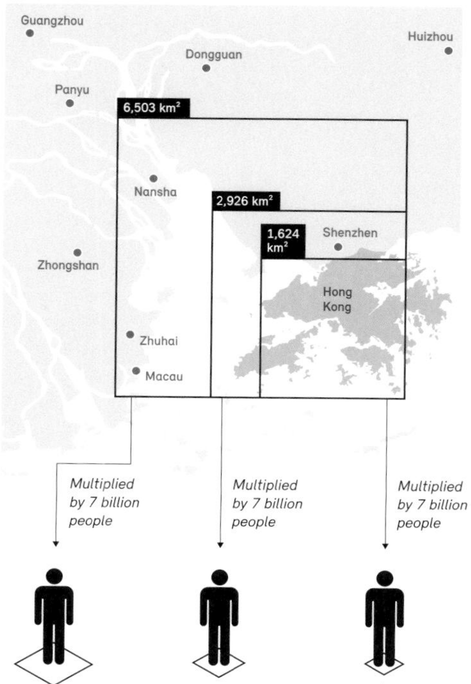

6,503 km²
2,926 km²
1,624 km²

Multiplied by 7 billion people

Loose crowd
An arm's length from the body of nearest neighbors.
0.929 square meter of space needed.

Multiplied by 7 billion people

Tight crowd
Standing shoulder to shoulder.
0.418 square meter of space needed.

Multiplied by 7 billion people

Packed crowd
Density similar to that of a mosh pit at a rock concert.
0.232 square meter of space needed.

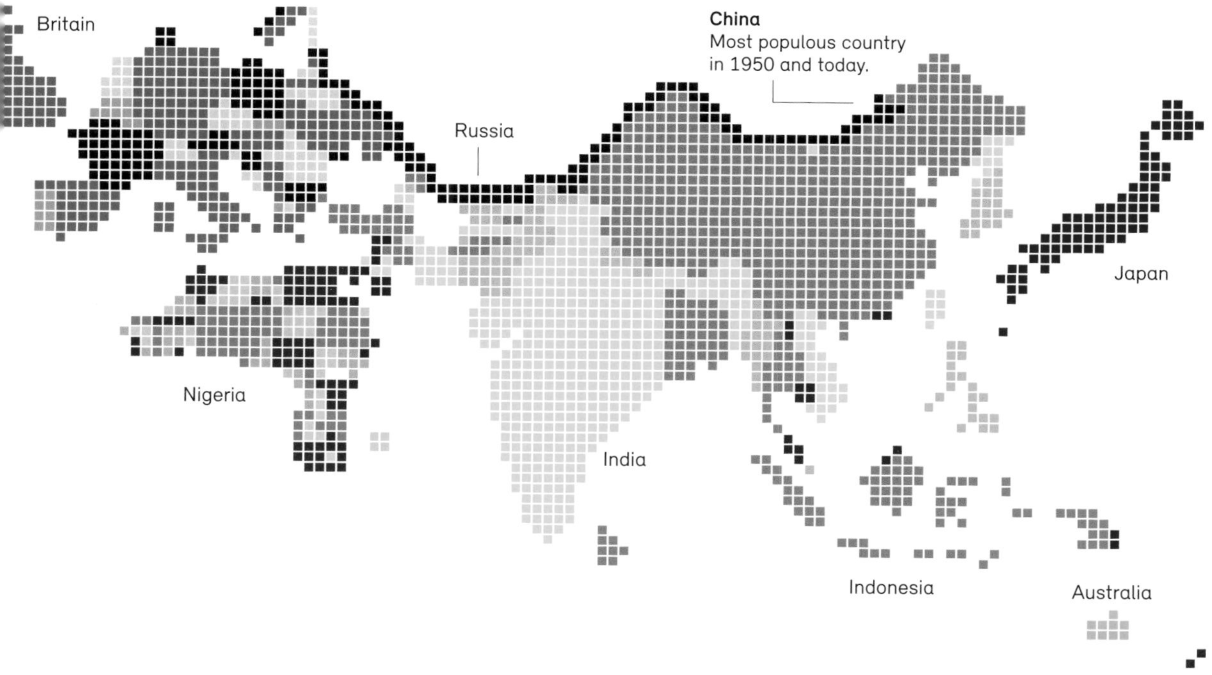

Britain

China
Most populous country in 1950 and today.

Russia

Nigeria

Japan

India

Indonesia

Australia

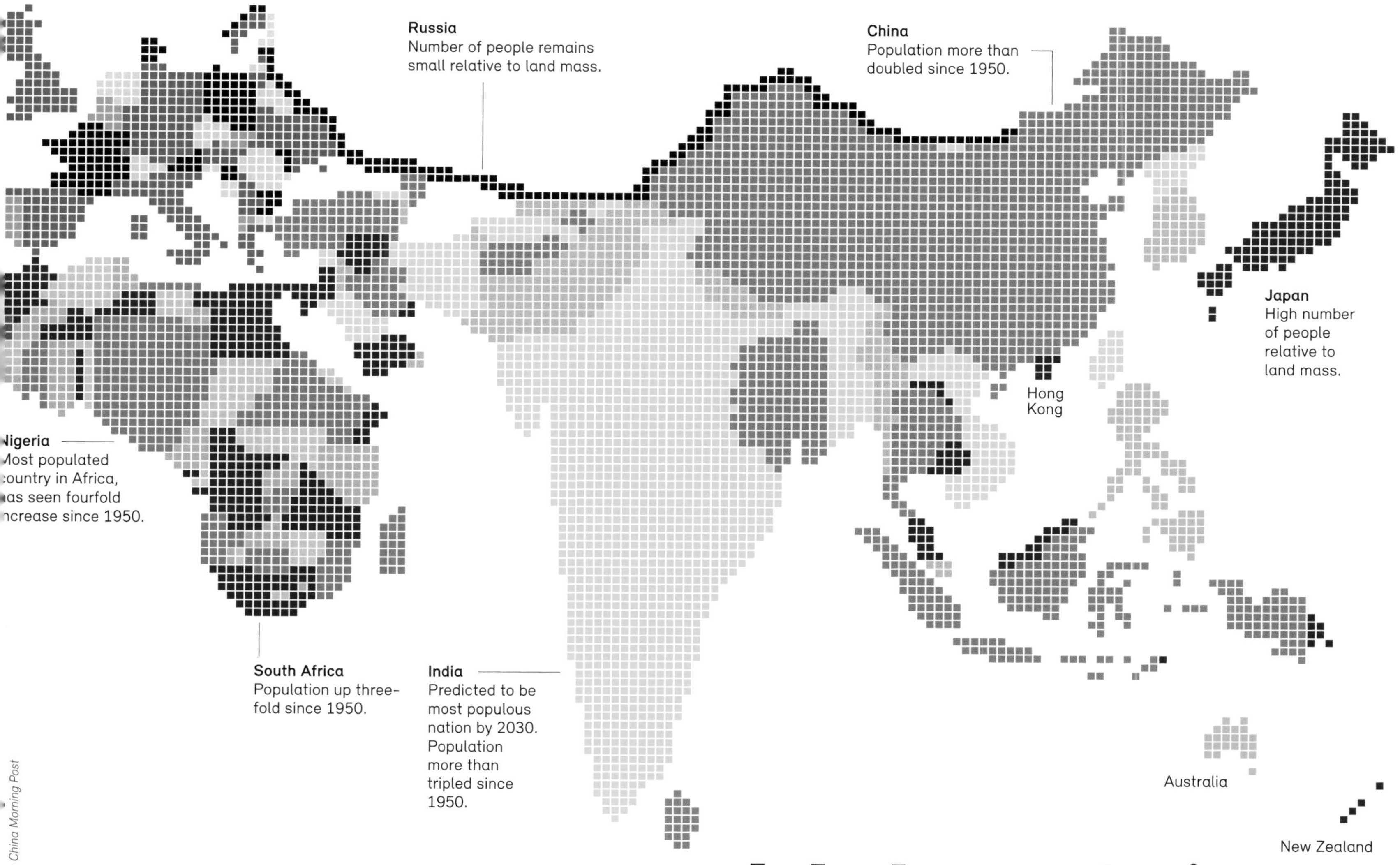

Russia
Number of people remains small relative to land mass.

China
Population more than doubled since 1950.

Japan
High number of people relative to land mass.

Hong Kong

Nigeria
Most populated country in Africa, has seen fourfold increase since 1950.

South Africa
Population up three-fold since 1950.

India
Predicted to be most populous nation by 2030. Population more than tripled since 1950.

Australia

New Zealand

Global Population

LIVING TOGETHER   41

Published: *South China Morning Post*

Bodies Movin'

Where do we want to live?

Since human existence began, man-made boundaries have never been able to hold populations in for long. Even in the millennia before high-speed long-distance travel, the human urge to expand and conquer was irresistible, and families trailed hundreds of miles along dusty roads that stretched endlessly across continents. Or else people uprooted and piled onto dangerous ships to seek fertile lands elsewhere. And yet never before have so many people lived so far from their homes than they do now. The reasons they migrate vary, but the main ones remain constant: curiosity, pursuit of happiness, and political or social oppression. This data flow merely shows the current major immigration patterns to and from various nations, with the world's itchy-footed masses represented by thick arching lines. The world's current top five migrant destinations are the United States, Russia, Germany, Saudi Arabia, and Canada.

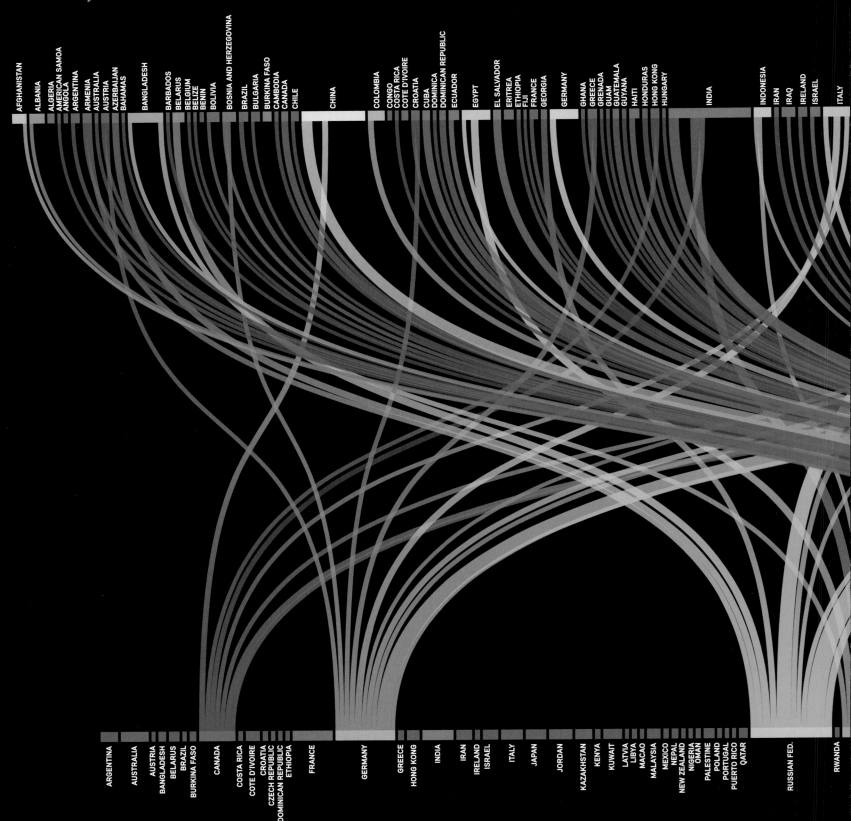

Design: MakingUse – Carlo Zapponi
Published: www.peoplemov.in

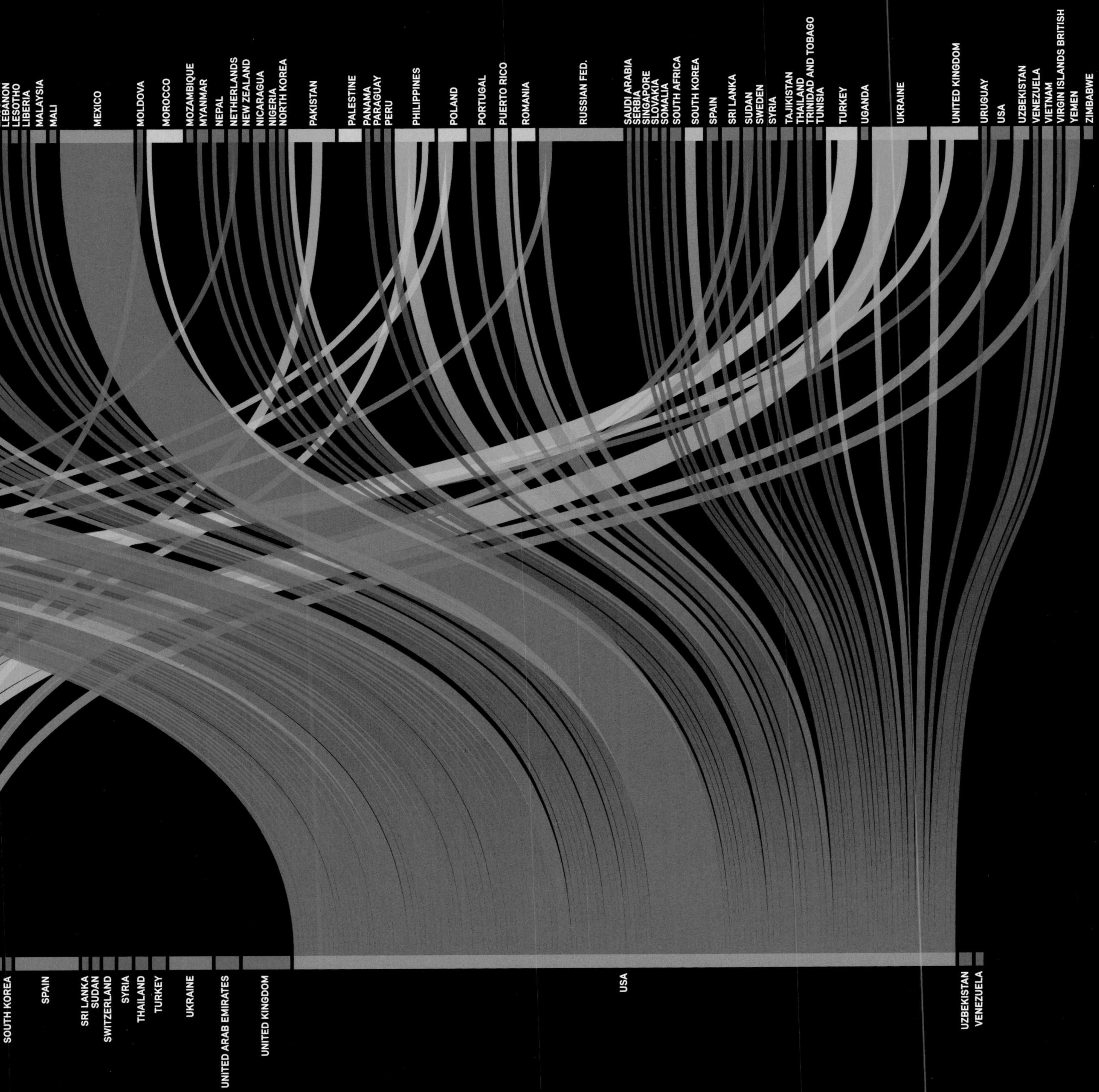
LEBANON
LESOTHO
LIBERIA
MALAYSIA
MALI
MEXICO
MOLDOVA
MOROCCO
MOZAMBIQUE
MYANMAR
NEPAL
NETHERLANDS
NEW ZEALAND
NICARAGUA
NIGERIA
NORTH KOREA
PAKISTAN
PALESTINE
PANAMA
PARAGUAY
PERU
PHILIPPINES
POLAND
PORTUGAL
PUERTO RICO
ROMANIA
RUSSIAN FED.
SAUDI ARABIA
SERBIA
SINGAPORE
SLOVAKIA
SOMALIA
SOUTH AFRICA
SOUTH KOREA
SPAIN
SRI LANKA
SUDAN
SWEDEN
SYRIA
TAJIKISTAN
THAILAND
TRINIDAD AND TOBAGO
TUNISIA
TURKEY
UGANDA
UKRAINE
UNITED KINGDOM
URUGUAY
USA
UZBEKISTAN
VENEZUELA
VIETNAM
VIRGIN ISLANDS BRITISH
YEMEN
ZIMBABWE
SOUTH KOREA
SPAIN
SRI LANKA
SUDAN
SWITZERLAND
SYRIA
THAILAND
TURKEY
UKRAINE
UNITED ARAB EMIRATES
UNITED KINGDOM
USA
UZBEKISTAN
VENEZUELA

Migration

Multiplying Metropolises

Humanity is congealing into megacities.

The first city to boast a million inhabitants was probably ancient Rome. But after the capital of the Western world moved east to Constantinople and the eventual collapse of the Empire, no city expanded to that size again until the nineteenth century, when London claimed the title. London was also the first city to boast five million inhabitants, the accepted benchmark we use to define the "megacity." In 1900, the British capital was the only one in the world — now, there are 63 megacities, and they are all displayed here on these pages.

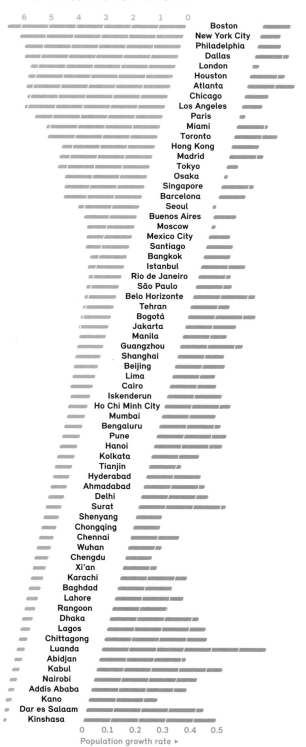

WEALTH VS. GROWTH
◂ GDP per capita 2006/2007 in $1,000, purchasing power parity (PPP) adjusted

Population growth rate ▸ 2005–2010 in percent

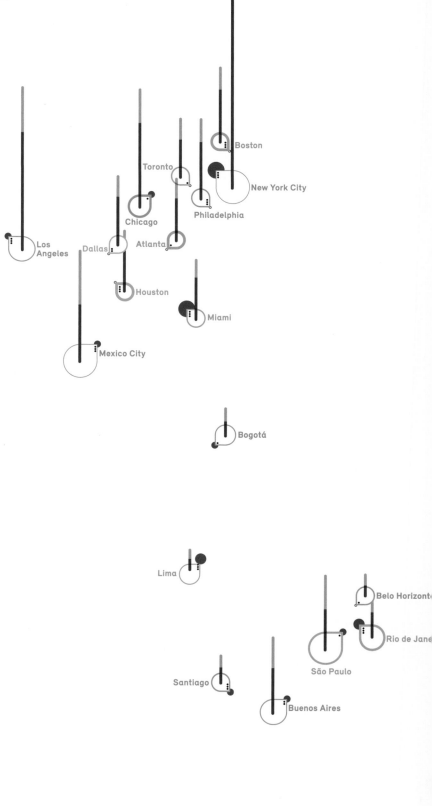

Agglomeration size 2007 2025 — 5 10 20 35 Mio.

Annual population growth 2005–2010 0 — 6

44

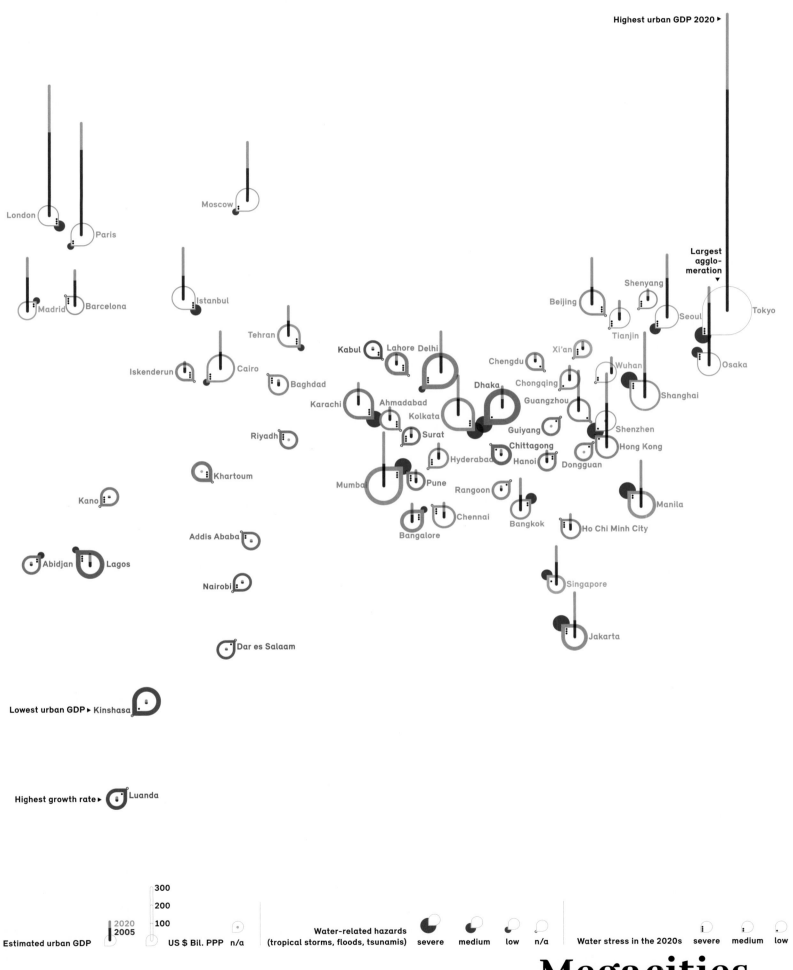

Megacities
LIVING TOGETHER

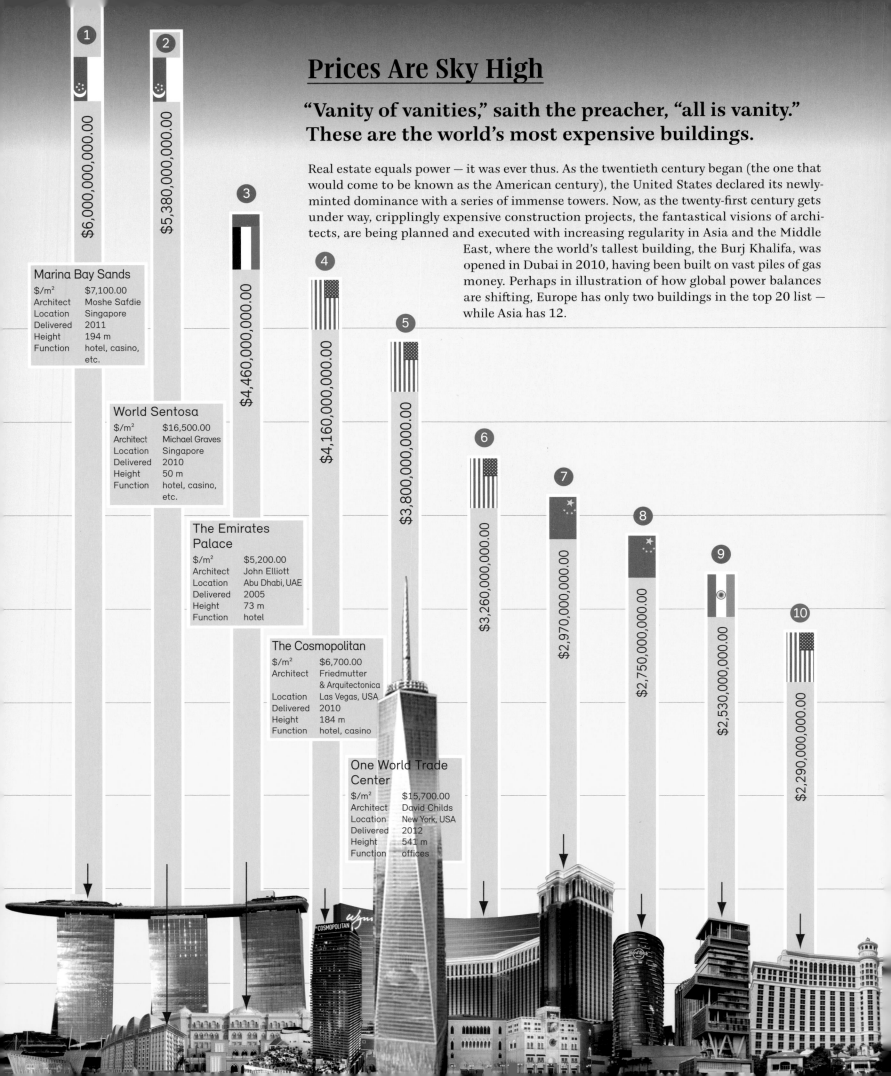

Prices Are Sky High

"Vanity of vanities," saith the preacher, "all is vanity." These are the world's most expensive buildings.

Real estate equals power — it was ever thus. As the twentieth century began (the one that would come to be known as the American century), the United States declared its newly-minted dominance with a series of immense towers. Now, as the twenty-first century gets under way, cripplingly expensive construction projects, the fantastical visions of architects, are being planned and executed with increasing regularity in Asia and the Middle East, where the world's tallest building, the Burj Khalifa, was opened in Dubai in 2010, having been built on vast piles of gas money. Perhaps in illustration of how global power balances are shifting, Europe has only two buildings in the top 20 list — while Asia has 12.

1 — $6,000,000,000.00

Marina Bay Sands
$/m²: $7,100.00
Architect: Moshe Safdie
Location: Singapore
Delivered: 2011
Height: 194 m
Function: hotel, casino, etc.

2 — $5,380,000,000.00

World Sentosa
$/m²: $16,500.00
Architect: Michael Graves
Location: Singapore
Delivered: 2010
Height: 50 m
Function: hotel, casino, etc.

3 — $4,460,000,000.00

The Emirates Palace
$/m²: $5,200.00
Architect: John Elliott
Location: Abu Dhabi, UAE
Delivered: 2005
Height: 73 m
Function: hotel

4 — $4,160,000,000.00

The Cosmopolitan
$/m²: $6,700.00
Architect: Friedmutter & Arquitectonica
Location: Las Vegas, USA
Delivered: 2010
Height: 184 m
Function: hotel, casino

5 — $3,800,000,000.00

One World Trade Center
$/m²: $15,700.00
Architect: David Childs
Location: New York, USA
Delivered: 2012
Height: 541 m
Function: offices

6 — $3,260,000,000.00

7 — $2,970,000,000.00

8 — $2,750,000,000.00

9 — $2,530,000,000.00

10 — $2,290,000,000.00

6 Wynn Resort
$/m² $3,700.00
Architect Butler/Ashworth & Jerde Partnership
Location Las Vegas, USA
Delivered 2005
Height 187 m
Function hotel, casino

7 Venetian Macau
$/m² $3,000.00
Architect Aedas
Location Macau, China
Delivered 2005
Height 225 m
Function hotel, casino

8 City of Dreams
$/m² $8,900.00
Architect Arquitectonica
Location Macau, China
Delivered 2009
Height 150 m
Function hotel, casino

9 Antilla
$/m² $68,000.00
Architect Perkins & Will
Location Mumbai, India
Delivered 2010
Height 173 m
Function private home

10 Bellagio Hotel
$/m² $5,100.00
Architect De Ruyter Butler & Atlandia
Location Las Vegas, USA
Delivered 1998
Height 151 m
Function hotel, casino

11 Princess Tower
$/m² $14,500.00
Architect E. Adnan Saffarini
Location Dubai, UAE
Delivered 2012
Height 414 m
Function residential

12 Bank of China
$/m² $15,500.00
Architect I.M. Pei & Partners
Location Hong Kong, China
Delivered 1990
Height 367 m
Function office

13 The Palazzo
$/m² $3,200.00
Architect HKS Inc.
Location Las Vegas, USA
Delivered 2007
Height 196 m
Function hotel, casino

14 Taipei 101
$/m² $10,300.00
Architect C.Y. Lee & Partners
Location Taipei, Taiwan
Delivered 2004
Height 509 m
Function retail, communication, restaurant

15 Wembley Stadium
$/m² no data
Architect Foster and Partners
Location London, UK
Delivered 2006
Height 133 m
Function stadium

16 Yankee Stadium
$/m² $13,300.00
Architect Populous
Location New York, USA
Delivered 2009
Height 61 m
Function stadium

17 ECB Headquater
$/m² $8,500.00
Architect Coop Himmelb(l)au
Location Frankfurt, Germany
Delivered 2013
Height 217 m
Function offices

18 Burj Khalifa
$/m² $5,000.00
Architect Adrian Smith (SOM)
Location Dubai, UAE
Delivered 2010
Height 828 m
Function hotel, restaurant, residential

19 World Finance Center
$/m² $3,600.00
Architect Kohn Pedersen Fox
Location Shanghai, China
Delivered 2008
Height 494 m
Function office, museum, hotel, etc.

20 Triple One
$/m² $4,300.00
Architect Renzo Piano
Location Yongsan, S. Korea
Delivered 2016
Height 620 m
Function office

● Asia: 12 out of top 20 buildings; $34.56 billion

● America: 6 out of top 20 buildings; $17.19 billion

● Europe: 2 out of top 20 buildings; $3.41 billion

Top 20 buildings = $57,000,000,000.00 = the GDP of Luxembourg

Skyscrapers

LIVING TOGETHER

Under the City of Love

The wonders of mining too much plaster of Paris.

It turns out it can be dangerous to build one of Europe's most ancient and beautiful capitals on sedimentary rock that happens to be an excellent multipurpose material. "Lescarrières de Paris" ("the quarries of Paris") were probably first dug in the thirteenth century, and now cross and re-cross the foundations of the city for some 280 kilometers. Most of the

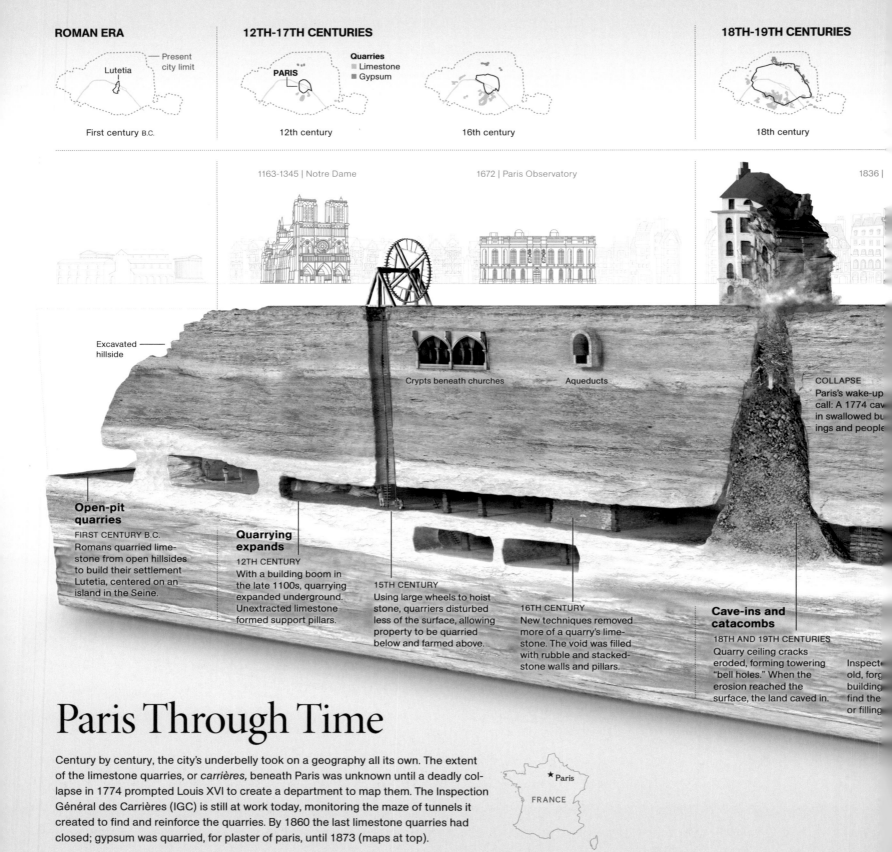

Paris Through Time

Century by century, the city's underbelly took on a geography all its own. The extent of the limestone quarries, or carrières, beneath Paris was unknown until a deadly collapse in 1774 prompted Louis XVI to create a department to map them. The Inspection Général des Carrières (IGC) is still at work today, monitoring the maze of tunnels it created to find and reinforce the quarries. By 1860 the last limestone quarries had closed; gypsum was quarried, for plaster of paris, until 1873 (maps at top).

tunnels have been blocked off from casual explorers, though 1.7 kilometers have become one of Paris's most famous tourist attractions — the collections of ancient bones stacked in the catacombs. The reason Paris lies on such a precarious spot is that in prehistoric times it was under the sea, part of what geologists call the Paris Basin. The sea finally receded toward the end of the Cretaceous period, around 66 million years ago — roughly the same time that the dinosaurs bowed out of history — and it left rich layers of sedimentary minerals. When humans first began excavating the rich seams in medieval times, they did not suspect that their toil would literally undermine the city centuries later.

Underworld

LIVING TOGETHER

Going Underground

Reclaim the streets — banish the automobile into tunnels.

Apart from vertical gardens and mobile trees, clearing the streets of the choking traffic jams would be a good way to make cities more pleasant to live in — making way for us all to roam free in expanding pedestrian and green areas. City authorities around the world have been noting the relief that traffic tunnels bring to a city for over a decade. Not only do they reduce CO_2 emissions, and pollution generally, they also reduce noise and so raise the quality of life for the townies.

More socialization
To redevelop urban space, areas could be set up to have lunch outdoors or to work with the computer via Wi-Fi. Even parking lots can be "regained" for play and relaxation areas. There are those who are already doing so, see the website parkinday.org.

Faster buses
With traffic underground, on the surface the streets narrow to make room for sidewalks, bike routes, and lanes for public transport only.

THE REBIRTH OF STREETS
If the lines of communication are relocated underground (below, a project that could accomplish this revolution), streets can be transformed into places of socialization (in the illustration on the right).

The fumes inside the tunnel are treated before being expelled.

30 – 40 meters below ground level

FROM BOSTON TO MILAN TEN "INVISIBLE" WORKS
Throughout the world the interest in underground urban infrastructures is growing. The State of New York, for example, will spend in the next few years 20 billion dollars in tunnels for subways and roads.

Boston — BIG DIG
YEAR	LENGTH	WIDTH
1991-2007	5.6 km	30.5 × 14 m

Big Dig is the unofficial name of the *Central Artery Tunnel Project CA/T*, a megaproject that has reconverted the Boston stretch of the Interstate highway 93 into a tunnel.

Kuala Lumpur — SMART TUNNEL
YEAR	LENGTH	DIAMETER
2003-07	4 km	13.2 m

The *Stormwater Management and Road Tunnel* is a 4 km road tunnel under which a drainage canal flows to prevent flooding.

AROUND THE WORLD BY SUBWAY

- Number of journeys per year
- Total length (km)
- 1900 Year of construction

The main metropolitan networks in the United States and around the world. Projects to amplify and extend are commonplace in many local authorities.

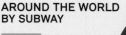

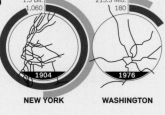

City	Journeys/yr	Length (km)	Year
NEW YORK	1.5 Bil.	1,060	1904
WASHINGTON	215.3 Mio.	180	1976
CHICAGO	190.3 Mio.	400	1947
BOSTON	187.5 Mio.	100	1897
SAN FRANCISCO	100 Mio.	170	1972
TOKYO	3 Bil.	320	1927
MOSCOW	2.5 Bil.	290	1935
SEOUL	1.7 Bil.	280	1974
MEXICO CITY	1.4 Bil.	200	1969
PARIS	1.4 Bil.	210	1900

CITIES DREAMT OF AND NEVER BUILT
The ideal city has always been at the center of interest of philosophers, artists, and urban planners. But it is with the Industrial Revolution that the redesign of urban areas becomes increasingly urgent.

Futurama
At the Universal Exposition in New York in 1939, the "Futurama" exhibition (sponsored by General Motors) imagined a future of megacities connected with the suburbs by an extensive network of fast highways.

Le Corbusier
The sketches of the Swiss Le Corbusier for a futuristic "Ville Contemporaine" envisaged, for traffic, fast highways positioned on different levels in densely populated cities.

Design: Francesco Franchi, Laura Cattaneo

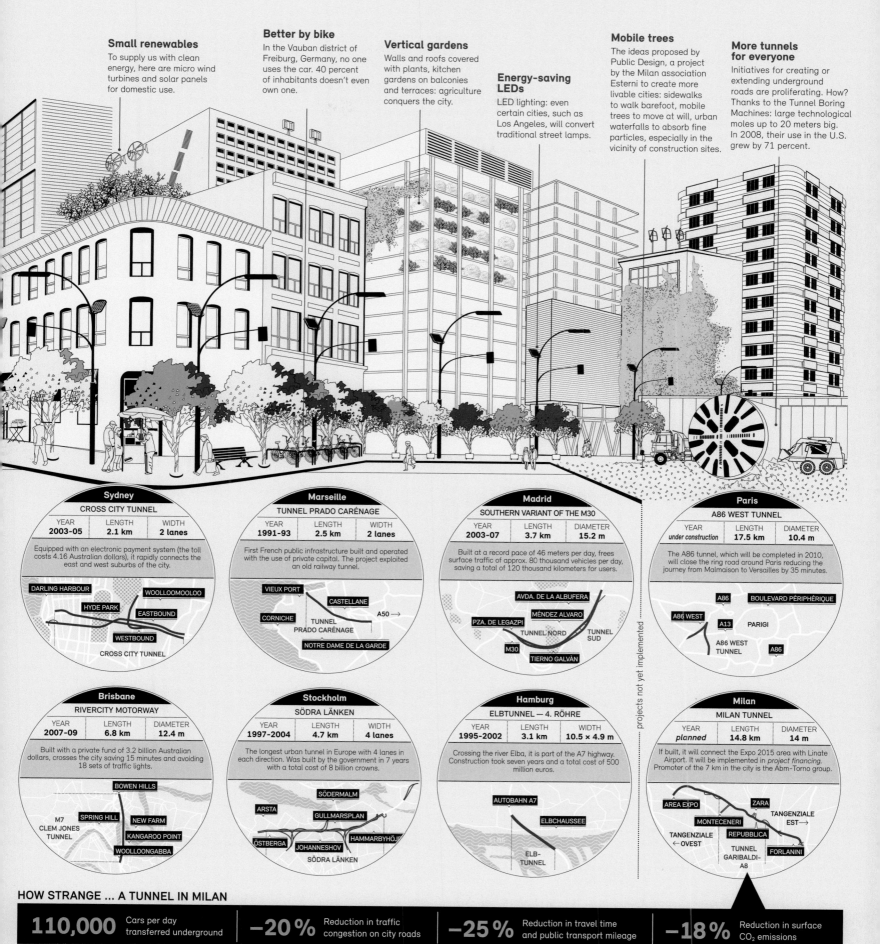

Small renewables
To supply us with clean energy, here are micro wind turbines and solar panels for domestic use.

Better by bike
In the Vauban district of Freiburg, Germany, no one uses the car. 40 percent of inhabitants doesn't even own one.

Vertical gardens
Walls and roofs covered with plants, kitchen gardens on balconies and terraces: agriculture conquers the city.

Energy-saving LEDs
LED lighting: even certain cities, such as Los Angeles, will convert traditional street lamps.

Mobile trees
The ideas proposed by Public Design, a project by the Milan association Esterni to create more livable cities: sidewalks to walk barefoot, mobile trees to move at will, urban waterfalls to absorb fine particles, especially in the vicinity of construction sites.

More tunnels for everyone
Initiatives for creating or extending underground roads are proliferating. How? Thanks to the Tunnel Boring Machines: large technological moles up to 20 meters big. In 2008, their use in the U.S. grew by 71 percent.

Sydney — CROSS CITY TUNNEL
YEAR 2003-05 | LENGTH 2.1 km | WIDTH 2 lanes
Equipped with an electronic payment system (the toll costs 4.16 Australian dollars), it rapidly connects the east and west suburbs of the city.

Marseille — TUNNEL PRADO CARÉNAGE
YEAR 1991-93 | LENGTH 2.5 km | WIDTH 2 lanes
First French public infrastructure built and operated with the use of private capital. The project exploited an old railway tunnel.

Madrid — SOUTHERN VARIANT OF THE M30
YEAR 2003-07 | LENGTH 3.7 km | DIAMETER 15.2 m
Built at a record pace of 46 meters per day, frees surface traffic of approx. 80 thousand vehicles per day, saving a total of 120 thousand kilometers for users.

Paris — A86 WEST TUNNEL
YEAR under construction | LENGTH 17.5 km | DIAMETER 10.4 m
The A86 tunnel, which will be completed in 2010, will close the ring road around Paris reducing the journey from Malmaison to Versailles by 35 minutes.

Brisbane — RIVERCITY MOTORWAY
YEAR 2007-09 | LENGTH 6.8 km | DIAMETER 12.4 m
Built with a private fund of 3.2 billion Australian dollars, crosses the city saving 15 minutes and avoiding 18 sets of traffic lights.

Stockholm — SÖDRA LÄNKEN
YEAR 1997-2004 | LENGTH 4.7 km | WIDTH 4 lanes
The longest urban tunnel in Europe with 4 lanes in each direction. Was built by the government in 7 years with a total cost of 8 billion crowns.

Hamburg — ELBTUNNEL – 4. RÖHRE
YEAR 1995-2002 | LENGTH 3.1 km | WIDTH 10.5 × 4.9 m
Crossing the river Elba, it is part of the A7 highway. Construction took seven years and a total cost of 500 million euros.

Milan — MILAN TUNNEL (projects not yet implemented)
YEAR planned | LENGTH 14.8 km | DIAMETER 14 m
If built, it will connect the Expo 2015 area with Linate Airport. It will be implemented in *project financing*. Promoter of the 7 km in the city is the Abm-Torno group.

HOW STRANGE ... A TUNNEL IN MILAN

| 110,000 Cars per day transferred underground | −20 % Reduction in traffic congestion on city roads | −25 % Reduction in travel time and public transport mileage | −18 % Reduction in surface CO_2 emissions |

Ron Herron
In the sixties, the British architect Ron Herron, from the avant-garde Archigram school, proposed the "Walking city" in which robotic structures roam the city to meet the needs and desires of the inhabitants.

Antonio Sant'Elia
The Italian architect Antonio Sant'Elia in 1914 imagined the "New Town," a project in which huge vertical conglomerates accommodate domestic and industrial structures, along with service and transport centers.

Buckminster Fuller
Richard Buckminster Fuller, an American architect (1895-1983), with the "Cloud Nine" project dreamed of putting cities into huge geodesic spheres floating above the clouds. The objective? Prevent exploitation of the land.

Metro Systems
LIVING TOGETHER

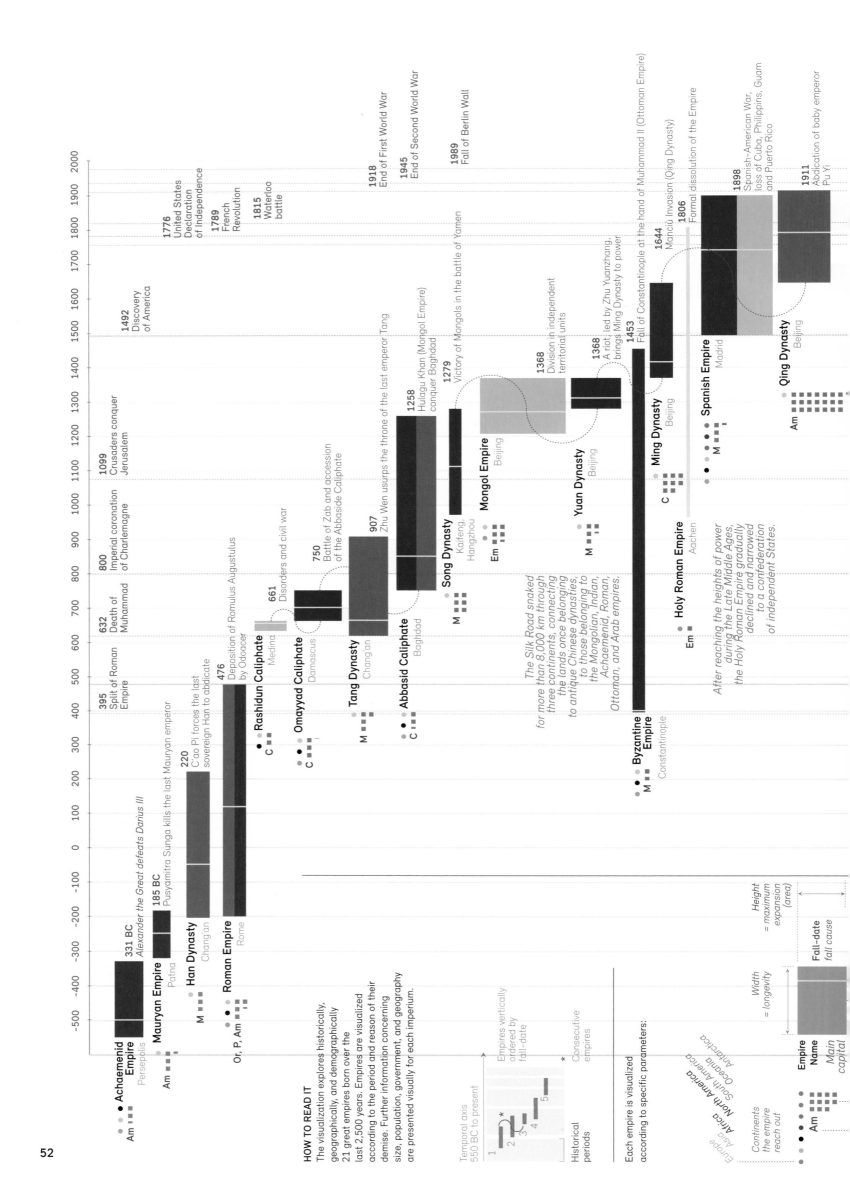

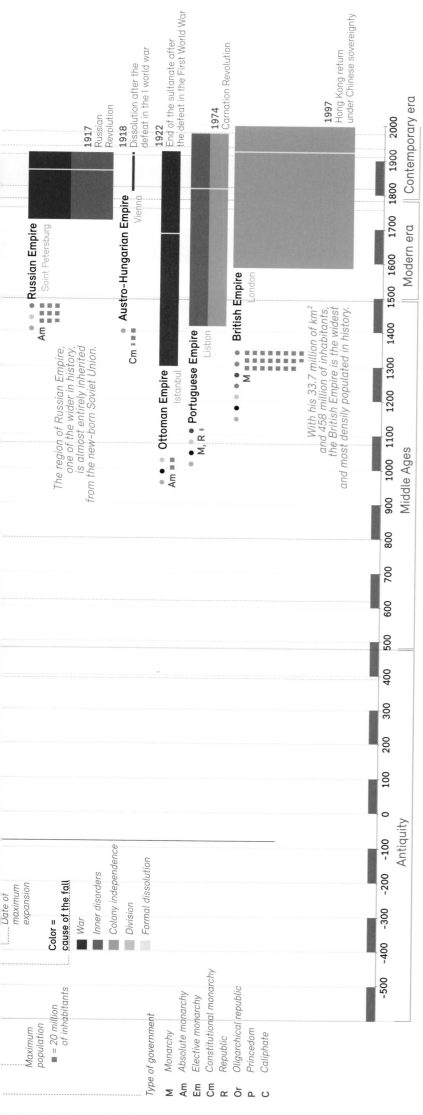

Age of Empires

Everybody wants to rule the world.

They all had their fifteen seconds of fame: the Romans and the Ottomans, the Chinese, the Britons, and a couple more. Empires have defined our history. They are the framework for progress and oppression. History tells us that emperors hardly ever leave their superior role voluntarily, but also that each empire has an expiration date. Right now we are witnessing this after the decline of the Soviet empire. China is stepping up to become the Asian counterpart to the American superpower, which could prove that even in the world of empires there is room for revival.

Empires

LIVING TOGETHER

Built-in Conflict

The United States Constitution designed a government whose key was complexity — where conflict and compromise were part of the structure.

Building a nation from scratch in times of global upheaval is not an easy thing to do. Often a nation state will pass through several other forms of government — usually involving some kind of dictatorship, either by monarchy or other means, before it gets to enjoy the fruits of a democratic republic. The United States was fortunate in that it managed to get through this part in a relatively short space of time. Attracting some of Europe's leading dissenters, free-thinkers, and revolutionaries, the thirteen colonies incubated a young, vibrant political culture early on. New ways to organize democracy, taxation, poor relief, and oversight were already being debated before the American Revolution even began. As a consequence, when the time came to draw up a constitution for a new republic, the law was already a highly professionalized profession — as the careers of men like John Adams and Thomas Jefferson showed — and the United States of America had the right men for the job. As a result, the U.S. government's founding document was a meticulously thought-out blueprint in which power was carefully balanced between different entities, laid out in detail in the graphics on this page. Though the rights of many people were completely ignored — including women, slaves, Native Americans — its ideals inspired others around the world.

THREE KEY IDEAS FOR UNDERSTANDING POLITICS

POLITICS IS CONFLICTUAL
Conflict and compromise are natural parts of politics.

Political conflict over issues like the national debt, abortion, and health care reflect disagreements among the American people and often require compromises within government.

POLITICAL PROCESS MATTERS
How political conflicts are resolved is important.

Elections determine who represents citizens in government. Rules and procedures determine who has power in Congress and other branches of government.

POLITICS IS EVERYWHERE
What happens in government affects our lives in countless ways.

Policies related to jobs and the economy, food safety and nutrition, student loans, and many other areas shape our everyday lives. We see political information in the news and encounter political situations in many areas of our lives.

VERSIONS OF FEDERALISM

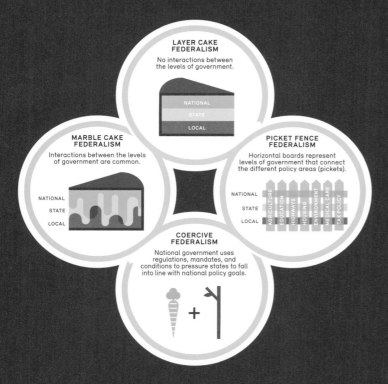

LAYER CAKE FEDERALISM
No interactions between the levels of government.

MARBLE CAKE FEDERALISM
Interactions between the levels of government are common.

PICKET FENCE FEDERALISM
Horizontal boards represent levels of government that connect the different policy areas (pickets).

COERCIVE FEDERALISM
National government uses regulations, mandates, and conditions to pressure states to fall into line with national policy goals.

CHECKS AND BALANCES

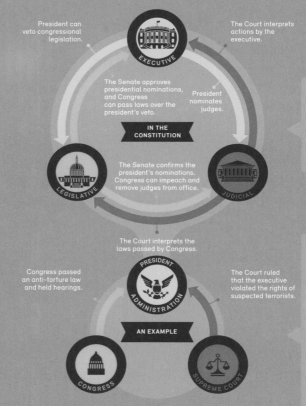

President can veto congressional legislation.

The Court interprets actions by the executive.

The Senate approves presidential nominations, and Congress can pass laws over the president's veto.

President nominates judges.

IN THE CONSTITUTION

The Senate confirms the president's nominations. Congress can impeach and remove judges from office.

The Court interprets the laws passed by Congress.

Congress passed an anti-torture law and held hearings.

The Court ruled that the executive violated the rights of suspected terrorists.

AN EXAMPLE

IN THE CONSTITUTION

If one branch tries to assert too much power, the other branches have certain key powers that allow them to fight back and restore the balance. (In addition to the powers noted in the diagram, Congress also can impeach the president and remove him or her from office.

AN EXAMPLE

During the war on terror, concerns arose that President Bush and the executive branch had assumed too much power— especially the unilateral power to disregard due process rights for suspected terrorists. In response, Congress checked the president by passing an anti-torture law and holding hearings to determine if the Department of Justice had acted illegally. The Supreme Court limited the president's power by ruling that the executive branch violated the rights and liberties of suspected terrorists, but the president and Congress responded to limit the scope of the Court's ruling.

THE PRESIDENT AS HEAD OF THE EXECUTIVE BRANCH

THE PRESIDENT

↓

THE WHITE HOUSE STAFF

↓

EXECUTIVE OFFICE OF THE PRESIDENT

White House Office
Office of Management & Budget
Council of Economic Advisers
National Security Council
Office of National Drug Control Policy
Office of the U.S. Trade Representative

Council on Environmental Quality
Office of Science & Tech. Policy
Office of Policy Development
Office of Administration
Vice President

INDEPENDENT AGENCIES & GOV. CORPORATIONS

Examples:
Federal Election Commission
Federal Trade Commission
Social Security Administration
National Transportation Safety Board

THE CABINET

Dept. of Housing & Urban Development
Dept. of the Interior
Dept. of Commerce
Dept. of Labor
Dept. of Education
Dept. of Transportation
Dept. of Energy
Dept. of Veterans Affairs
Dept. of Justice
Dept. of Defense
Dept. of State
Dept. of Homeland Security
Dept. of Health & Human Services
Dept. of the Treasury
Dept. of Agriculture

NOMINATING PRESIDENTIAL CANDIDATES

OPEN PRIMARIES — Open to voters from any political party and independents

CLOSED PRIMARIES — Only voters registered with a party vote

CAUCUSES — Party members meet in groups to select delegates

↓

SELECT DELEGATES TO NATIONAL CONVENTION

Republican Party — States can divide delegates or give all to the winning candidate.

Democratic Party — The state's delegates are divided up proportionately.

↓

NATIONAL NOMINATING CONVENTIONS

Delegates from all states attend the national convention, where they vote for the party's presidential and vice presidential nominees, based on the primary and caucus results. Superdelegates—important party leaders—also vote at the convention.

THE BUDGET PROCESS

The president submits budget to Congress. — **1ST MONDAY IN FEB.**

FEB. 15TH — CBO issues budget and economic outlook report.

Other committees with budgetary responsibilities submit "views and estimates" to budget committees. — **WITHIN SIX WEEKS OF PRESIDENT'S SUBMISSION**

House Budget Committee creates its budget resolution and the House votes on it. — **EARLY APRIL** — Senate Budget Committee creates its budget resolution and the Senate votes on it.

Budget Conference Committee reconciles House and Senate versions of the budget resolution.

House votes on conference version. — **BY APRIL 15TH** — Senate votes on conference version.

APPROPRIATIONS

After both houses approve the budget resolution, appropriations committees draft legislation authorizing expenditures to the relevant agencies. Each appropriations bill must be passed by both houses and signed into law by the president. If this process is not completed by October 1st, and no temporary measure (a "continuing resolution") is in place, the government will shut down.

Start of the fiscal year. — **OCT 1ST**

THE POLICY PROCESS

PROBLEM RECOGNITION AND DEFINITION — A large proportion of elderly living in poverty.

AGENDA SETTING — As poverty among elderly worsened during the Great Depression, solving the problem became a priority as part of FDR's New Deal agenda.

DELIBERATION AND FORMULATION — The proposal for Social Security came from FDR's President's Committee on Economic Security.

ENACTMENT — Social Security was passed by Congress and signed into law by the president in 1935.

IMPLEMENTATION — The Social Security Administration implements the policy.

EVALUATION — Social Security has been evaluated at various times, most recently by presidential commissions on reform.

POSSIBLE MODIFICATION, TERMINATION, OR EXPANSION OF POLICY — Social Security has been modified and expanded several times, most recently with the 1983 amendments.

The American Way

LIVING TOGETHER 55

The Highest Office

Where the most powerful person in the world hangs their hat.

Building a new seat of government is a careful balancing act. On the one hand, you want to project your nation's power and standing. Then again, it's meant to be the office and home of a servant of the people. You want an aura of strength, a home fit for a man or woman of the people. If you are creating a new seat of government, you probably wouldn't necessarily put it in a cramped side street like London's Downing Street. 1600 Pennsylvania Avenue is probably the most famous home address in the world.

The residence and workplace of the American President is a neoclassical mansion, popular among southern landowners of the time. It was originally referred to as "The President's Palace" or "President's Mansion" before it gained its more commonly used nickname.

Area of the White House

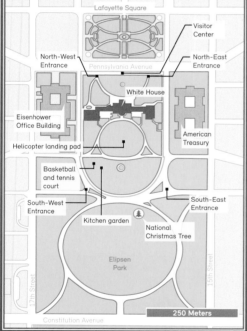

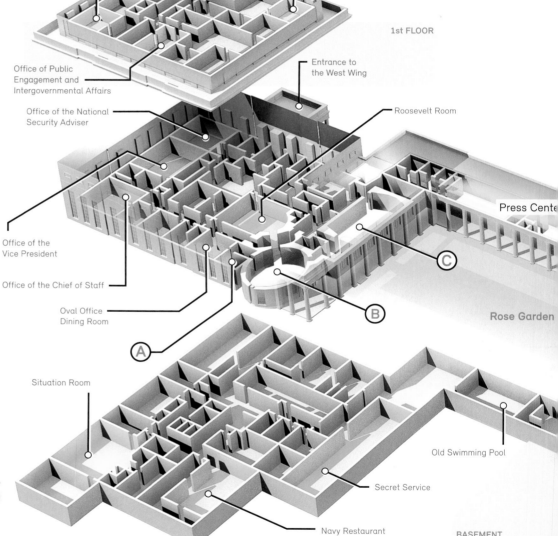

West Wing — The real center of power. In addition to the President, all of the important executive staff are based here.

Profile – The White House

Official seat and residence of the President of the USA
Address: Washington, D.C.
1600 Pennsylvania Avenue
USA

Rooms	132
Bathrooms	35
Doors	412
Windows	147
Staircases	8
Elevators	3

Height: 21.33 m
Total floorspace: 16,764 m²
Built: 1800–1817
Designed by: James Hoban (1762–1831)

Ⓐ The President's Room
In the past, the room was used as a workplace for secretaries or the President himself. Today the room is used as a dining room.

Ⓑ The Oval Office
Probably the most famous room of the White House. The **Oval Office**, which is located in the West Wing, is the main workplace of the President. The windows are made of bullet-proof glass. Each President may redecorate the room as he wishes.

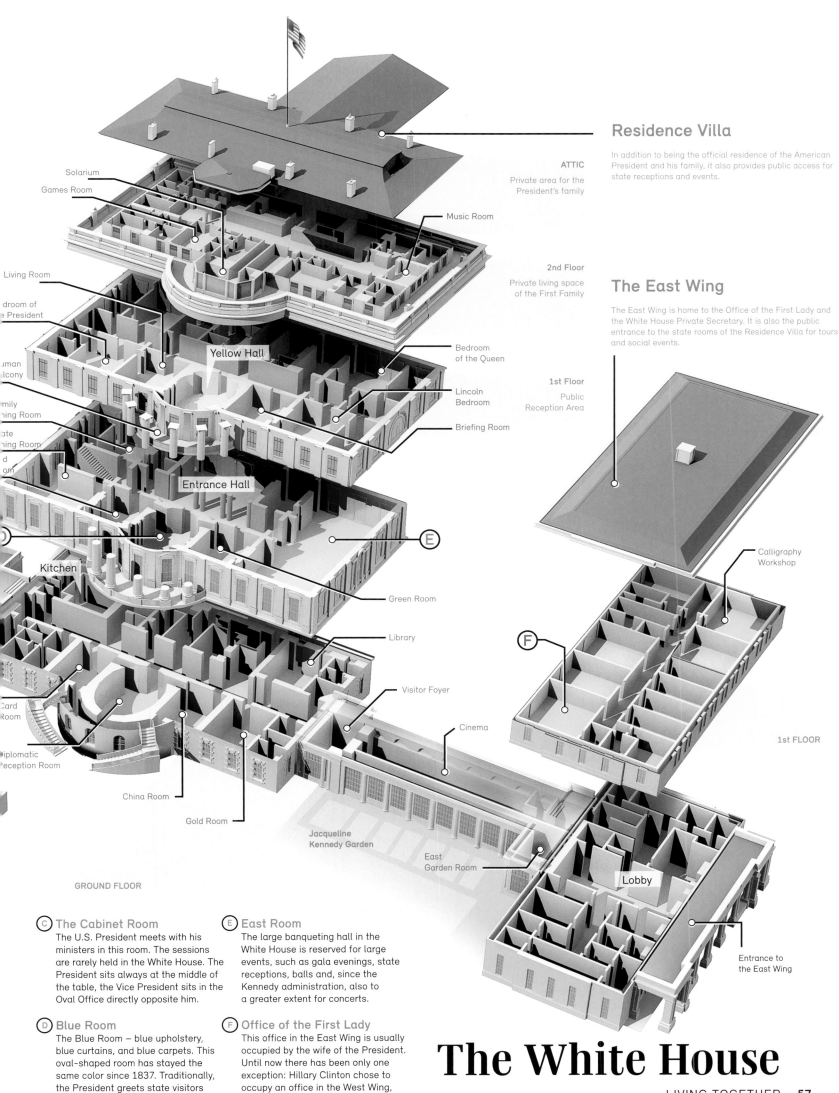

The White House

LIVING TOGETHER

Residence Villa

In addition to being the official residence of the American President and his family, it also provides public access for state receptions and events.

ATTIC — Private area for the President's family

2nd Floor — Private living space of the First Family

1st Floor — Public Reception Area

The East Wing

The East Wing is home to the Office of the First Lady and the White House Private Secretary. It is also the public entrance to the state rooms of the Residence Villa for tours and social events.

C The Cabinet Room
The U.S. President meets with his ministers in this room. The sessions are rarely held in the White House. The President sits always at the middle of the table, the Vice President sits in the Oval Office directly opposite him.

D Blue Room
The Blue Room – blue upholstery, blue curtains, and blue carpets. This oval-shaped room has stayed the same color since 1837. Traditionally, the President greets state visitors and congress representatives here.

E East Room
The large banqueting hall in the White House is reserved for large events, such as gala evenings, state receptions, balls and, since the Kennedy administration, also to a greater extent for concerts.

F Office of the First Lady
This office in the East Wing is usually occupied by the wife of the President. Until now there has been only one exception: Hillary Clinton chose to occupy an office in the West Wing, the power center of the White House.

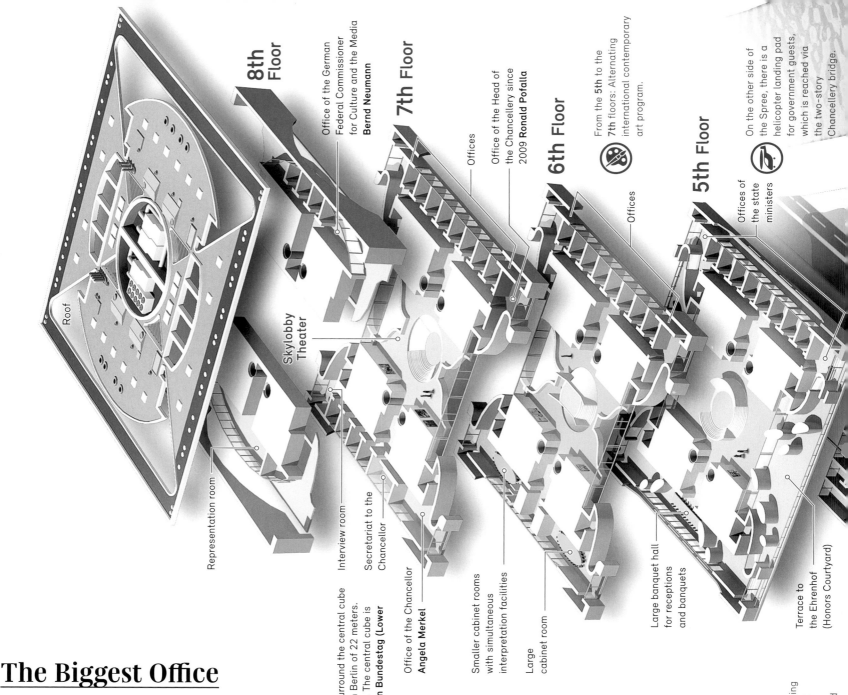

8th Floor
- Roof
- Representation room
- Office of the German Federal Commissioner for Culture and the Media **Bernd Neumann**

7th Floor
- Skylobby Theater
- Interview room
- Secretariat to the Chancellor
- Offices
- Office of the Head of the Chancellery since 2009 **Ronald Pofalla**
- Office of the Chancellor **Angela Merkel**

6th Floor
- Smaller cabinet rooms with simultaneous interpretation facilities
- Large cabinet room
- Offices
- From the 5th to the 7th floors: Alternating international contemporary art program.

5th Floor
- Large banquet hall for receptions and banquets
- Offices of the state ministers
- Terrace to the Ehrenhof (Honors Courtyard)
- On the other side of the Spree, there is a helicopter landing pad for government guests, which is reached via the two-story Chancellery bridge.

The Biggest Office

An imposing house for the most powerful woman in the world.

It's not often that one of the world's major economies needs to build an entirely new base for its head of government from scratch. When the Germans conceived a new office for a chancellor in their newly unified capital in the 1990s, it's safe to say they went big. Located slap bang in the center of Berlin, the Chancellery is the world's largest seat of government, and it was commissioned, designed, and partially constructed during the tenure of the long-serving Chancellor Helmut Kohl — himself of pretty colossal dimensions. On its completion in 2001, accompanied by mutterings about hubris, Gerhard Schröder became the first German chancellor to move in. Though he couldn't match Kohl's girth, Schröder was a larger-than-life leader, a super-confident embodiment of modern, prosperous, unified Germany, much like his new, ultra-modern office. A more austere public figure, Angela Merkel currently occupies the space.

The new building on the banks of the Spree

The two long lines of the side wings in the dimensions of the **Ribbon of Government** surround the central cube of the Chancellery. At 18 meters in height, the building is in line with building height in Berlin of 22 meters. The winter gardens and offices for approximately 450 employees alternate within this. The central cube is 36 meters high and is, therefore, only four meters lower than the cupola of the **German Bundestag (Lower House of Parliament)**.

The Ribbon of Government
The arrangement of the buildings in the government quarter represent a symbolic bridge between the formerly divided halves of the city.

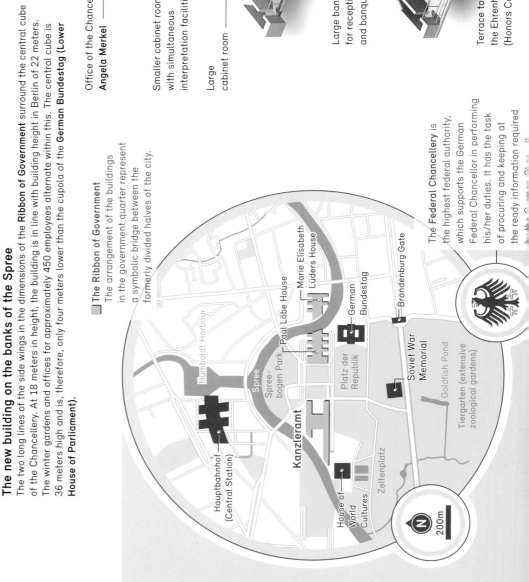

The **Federal Chancellery** is the highest federal authority, which supports the German Federal Chancellor in performing his/her duties. It has the task of procuring and keeping at the ready information required

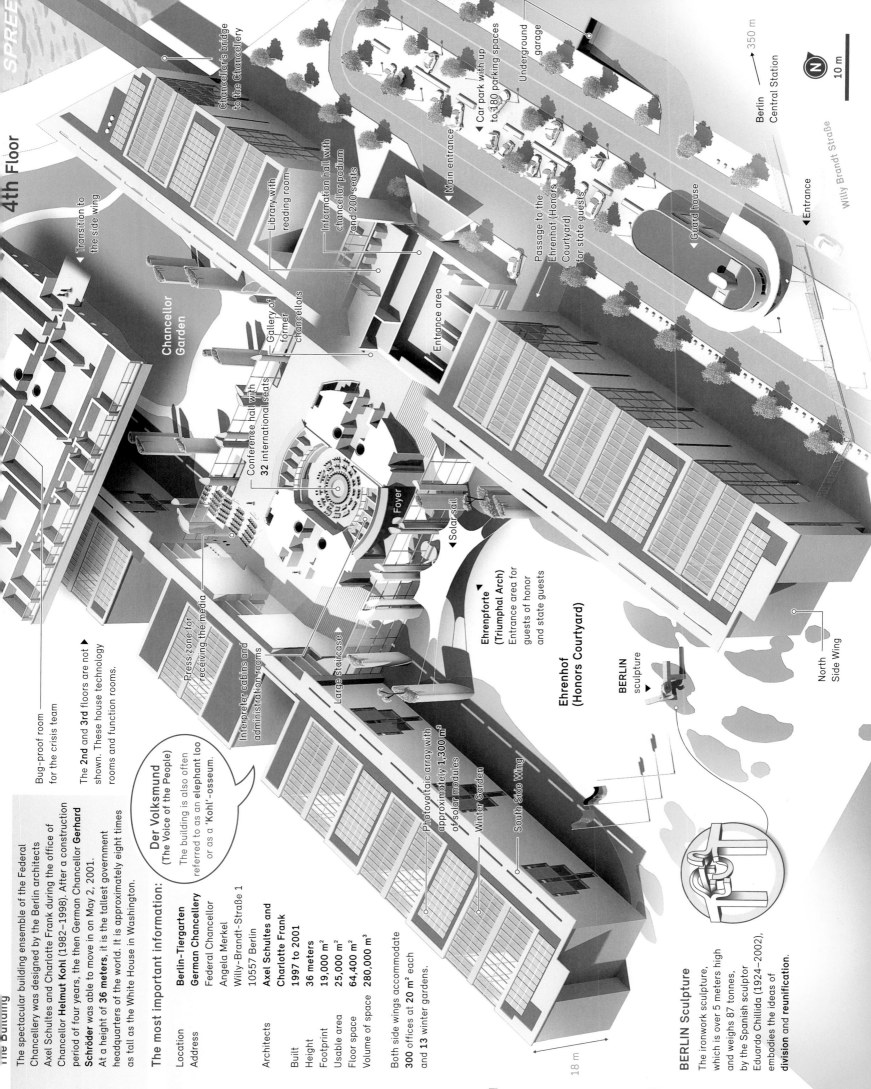

The Kanzleramt

LIVING TOGETHER

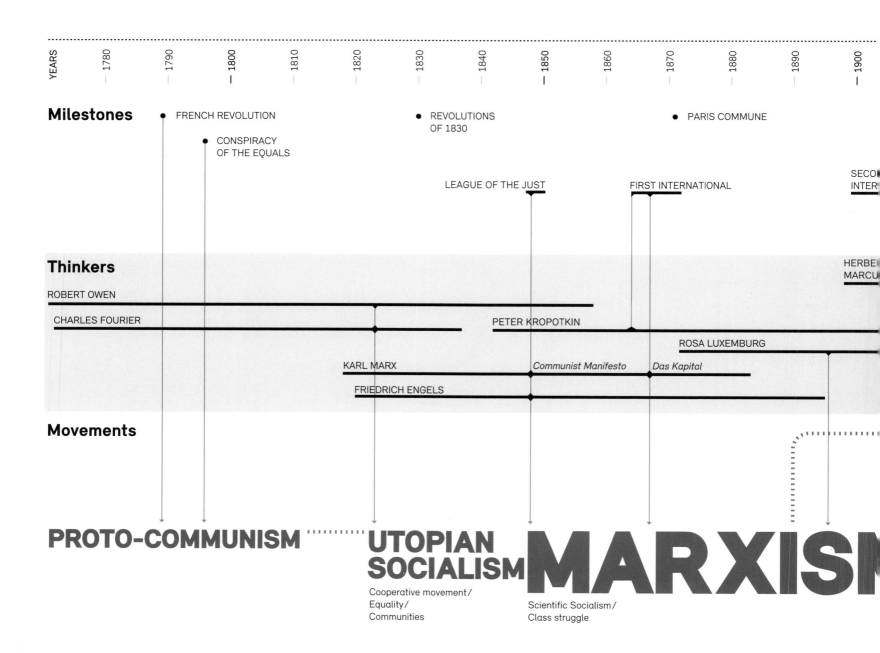

One Party, Many Movements

How communism developed into a world ideology — and is still part of our lives.

In a world where the government offers small handouts for the poor and large bailouts for rich companies, it's odd that many declare socialist ideals out of date. Both Russia and China have largely abandoned the total economic control that marked much of their twentieth-century history, but it would be naïve to think those governments don't still play a major role in running their economies. In the West, meanwhile, the 2008 financial crisis left many nations forced into guaranteeing debts and providing financial injections for banks, and discussing legislation to rein in the worst excesses of market trading. It might not be the kind of socialist communes that the early idealists imagined, but we don't live in a no-holds-barred free market either. This genealogy shows that almost every economy in the world has — to a greater or lesser extent — some vestige of governmental control left in it, even if corporate lobbyists clearly still have too much of an influence on political decisionmaking.

The Left Wing

The Nuclear Family

Because you never know when you might need to destroy the world.

No one knows exactly how many nuclear weapons there are in the world, because, as many eminent scientists and peace activists have pointed out, the major nuclear nations — the United States, Russia, and those belonging to NATO — like to keep their nuke cards close to their chest.

This graphic is an attempt to tally up the number of immediately available nuclear weapons — the ones that are hanging over the world's collective head at any one time. The size of the real arsenal, including those weapons currently decommissioned that could be armed again, is anyone's guess.

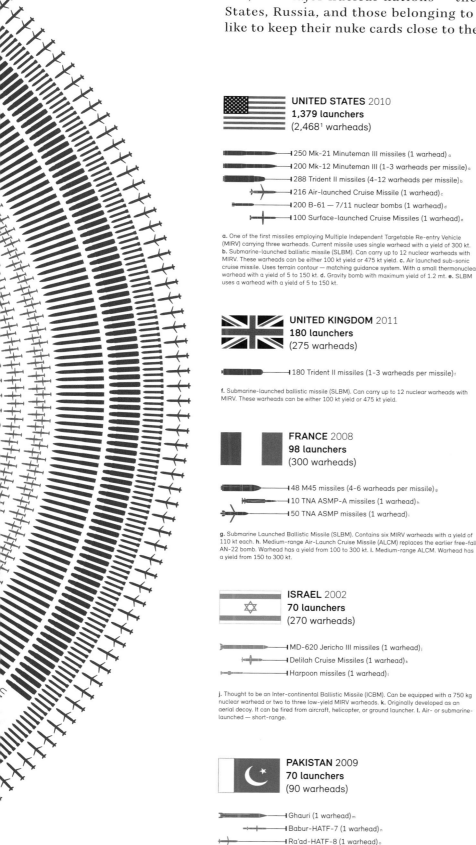

UNITED STATES 2010
1,379 launchers
(2,468[1] warheads)

- 250 Mk-21 Minuteman III missiles (1 warhead) a
- 200 Mk-12 Minuteman III (1-3 warheads per missile) a
- 288 Trident II missiles (4-12 warheads per missile) b
- 216 Air-launched Cruise Missile (1 warhead) c
- 200 B-61 — 7/11 nuclear bombs (1 warhead) d
- 100 Surface-launched Cruise Missiles (1 warhead) e

a. One of the first missiles employing Multiple Independent Targetable Re-entry Vehicle (MIRV) carrying three warheads. Current missile uses single warhead with a yield of 300 kt. **b.** Submarine-launched ballistic missile (SLBM). Can carry up to 12 nuclear warheads with MIRV. These warheads can be either 100 kt yield or 475 kt yield. **c.** Air launched sub-sonic cruise missile. Uses terrain contour — matching guidance system. With a small thermonuclear warhead with a yield of 5 to 150 kt. **d.** Gravity bomb with maximum yield of 1.2 mt. **e.** SLBM uses a warhead with a yield of 5 to 150 kt.

UNITED KINGDOM 2011
180 launchers
(275 warheads)

- 180 Trident II missiles (1-3 warheads per missile) f

f. Submarine-launched ballistic missile (SLBM). Can carry up to 12 nuclear warheads with MIRV. These warheads can be either 100 kt yield or 475 kt yield.

FRANCE 2008
98 launchers
(300 warheads)

- 48 M45 missiles (4-6 warheads per missile) g
- 10 TNA ASMP-A missiles (1 warhead) h
- 50 TNA ASMP missiles (1 warhead) i

g. Submarine Launched Ballistic Missile (SLBM). Contains six MIRV warheads with a yield of 110 kt each. **h.** Medium-range Air-Launch Cruise Missile (ALCM) replaces the earlier free-fall AN-22 bomb. Warhead has a yield from 100 to 300 kt. **i.** Medium-range ALCM. Warhead has a yield from 150 to 300 kt.

ISRAEL 2002
70 launchers
(270 warheads)

- MD-620 Jericho III missiles (1 warhead) j
- Delilah Cruise Missiles (1 warhead) k
- Harpoon missiles (1 warhead) l

j. Thought to be an Inter-continental Ballistic Missile (ICBM). Can be equipped with a 750 kg nuclear warhead or two to three low-yield MIRV warheads. **k.** Originally developed as an aerial decoy. It can be fired from aircraft, helicopter, or ground launcher. **l.** Air- or submarine-launched — short-range.

PAKISTAN 2009
70 launchers
(90 warheads)

- Ghauri (1 warhead) m
- Babur-HATF-7 (1 warhead) n
- Ra'ad-HATF-8 (1 warhead) o

m. Medium-range ballistic missile. Fired from transporter-erected launcher. **n.** Medium-range ground-launched cruise missile. Fired from transporter-erected launcher. **o.** Air-launched cruise missile. Could be launched at sea-based targets as well as land.

RUSSIA 2012
1,286 launchers
(2,430[2] warheads)

- 50 SS-18 M6 Satan (10 warheads per missile) p
- 48 SS-19 M3 Stiletto (6 warheads per missile) q
- 135 SS-25 Sickle (1 warhead per missile) r
- 89 SS-27 Mod. 1/2 (1-6 warheads per missile) s
- 48 SS-N-18 M1 Stingray (1-7 warheads per missile) t
- 96 SS-N-23 M1 (1-4 warheads per missile) u
- 32 SS-N-32 (6-10 warheads per missile) v
- 168 AS-15A Air-launched Cruise Missiles (1 warhead) w
- 496 AS-15B Air-launched Cruise Missiles (1 warhead) x
- 156 AS-16A Surface-launched Cruise Missiles (1 warhead) y

p. Heaviest intercontinental ballistic missile (ICBM) in the world. 10 MIRV warheads with a yield of 550-750 kt. **q.** Has an inertial guidance system. MIRV warheads with a blast yield of up to 5 mt each. **r.** Single 800 kt warhead. Road mobile launch platform. **s.** Road mobile launch platform. **t.** Submarine Launched Ballistic Missile (SLBM). Up to seven warheads with a blast yield of 0.45 mt. **u.** SLBM. Up to four warheads per missile. **v.** SLBM. Up to four warheads per missile. **w.** ALCM. 200 kt yield. **x.** Upgraded ALCM. AS-15A with extra fuel tanks. 200 kt yield. **y.** ALCM. Fastest air-launched missile.

CHINA 2011
138 launchers
(178 warheads)

- 16 DF-3A (1 warhead) z
- 12 DF-4 (1 warhead) aa
- 20 DF-5A (1 warhead) bb
- 60 DF-21 (1 warhead) cc
- 20 DF-31 (1 warhead) dd
- JL-1 (1 warhead) ee
- JL-2 (1-4 warheads) ff
- H-6M Cruise missiles (1 warhead) gg
- DH-10 Cruise missile (1 warhead) hh

z. Considered China's first domestic intermediate-range ballistic missile. An improved version with a conventional warhead was exported to Saudi Arabia. **aa.** China's first two-stage missile. Developed to provide strike capability against Moscow and Guam. **bb.** Designed to intercept ballistic missiles and satellites. **cc.** China's first medium-range ballistic missile. Used to carry out China's first nuclear ballistic missile test. **dd.** Road mobile lauched ICBM. Can carry a single one megaton warhead or up to three 20-150 kt MIRV warheads. **ee.** China's first Submarine-launched ballistic missile (SLBM). Carries a single 200-300 kt warhead. **ff.** China's second-generation SLBM. Can be equipped with a single 250-1,000 kt warhead or three or four 90 kt MIRV warheads. **gg.** Air-launched Cruise Missile. **hh.** Subsonic cruise missile. It is believed it carries a 90 kt warhead.

INDIA 2008
50 launchers
(60 warheads)

- Agni III (1 warhead) ii
- Prithvi (1 warhead) jj
- Brahma (1 warhead) kk

ii. Intermediate-range ballistic missile. Most accurate ballistic missile in its range category. **jj.** Short-range ballistic missile. First missile developed under India's Integrated Guided Missile Development Program. **kk.** Supersonic cruise missile. Fastest cruise missile in operation in the world.

1. As of January 2010 the United States maintained an arsenal of 2,468 operational warheads. The arsenal consists of roughly 1,968 strategic warheads deployed on 798 strategic delivery vehicles and 500 nonstrategic warheads. In addition, approximately 2,600 warheads are held in reserve. Another several thousand warheads designated retired — probably 3,500-4,500 — are awaiting dismantlement.

2. It is estimated that as of early 2012 Russia had 2,430 nuclear warheads to its ICBMs, SLBMs, and heavy bombers. Russia also keeps an inventory of 2,000 non-strategic warheads for potential use by ships, aircraft, and air defence forces. An additional 5,500 already retired strategic and non-strategic warheads may be awaiting dismantlement, for a total inventory of 10,000 nuclear warheads. 3. Israel has never confirmed it possesses nuclear weapons.

Nuclear Weapons

LIVING TOGETHER

Death Toll

Served up for your delectation, a century of world conflict on one kitchen table.

Pictured here are less than a quarter of all the deaths that have happened since 1915, often caused by a fatal interpretation of the idea of nations and identity or by the wrong delineation of borders. The casualties represented her are some of the least known in the west. These are the deep, vast lakes of blood spilt because of hatred, nationalist hatred, and irrational hatred in general. And it was spilt because of the illusion — often peddled by someone with personal gain in mind — that we are somehow different. As you can see, all the blood looks pretty similar.

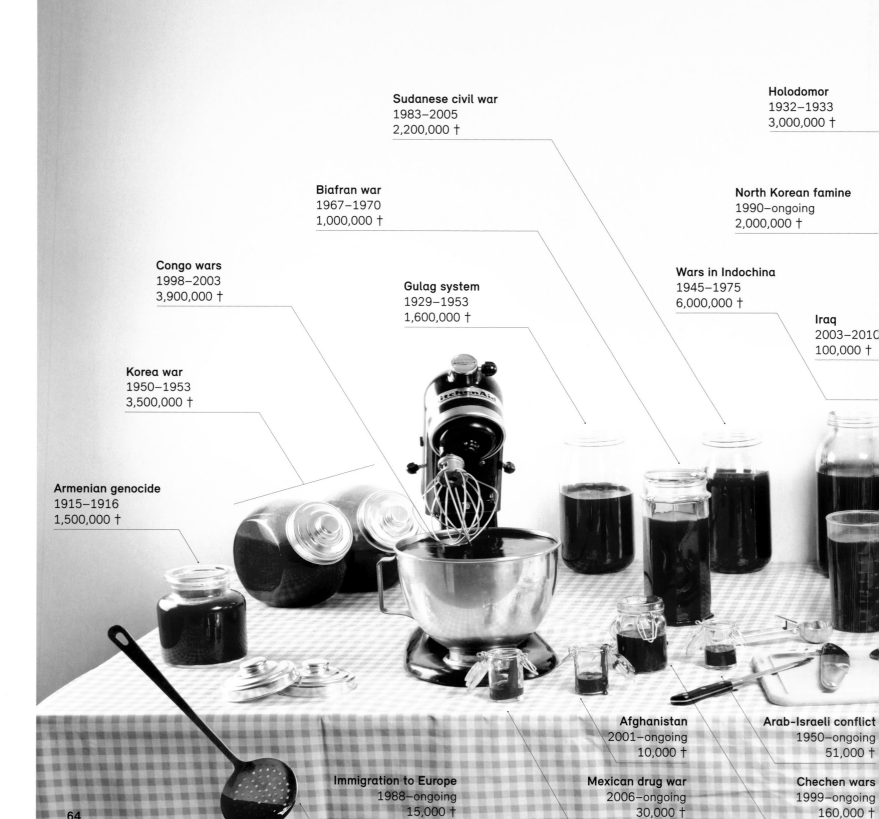

Sudanese civil war
1983–2005
2,200,000 †

Holodomor
1932–1933
3,000,000 †

Biafran war
1967–1970
1,000,000 †

North Korean famine
1990–ongoing
2,000,000 †

Congo wars
1998–2003
3,900,000 †

Gulag system
1929–1953
1,600,000 †

Wars in Indochina
1945–1975
6,000,000 †

Iraq
2003–2010
100,000 †

Korea war
1950–1953
3,500,000 †

Armenian genocide
1915–1916
1,500,000 †

Afghanistan
2001–ongoing
10,000 †

Arab-Israeli conflict
1950–ongoing
51,000 †

Immigration to Europe
1988–ongoing
15,000 †

Mexican drug war
2006–ongoing
30,000 †

Chechen wars
1999–ongoing
160,000 †

The Tree of Faith

How the world's major religions sprouted and grew from one another.

Mapping a genealogy of religions is what you'd call a Biblical or perhaps Qu'ranic task — not only has each major religion splintered into innumerable sects, schools, and confessions over the centuries, but all these faiths have borrowed or stolen so many features from one another so that it's hard to even delineate where one ends and another begins. This graphic presents a somewhat simplified family tree of the world's most widespread religions today, labeled by their numbers of adherents. It's a bewildering kaleidoscope of belief.

The so-called "three doctrines" complement and influence each other.

SHINTOISM up to 108 million (up to 1.5%)

BUDDHISTS approx. 474 million (6.7%)

TAOISTS approx. 385 million (5.5%)

CONFUCIANS approx. 6 million (0.1%)

ATHEISTS approx. 136 million (1.9%)

Atheism
Atheists do not believe in supernatural beings or in a life after death. For them, the only thing which exists is the natural world which we can perceive. Agnostics, on the other hand, hold that the question of the existence of God is basically indeterminable.

Shintoism
Various forms of belief are combined to create this nature religion, which is practiced almost exclusively in Japan. According to the creation myth, the primordial pair of gods, Izanagi and Izanami, created the Japanese islands as well as all other divinities. Adherents of Shinto revere a limited number of divinities — the Kami — which may take the form of objects, animals, people, or abstract beings.

Feast days: Tanabata (Star Festival), Kodomo no Hi (Children's Day)
Important texts: Kojiki, Nihon Shoki

Buddhism
The Four Noble Truths form the basis of the teachings of Siddhartha Gautama, called the Buddha. These are the ubiquity of suffering, the cause of suffering, the cessation of suffering, and the correct path which leads to the cessation of suffering. Many Buddhists believe in a cycle of birth and rebirth. Following the teachings of the Buddha, they seek to become enlightened.

Feast days: Vesak (birth, enlightenment and death of the Buddha), Asalha Puja (first sermon)

Taoism
The adherents of this centuries-old Chinese philosophical tradition attempt to attain health and wisdom and, therefore, ultimately immortality. To this end, they observe various doctrines such as the interplay of opposites (yin and yang) and practice techniques such as meditation or tai chi. A central tenet is the unity of all things. The supreme cosmic principle is the Tao — the path — with which everything begins and to which everything returns.

Feast days: Chunjie (New Year's Festival), many birthdays of gods and goddesses
Important text: Tao Te Ching

Confucianism
This Chinese doctrine of state and society is attributed to the philosopher Confucius, whose real name was Kong Qiu. Confucians attempt to attain an order in things. This can be achieved by reverence for the ancestors and respect for other people, an example being the veneration of children for their parents. Whoever acts within this order in a loyal, honest, and just manner attains true humanity as the sum of all virtues.

Feast day: Confucius' birthday
Important texts: Lunyu, the Confucian canon

Religion

LIVING TOGETHER

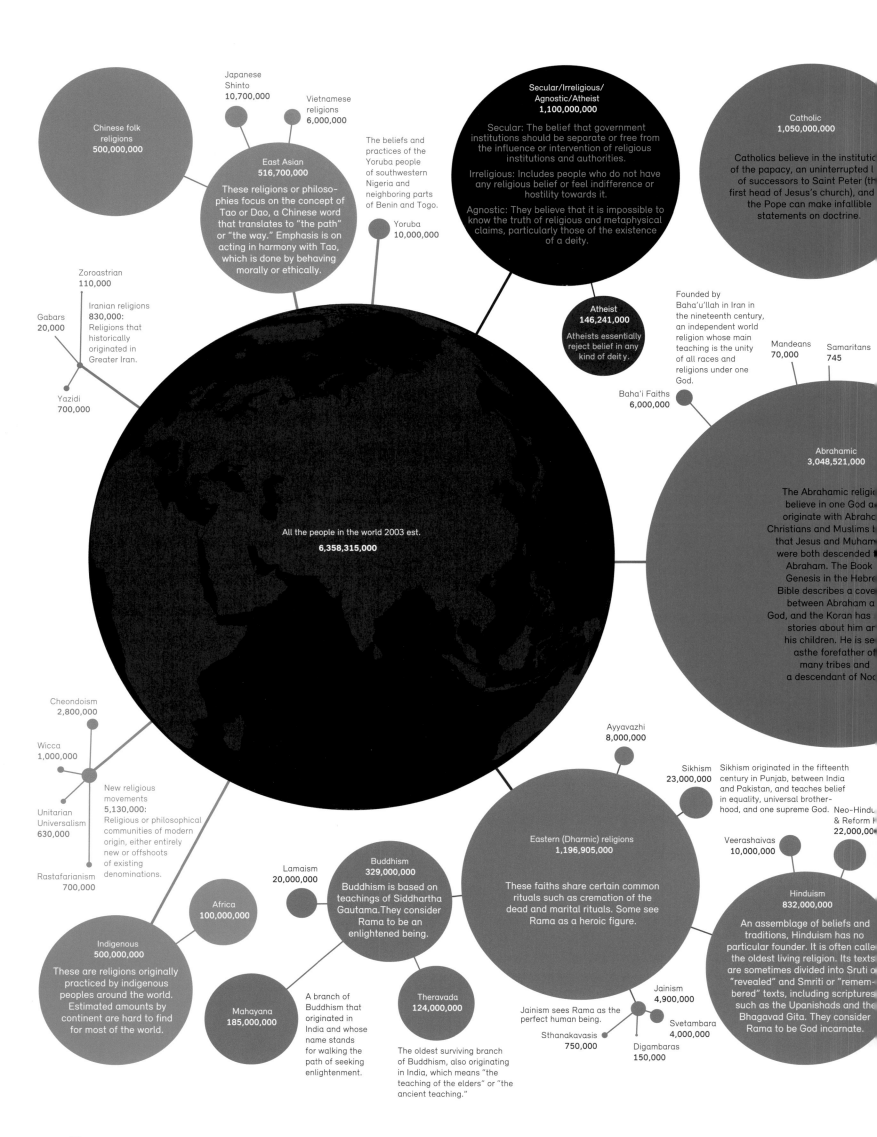

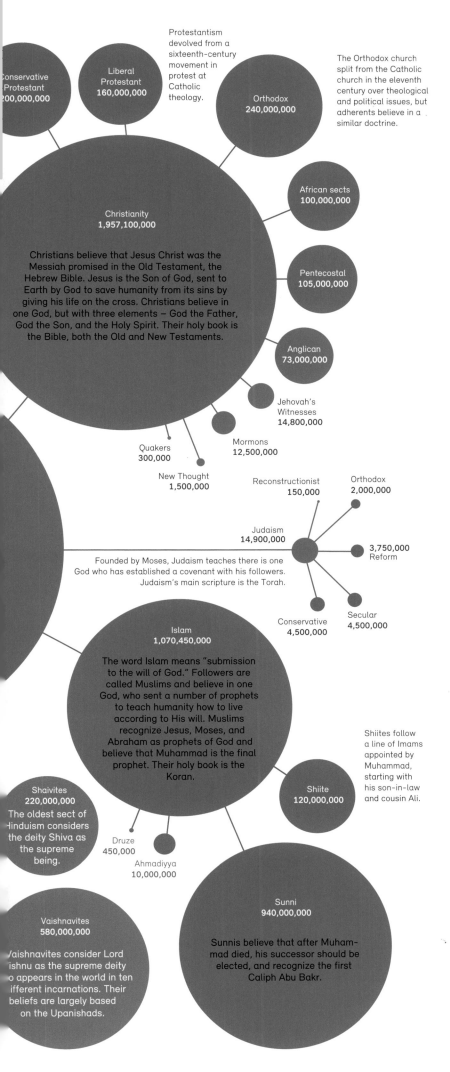

The Children of God

Finding your true faith can be a lifelong journey.

Even though the question "do you believe in God?" may sound simple, finding an answer can be a complex challenge. Within the universe of faith there are a number of planets that provide home to many different ideas, ideologies, and variations. Your place of birth and your family tradition might have the greatest impact on your religious orientation. But over the course of a lifetime your beliefs might develop further. Some believers turn from moderate to fundamentalist, others from dedicated to atheist. Some try other spiritual routes, but the need and search for faith remains a universal. We just don't seem to be able to shake our want for some all-powerful, incomprehensible being in the sky to look after us and tell us what to do.

GLOBAL RELIGIOUS FERVOR 2006-2008
COUNTRIES AND THEIR RELIGIOSITY FROM A PEW SURVEY

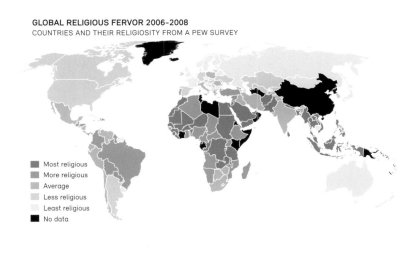

- Most religious
- More religious
- Average
- Less religious
- Least religious
- No data

ABRAHAMIC AND DHARMIC RELIGIONS
PERCENTAGE OF COUNTRY FOLLOWING THAT DOCTRINE

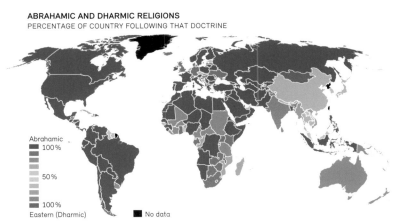

Abrahamic 100% / 50% / 100% Eastern (Dharmic)   ■ No data

Religion

LIVING TOGETHER   69

Different, but the Same

Many religions are based on each other.

Many religious rituals, such as Christmas, were derived from those of other cultures. And the same goes for sacred texts. This infographic artwork shows the most common characters from the holy books of the five world religions in one world-spanning arc — and their interconnections. Arranged alphabetically along the bottom are the 41 most frequently mentioned characters from a total of 2,903,611 words collected in English translations of the holy scriptures of Christianity, Islam, Judaism, Hinduism, and Buddhism. (Caveat: while the first three have just one book each, the holy texts of the latter two religions are very scattered, and some actually remain untranslated.) The size of the colored arcs represents the frequency of the word counts for each word, while the gray arcs above symbolize the similarities among the activities assigned to a character pair, defined as the verbs below each name. Their thickness is determined by an algorithm — the wider and darker the gray arc, the closer the similarity. The frequency of each verb determines its font size. Oh, and the character You is you.

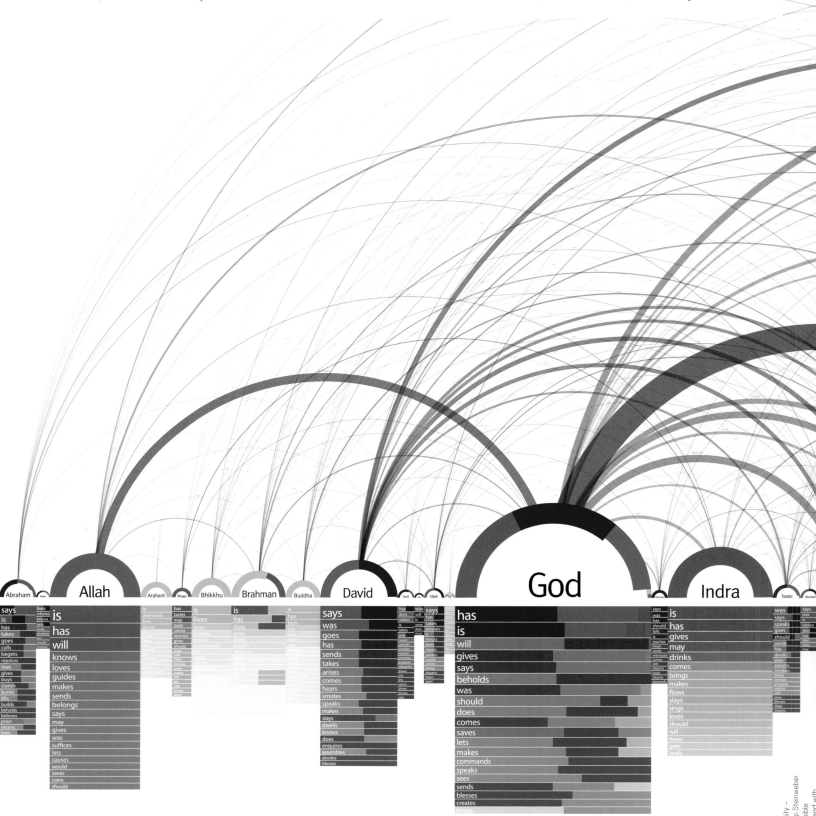

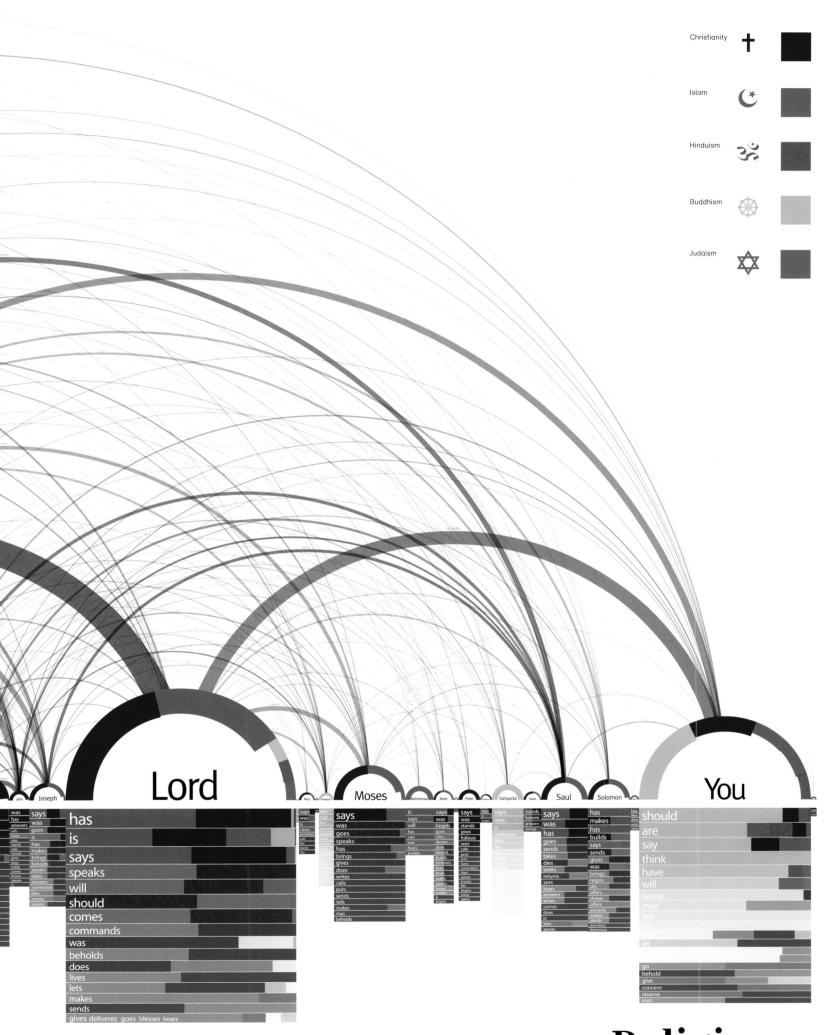

Religion
LIVING TOGETHER

JERUSALEM'S HOLY GROUND

"The air over Jerusalem is saturated with prayers and dreams," wrote poet Yehuda Amichai—and they gather like a storm over the plateau called Mount Moriah. To Jews this is the Temple Mount, their holiest site, where the Western Wall survives from the Second Temple era. To Christians it is ground that Jesus walked. To Muslims it is the Haram al-Sharif, the Noble Sanctuary, where the golden Dome of the Rock shelters a scarred outcrop of limestone. The meaning of that rock—linked to Muhammad and his mystical journey to heaven, to Solomon and the First Temple, to David and the Ark of the Covenant, to Abraham and the near sacrifice of his son—is at the heart of what makes this such sacred, and deeply contested, ground.

CA 960 B.C. | SOLOMON'S TEMPLE

At the direction of God, the Bible recounts, Israelite King David built an altar on or near the rock about 1000 B.C. King Solomon, David's son, fulfilled his father's ambition to erect a Temple to the Lord, dedicating it about 960 B.C. It was destroyed in 586 B.C., when King Nebuchadrezzar sacked Jerusalem and exiled the Jews to Babylon. The Ark of the Covenant, previously enshrined in the Temple's Holy of Holies, vanished from the historical record.

CA 10 B.C. | HEROD'S TEMPLE

With monumental vision, Herod the Great, appointed by Rome as King of Judaea, doubled the size of the Temple Mount. The grand limestone temple he dedicated about 10 B.C. was a renovation of the Second Temple, built 500 years earlier when the Jews returned from Babylonian exile. Jesus taught on the Temple Mount in the week before his death, about A.D. 30, and argued with the priests. In A.D. 70, 6,000 Jews died on the Temple Mount as the Roman army crushed a revolt, torching the Temple and demolishing the complex.

THE CITY THROUGH TIME

Politics and faith have always driven the story of Jerusalem. Watered by a spring on the edge of a desert, its naturally defensible terrain (above) was home to the Jebusites, related to the Canaanites, when King David captured it about 1000 B.C. He made it his capital—neutral ground for the 12 tribes of Israel, whom he had united as one kingdom.

DAVID & SOLOMON ca 1010–970 B.C. Reign of King David. The captured Jebusite settlement becomes the City of David (map left).

ca 970–931 Reign of King Solomon, David's son, who dedicates the First Temple ca 960. After his death, Israel and Judah become separate kingdoms, with Jerusalem as the capital of Judah.

THE LOST TRIBES ca 720–701 Jerusalem's population soars from about 1,000 to 15,000 as refugees move south into Judah from Israel. Neo-Assyrians had conquered the northern kingdom and scattered its ten Jewish tribes, giving rise to the legend of the Ten Lost Tribes of Israel.

BABYLONIAN EXILE & PERSIAN EMPIRE 586 Nebuchadrezzar destroys Solomon's Temple and exiles the Jews to Babylon.

538 Cyrus the Great releases the Jews, now subjects of the Persian Empire; allows building of the Second Temple, dedicated ca 516.

HELLENISTIC PERIOD 332 Alexander the Great sweeps Jerusalem into his empire. After his death in 323, his general Ptolemy I wins the city for his Egyptian kingdom. Other Alexander successors, the Syrian Seleucids, take control from 198–167.

MACCABEAN REVOLT 167–141 Jewish rebellion against the Seleucids. Judah Maccabee reconsecrates the Temple in 164 (an act celebrated as Hanukkah).

INDEPENDENT JEWISH KINGDOM OF JUDAEA 141–63 The Hasmonaeans (family name of the Maccabees) rule as kings and priests. In 63, Roman general Pompey is invited to help settle a disputed succession to the throne and ends up occupying Judea.

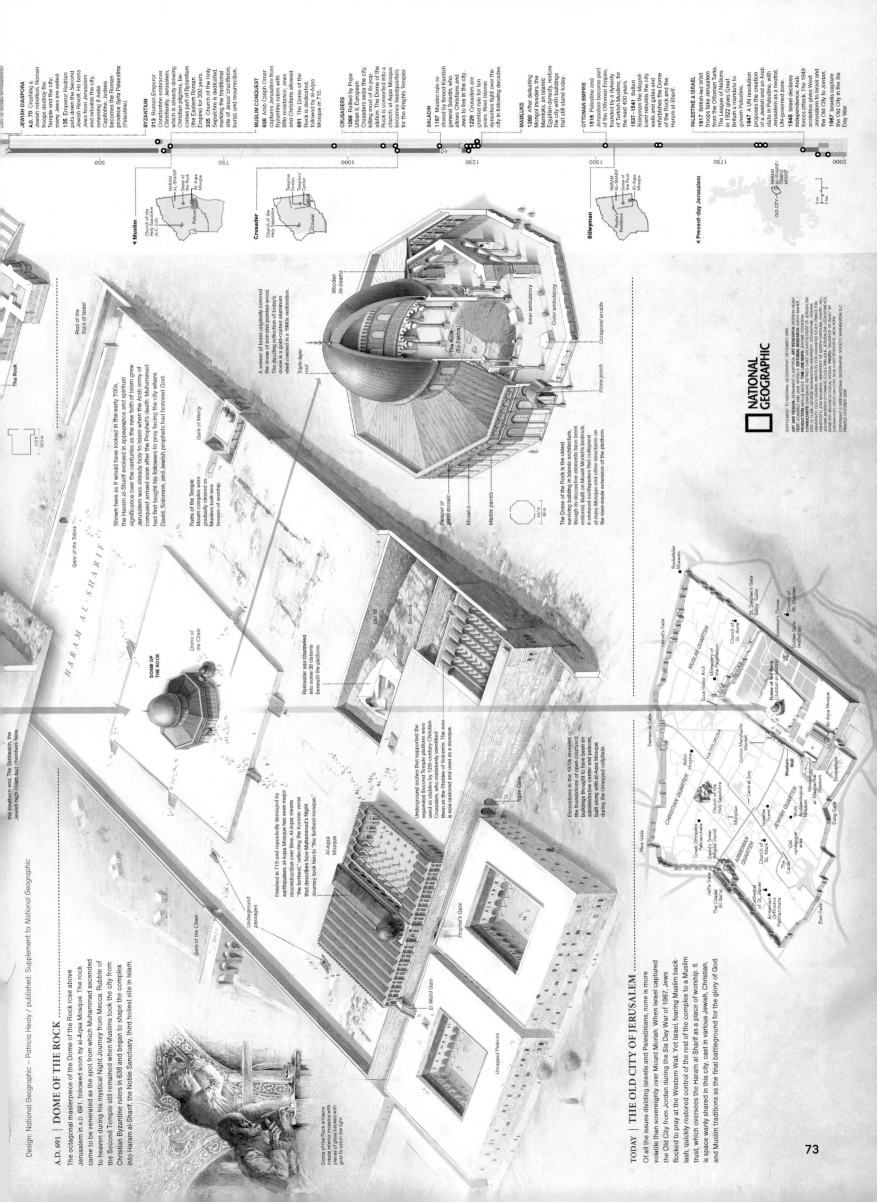

Pope Pickers

The election of a new leader of the Catholic Church is far from a straightforward process.

The Catholic Church has its own special way of doing everything. While other major organizations might distribute press releases to the media and hold conferences for reporters to announce a new leader, the Holy See prefers to send some chemically-infused smoke up a chimney. When that white smoke finally emerges from the Sistine Chapel, not only do 1.2 billion people have a new spiritual leader, but it marks the end of a complex ritual that dates back centuries. Currently, "by scrutiny" (*per scrutinium*) is the only method used for the papal election. In the past, there were other procedures, such as "by acclamation" (*per inspirationem*), when an evident favorite was unanimously declared *quasi afflati Spiritu Sancto*, "as if inspired by the Holy Spirit." And then there was "by compromise" (*per compromissum*), when a deadlocked Conclave would delegate the election to a committee whose choice they all agreed to abide by. The last Conclave, in March 2013, was unique of course. Not because the previous pope Benedict XVI was still alive (that had happened at least five times before, the last time in the fifteenth century), but because this was the first time an ex-pope could watch the holy smoke emerge from the chimney on TV.

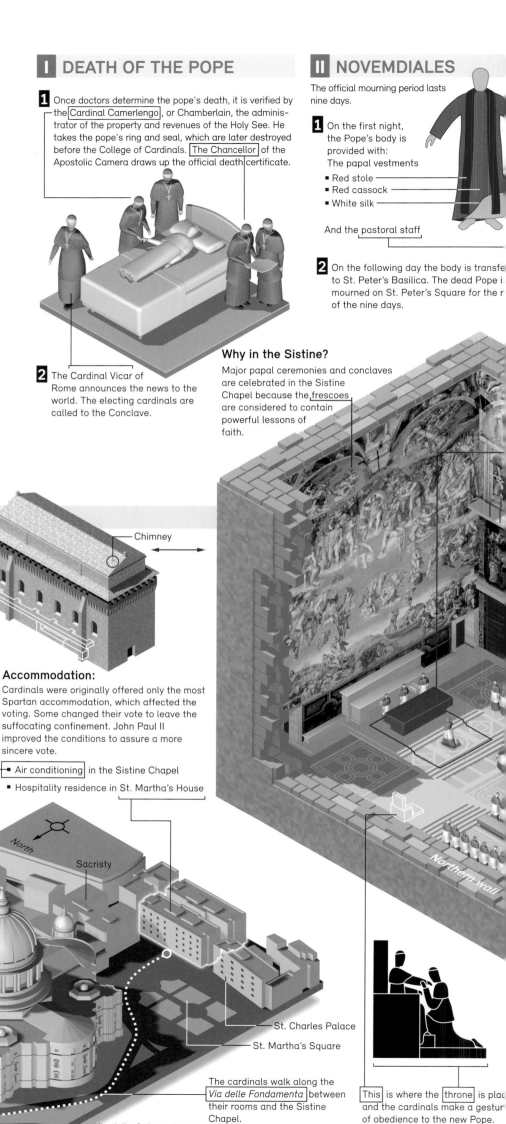

I DEATH OF THE POPE

1 Once doctors determine the pope's death, it is verified by the Cardinal Camerlengo, or Chamberlain, the administrator of the property and revenues of the Holy See. He takes the pope's ring and seal, which are later destroyed before the College of Cardinals. The Chancellor of the Apostolic Camera draws up the official death certificate.

2 The Cardinal Vicar of Rome announces the news to the world. The electing cardinals are called to the Conclave.

II NOVEMDIALES

The official mourning period lasts nine days.

1 On the first night, the Pope's body is provided with:
The papal vestments
- Red stole
- Red cassock
- White silk

And the pastoral staff

2 On the following day the body is transfe[rred] to St. Peter's Basilica. The dead Pope i[s] mourned on St. Peter's Square for the r[est] of the nine days.

Why in the Sistine?
Major papal ceremonies and conclaves are celebrated in the Sistine Chapel because the frescoes are considered to contain powerful lessons of faith.

At the Vatican

- Building
- Gardens

St. Peter's Square
Swiss guard's barracks
Pope's residence
The Sistine Chapel
Chimney
St. Peter's Basilica

Meters 0 100

The wall doubles as the border frontier to the Vatican.

Accommodation:
Cardinals were originally offered only the most Spartan accommodation, which affected the voting. Some changed their vote to leave the suffocating confinement. John Paul II improved the conditions to assure a more sincere vote.
- Air conditioning in the Sistine Chapel
- Hospitality residence in St. Martha's House

Paul VI Audience Hall
Sacristy
St. Peter's Basilica
Belvedere Courtyard
Borgia Tower
State Secretary
St. Charles Palace
St. Martha's Square
Piazza del Forno

The cardinals walk along the *Via delle Fondamenta* between their rooms and the Sistine Chapel.

Via delle Gobernatorato

Northern wall

This is where the throne is plac[ed] and the cardinals make a gestur[e] of obedience to the new Pope.

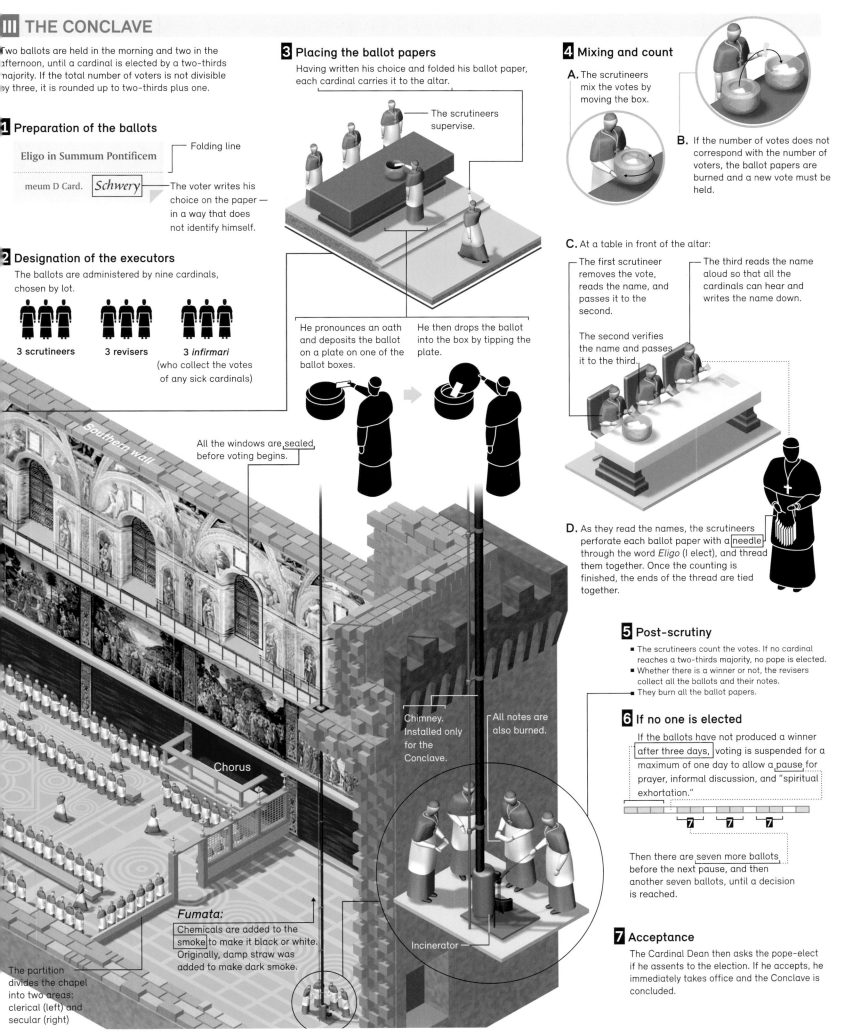

Picking the Pope

Holy Duties

Every Muslim is supposed to perform the Hajj pilgrimage once — though deputies are also allowed.

The Hajj (Arabic for "pilgrimage") is one of the five pillars of Islam, and is meant to demonstrate the solidarity of the Muslim people and their submission to God, but its roots go much deeper into the origins of all three of the world's major monotheistic faiths. The annual trip to Mecca may have originated thousands of years before the Prophet Muhammad made the journey, at the time of Abraham — thought to be 2,000 years BCE. For many centuries, people of several faiths — including Christian and pagan — subsequently performed the pilgrimage. The Hajj is partially based on the story of Abraham's second wife Hagar, abandoned in the desert with her son Ishmael, as told in the Book of Genesis. Part of the present-day ritual still recreates the desperation of that lonely single mother dashing between two hills in search of food and shelter for her baby. Nowadays the trip has become complex and highly ritualized, and attracts around three million people in one stunning week-long spectacle. Civil rights activist Malcolm X wrote that the Hajj demonstrated that Islam had "erased the race problem" from its society, while a Harvard study suggested that those who have performed the pilgrimage show more belief in peace, equality, and harmony — possibly because of exposure to the many different nationalities that attend every year.

The Hajj can be performed in three different ways, one of which must be chosen by the pilgrim at the beginning of his journey. First of all there is the elaborate Hajj al-Tamattu', which is considered to be the "best" way to perform the Hajj since the Umrah, or "small pilgrimage" is included in the ritual. The pilgrim must exit the state of consecration on the evening of the first day and repeat the ritual of entrance. The second way is called the Hajj al-Qiran, which merges the Hajj with the Umrah. In this case the pilgrim waives the exit and entrance of the state of consecration. Last of all there is the Hajj al-Ifrad during which only the Hajj is performed, and no animal needs to be sacrificed.

ARRIVAL AT THE HAJJ

Over two million pilgrims attend the Hajj every time, most of whom traveled to Jeddah by plane.
The "Haramain High Speed Rail" is currently under construction to ensure fast and safe transportation to the holy city of Mecca, and to improve the connection to Medina (the second holiest city of Islam).

GENERAL MAP

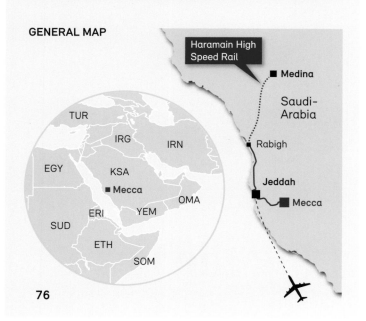

KAABA — THE HOUSE OF GOD

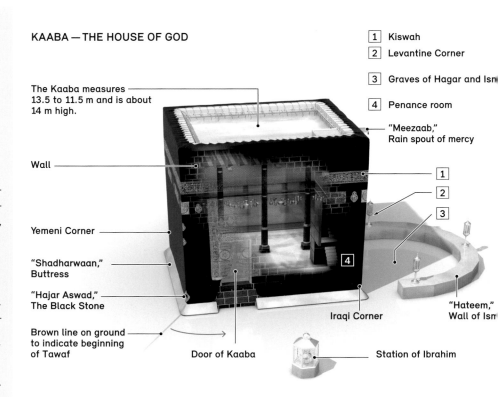

SEATING DURING PRAYER AT THE AL-HARAM-MOSQUE

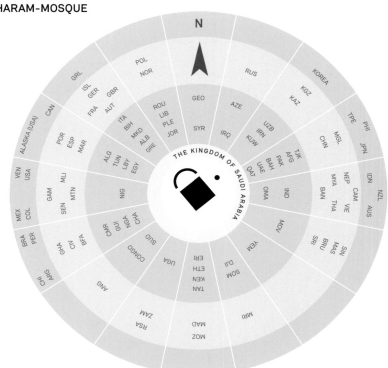

AL-HARAM-MOSQUE

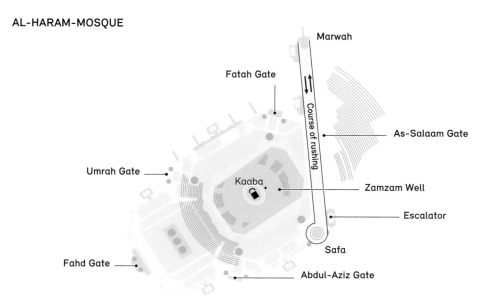

JAMARAAT BRIDGE

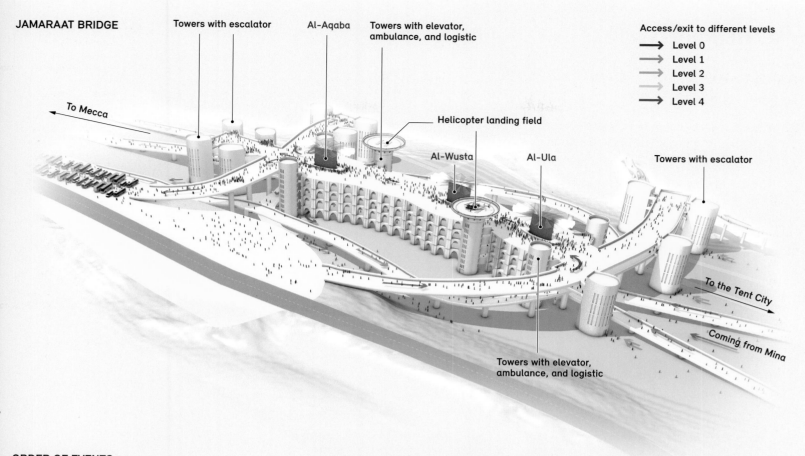

ORDER OF EVENTS

A preset series of actions must be followed to do the pilgrimage correctly. These start on the 8th day of the last month of the Islamic calender:

Dhul Hajj (day 8)
- Entering the consecration/putting on the pilgrim's garment
- Arrival Tawaf (circling the Kaaba seven times)
- Rushing back and forth between the rocks Safa and Marwah seven times
- Spending the night at Mina

Dhul Hajj (day 9)
- Lingering and praying close to Mount Arafat until sunset
- Collecting 70 small pebbles for the stoning ceremony from mountain Muzdalifah
- Spending the night in the open air close to Muzdalifah

Dhul Hajj (day 10)
- Stoning of the pillar Al-Aqaba from the Jamaraat Bridge
- Sacrificing an animal
- Shaving the head (men) or cutting the hair (women)
- Returning to Mecca for Tawaf al-Hajj
- Again rushing back and forth between the rocks Safa and Marwah seven times
- Spending the night at Mina

Dhul Hajj (day 11 to 13)
- Staying in Mina up to three days for stoning all pillars three times
- Farewell Tawaf

THE PILGRIMAGE ROUTE

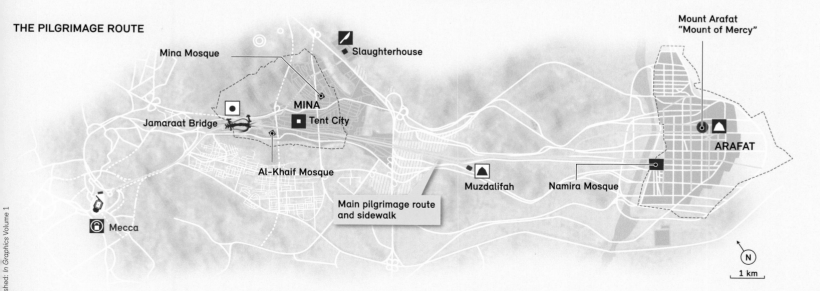

The Hajj

LIVING TOGETHER

Holiday Distribution

When does the world celebrate?

As long as you have a canny travel plan and a few free air miles, this graphic can help you exploit the world's feast day potential. That is, if you ever felt like spending the year engaged in intercontinental hopping from national holiday to religious celebration in a crazy effort to avoid doing any work. But unfortunately, while some dates are manifestly over subscribed, like October 1 and September 15, there are still plenty of dry patches in between where you may need to put in a few desk hours.

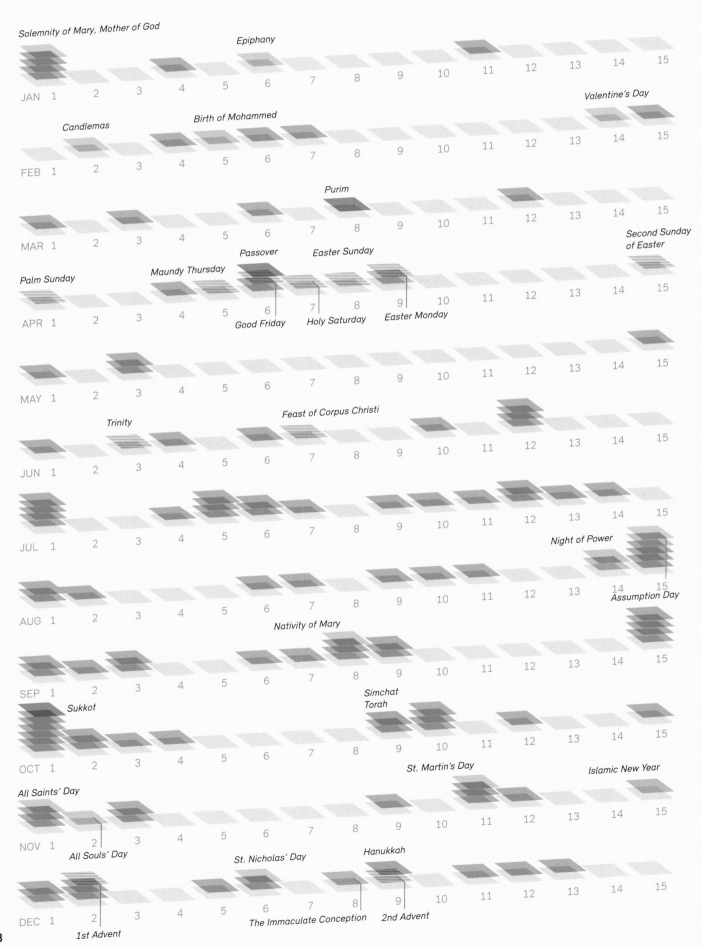

The Key Is Choice

It's no accident that the richest countries have the most emancipated women.

"Gender equality is not only morally right, it is pivotal to human progress and sustainable development. Gender equality will not only empower women to overcome poverty, but also their children, families, communities, and countries." — UNESCO. The data proves that a woman's right to choose when she wants a child is directly related to a nation's prosperity. But what social factors actually block those rights? Here, 20 countries have been chosen from the various points on the scale of the "unmet need for family planning" — a factor based on surveys of the number of women who have expressed a wish either to not have any more children or to delay the birth of their next child. These countries have been broken down into seven systems — health, government, religion, culture, society, media, and development — in an attempt to determine which one might be blocking the empowerment of women. Below, two of these countries, Italy and China, have been further broken down, because they showed particularly conspicuous peaks in the prominent influence of three respective factors: religion, media, and government.

I CAN CHOOSE, RIGHT?
1/3 WELL, IT DEPENDS ON WHERE YOU WERE BORN

37.6% - 45%
30.1% - 37.5%
22.6% - 30%
15.1% - 22.5%
7.6% - 15%
0% - 7.5%
No data

The unmet need for family planning is the percentage of women of reproductive age, who are married or in a union, fecund and sexually active, but not using any method of contraception, and reporting not wanting any more children or wanting to delay the birth of their next child. This data used in the above representation highli[ghts] geographic trends and contrasts, showi[ng] the fact that not everyone around the world can freely acce[ss] family planning facili[ties].

2/3 BECAUSE YOUR SOCIAL STRUCTURE AFFECTS YOUR CHOICE

RELIGIOUS SYSTEM — The collection of beliefs, values and ethics that relate to people's daily lifestyle. MEASURED BY: Religion type and rules.

CULTURAL SYSTEM — The collection of education, ideals, traditions, personal background, and national aspirations. MEASURED BY: Literacy index.

SOCIAL SYSTEM — The interaction and the roles of people living within the same environment. MEASURED BY: Gender equality index.

MEDIA SYSTEM — The set of storage and transmission channels, tools, and regulations used to deliver information. MEASURED BY: Press freedom index.

DEVELOPMENT SYSTE[M] — The intersection between living standard, industri[al] base, and safe environ[ment] of a certain nation. MEASURED BY: Development index.

GOVERNMENT SYSTEM — The bureaucracy and the administrative group of people with authority to govern a political state. MEASURED BY: Democracy index.

HEALTH SYSTEM — The organization of people, institutions, resources that deliver health care services. MEASURED BY: Life expectancy index.

10 Indexes values
AVG %
COUNTRY NAME Female population (15-65)
*=2 Mio.

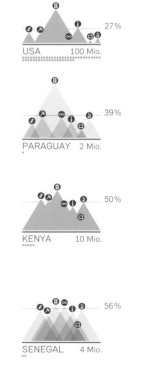

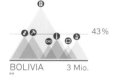
USA 100 Mio. 27%

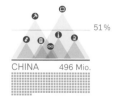
PARAGUAY 2 Mio. 39%

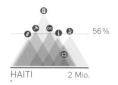
SENEGAL 4 Mio. 56%

FRANCE 20 Mio. 28%

BOLIVIA 3 Mio. 43%

HAITI 2 Mio. 56%

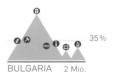
ITALY 19 Mio. 30%

THAILAND 22 Mio. 45%

KENYA 10 Mio. 50%
CHINA 496 Mio. 51%
INDIA 396 Mio. 52%

PAKISTAN 54 Mio. 62%

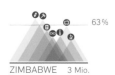
BULGARIA 2 Mio. 35%

PHILIPPINES 30 Mio. 46%

EGYPT 26 Mio. 53%

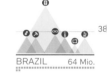
ZIMBABWE 3 Mio. 63%

BRAZIL 64 Mio. 38%

TURKEY 24 Mio. 50%

UGANDA 8 Mio. 55%

YEMEN 6 Mio. 71%

Design: DensityDesign Research L[ab]

3/3 AND ITS MECHANISM CAN EVEN DENY YOUR FAMILY PLANNING RIGHTS

The Universal Declaration of Human Rights is a declaration adopted by the United Nations General Assembly on December 10, 1948. Six of the thirty-one articles were chosen according to their pertinence with family planning, and the relation of relevance between these six articles and the seven systems of the social structure has been shown for both Italy and China.

03	RIGHT TO PRACTICE FAMILY PLANNING IN FREEDOM AND SECURITY
12	RIGHT TO FOUND A FAMILY WITHOUT EXTERNAL INFLUENCES
16	RIGHT TO FOUND A FAMILY REGARDLESS OF RACE, RELIGION, OR CITIZENSHIP
25	RIGHT TO PROTECT THE ACTUAL STANDARD OF LIVING BY DOING FAMILY PLANNING
26	RIGHT TO ACCESS TO DIFFERENT TECHNOLOGIES REGARDLESS OF THE STANDARD OF LIVING
27	RIGHT TO BE INFORMED ABOUT THE TECHNOLOGIES FOR FAMILY PLANNING

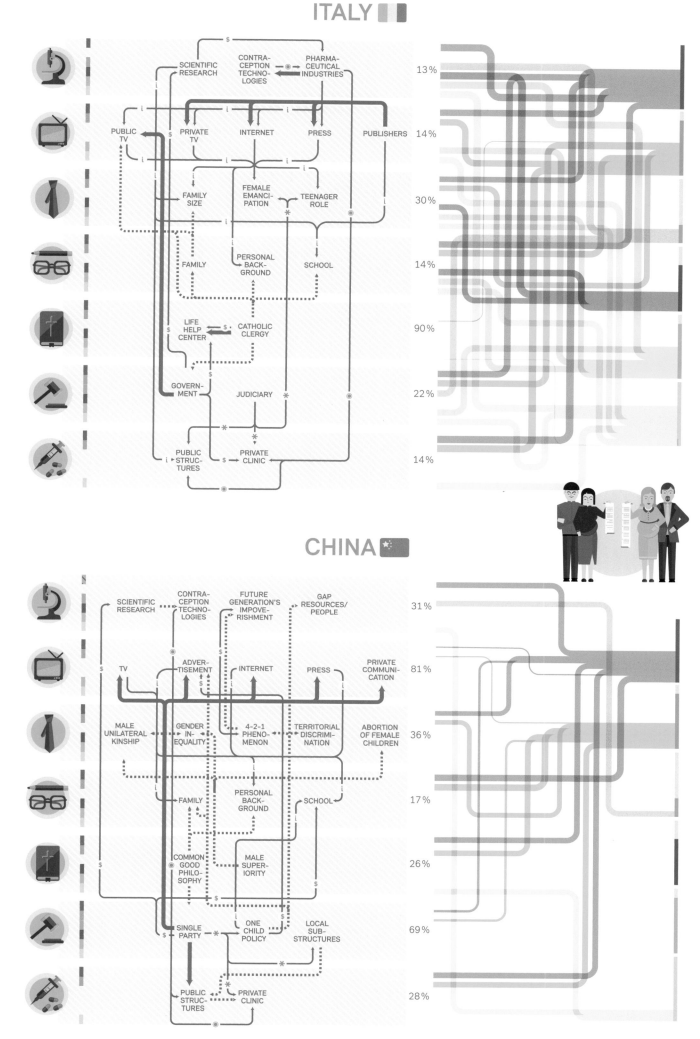

Family Planning
LIVING TOGETHER

The Knock-on Effect

How development aid could make us all richer.

In 1994, the International Conference on Population and Development in Cairo identified one key factor to help us meet the United Nations' much-vaunted Millennium Development Goals for reducing poverty, inequality, and hunger — family planning. The idea is presented here as a Newton's Cradle, much like the one possibly sitting on the desk of a few international investment bankers. Money from the development banks provides the energy from the left, and the momentum passes on through various levels in various forms — education and counseling, access to contraception — until they reach local communities. With any luck, the changes that that energy brings to people in the developing world will be passed back along the chain of ball-bearings — both in wealth that can be invested in the world economy and in information on how to improve matters further. Of course, as Isaac Newton could have pointed out, some energy is lost each step of the way, and environmental factors (local religious beliefs and cultural norms) add more friction, reducing the momentum.

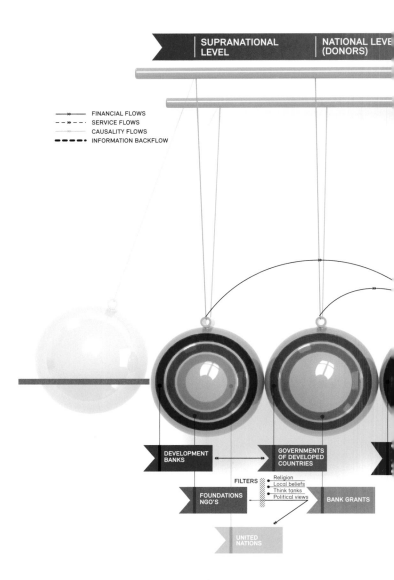

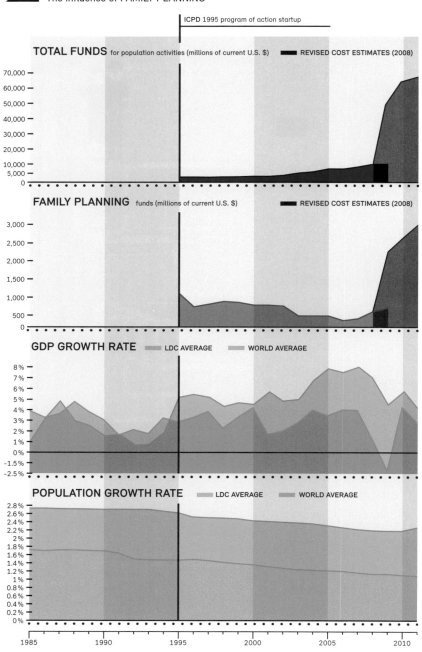

WOMAN CONDITION
Focus on INDIA and SUB-SAHARAN AFRICA

- ◆ World
- ◆ India
- ◆ Sub-saharan Africa

> Particular attention is to be given to the socio-economic improvement of poor women in developed and developing countries.
> As women are generally the poorest of the poor and at the same time key actors in the development process, eliminating social, cultural, political, and economic discrimination against women is a prerequisite of eradicating poverty, promoting sustained economic growth in the context of sustainable development, ensuring quality family planning and reproductive health services, and achieving balance between population and available resources and sustainable patterns of consumption and production.
>
> ICPD
> International Conference on Population and Development

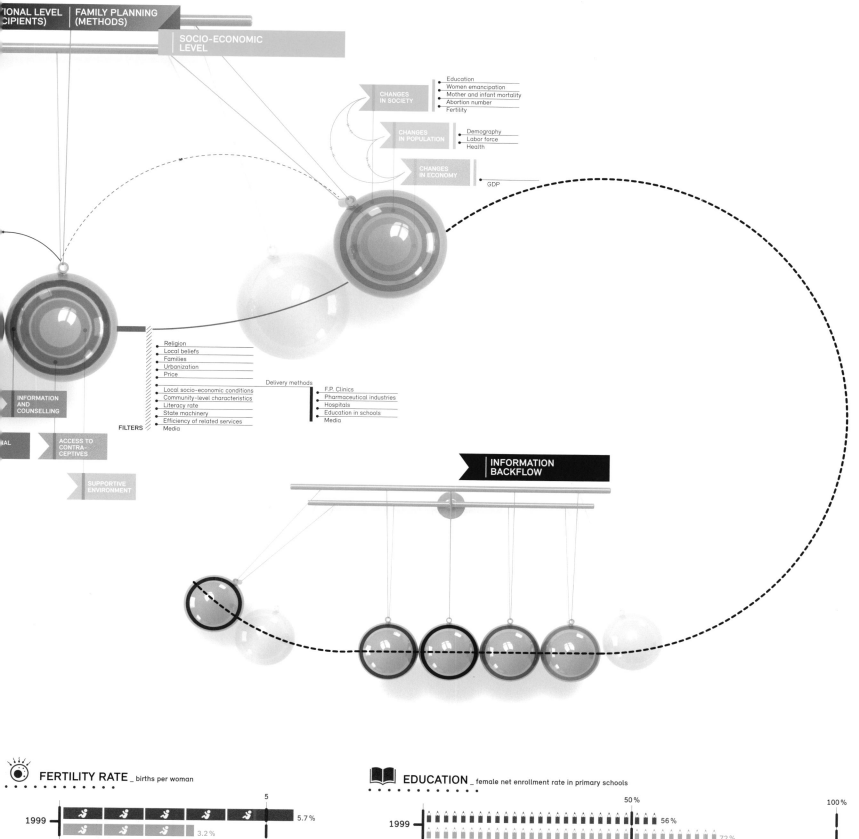

It's a Woman's World

LIVING TOGETHER

THE DAYS THE EARTH STOOD STILL

You remember. You remember what you did on 9/11. Your parents or grandparents remember when Neil Armstrong stepped on the moon, the day the Second World War ended, and the day the Berlin Wall fell. Even if you didn't see them happen, even if you were asleep or watching a movie or on holiday or not yet born, or for some other reason not close to a TV, a radio, a mobile phone, or the Internet, you remember them because they are the moments when the world seems to stop spinning, or spins the other way, or shifts in its orbit because the events that will go down in History with a capital H have occurred, and the world around you changed thereafter. This is a selection of some of those essential moments when the entire planet seemed to gasp as one.

Some were events that cut deep and made us feel suddenly, dizzyingly afraid. The assassination of JFK. The sinking of the famously unsinkable. The meltdown of a nuclear power station. Others were moments of remarkable achievement that changed our understanding of what's possible. A man on the moon, for example. An airlift dropping supplies behind supposedly impenetrable lines. Or finding the world's most wanted man who was hidden in a compound far from sight. The impact of these events reached and continues to reach far beyond those involved, to become part of our consciousness, a part of how we understand ourselves and the world.

We have chosen graphics that represent situations you know, telling the stories behind them that you probably don't. More than anything, these graphics are intended to explain the "how." But rarely can they succinctly answer the follow-up question, whose detailed response would require an explanation longer than this book: "Why?"

The Sinking of The Unsinkable

How an iceberg, some overconfident engineers, and poor safety measures led to the most famous shipwreck of the twentieth century.

Shortly before the sinking, smoke was seen coming out of the rear funnel, which was not a true fume outlet, but a ventilation shaft. The water pressure inside the ship forced smoke out from the engine rooms, away from the ship.

And as the smart ship grew
In stature, grace, and hue,
In shadowy silent distance grew the Iceberg too.

In his spine-shivering poem "The Convergence of the Twain," the English poet Thomas Hardy saw it as destiny for the ship and the iceberg to meet in their terrible cataclysm. A more ancient metaphor fits the awful night of April 15, 1912 too: the old mythical clash between Nature and the Hubris of Man, for instance. The battle was more drawn out than some suppose: after the initial raking punches from that chunk of northern ice, it took a full two hours for the Atlantic Ocean to finally vanquish what was supposedly the greatest ship ever built, and make its point about human vanity. In purely human terms, the 1,514 lost lives don't make it the greatest disaster ever. It is not even the greatest maritime disaster of the twentieth century — in 1945, around 10,000 German civilian refugees and soldiers were killed in the Baltic Sea after their overcrowded ship, the Wilhelm Gustloff, was sunk by a Soviet submarine. But the Titanic's symbolic power could not be greater — it contained a cross-section of the society of the day, and its destruction anticipated in microcosm the utter devastation of European society that would begin just two years later with the First World War and all its terrible consequences. In many ways, the sinking of the Titanic marked the metaphorical beginning of the century.

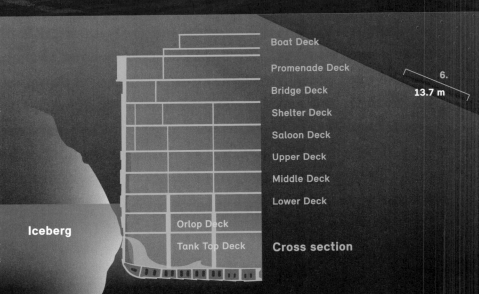

Iceberg

Boat Deck
Promenade Deck
Bridge Deck — 6. 13.7 m
Shelter Deck
Saloon Deck
Upper Deck
Middle Deck
Lower Deck
Orlop Deck
Tank Top Deck

Cross section

How did the Titanic sink?

11:40 p.m.
The starboard bow of the Titanic under full steam, collides with an iceberg of about 300,000 tonnes.

12:40 a.m.
Within an hour, up to 25,000 tonnes of water flooded the hull. Five forward compartments are flooded and the tilt of the ship is about five degrees.

01:40 a.m.
During the second hour after the collision, only about 6,000 tonnes of water flooded the ship. Slowly but surely the non-waterproof parts of the ship, like portholes, ventilation shafts, and hatchways sink below the water line. This hastens the sinking.

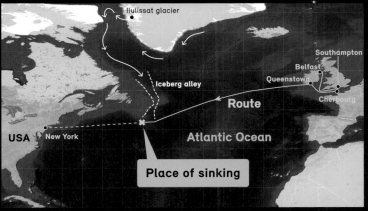

About 300 nautical miles before Newfoundland (Canada) the Titanic sank. In that year, an abnormal amount of icebergs were reported unusually south.

LIFE BOATS

The 20 life boats provided space for nearly 1,200 people, but there were about 2,200 passengers on board. The ship was actually allowed to carry 3,300 passengers. But even so, only 60 percent of the given capacity was used. The crew was not well-trained and the rules were not enforced for such eventualities.

1.18 m²
Total leak area

THE WRECK

The Titanic sank to a depth of 3,803 meters. For 73 years no light reached the wreck until it was located in 1985. The bow lies in relatively good condition about 600 meters from the heavily damaged stern, with much debris in between. It is estimated that about 5,500 artifacts have been recovered to date. By the way: a private dive cruise to the wreck is available for about €30,000.

02:20 a.m.
The stern rose out of the water to an angle of between 15° and 20° and broke apart due to the high stresses. But this is not the only possible scenario. How it really happened is not clear. We show two variations: A and B.

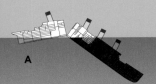

Titanic

THE DAYS THE EARTH STOOD STILL 87

RMS TITANIC

TECHNICAL DATA	
Length	269.04 m
Width	28.19 m
Draft	10.54 m
Displacement	53,147 t
Speed	21 kn (39 km/h)
Machine power	51,000 PS
Crew	897
Approved passengers	2,400
First Class	750
Second Class	550
Third Class	1,100
Coal	5,344 t

3 CLASSES OF SOCIETY

The division into three classes was common at that time. Titanic provided accommodation for 1,100 3rd-class and 550 2nd-class passengers. The 1st-class passengers, a maximum of 750, had access to the largest part of the ship's interior. Between the first and second funnel was the magnificent stairway. The 1st-class facilities offered salons, cafés, a Turkish bath, a dining room, and a restaurant. The 3rd-class conditions were relatively comfortable for the time, providing cabins with 6, 4, or 2 beds instead of the then common dormitories.

CONSTRUCTION AND BUILDING

The keel of Titanic was laid in 1909 in Belfast as the second ship of the Olympic-class. Before the shipyard, Harland & Wolff started building the Olympic. On May 31, 1911 Titanic slid into water. A launch ceremony was not held though it was usual at White Star Line. For its status as a Royal Mail Ship (RMS) it had a post office aboard. The total building cost was about £112 million (after taking inflation into account, and with the exchange rate of mid-November 2011 this would have been €131 million).

THE ENGINE

Titanic was designed for a cruising speed of 21 knots (about 39 km/h). The engines provided more than 50,000 HP power and needed about 630 tonnes of coal a day, with an on-board capacity of 6,700 tonnes. 159 furnaces could heat 29 boilers that provided steam to the turbines driving the three propeller shafts. Additionally, four generators provided electrical power to the ship.

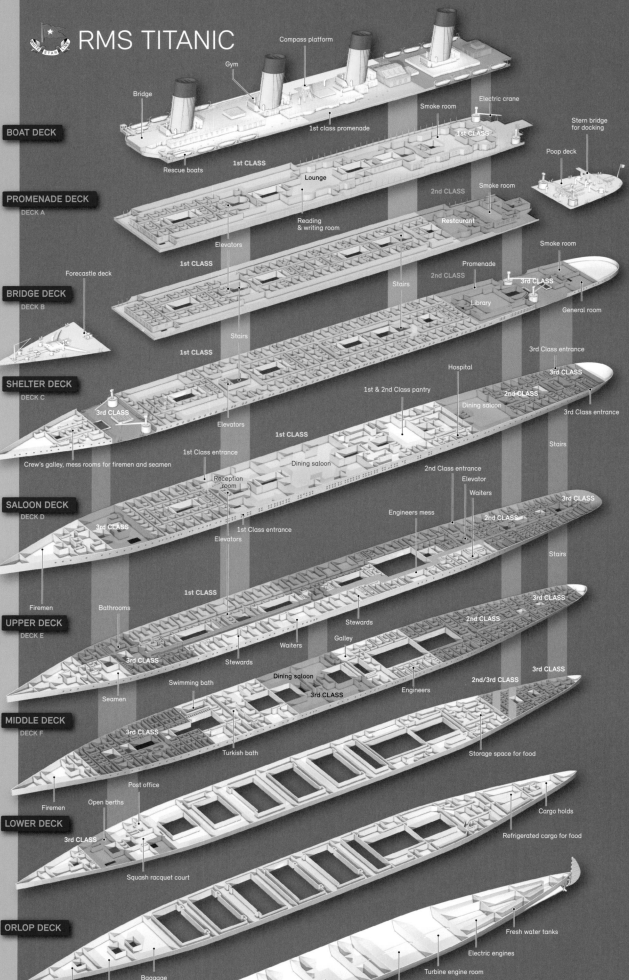

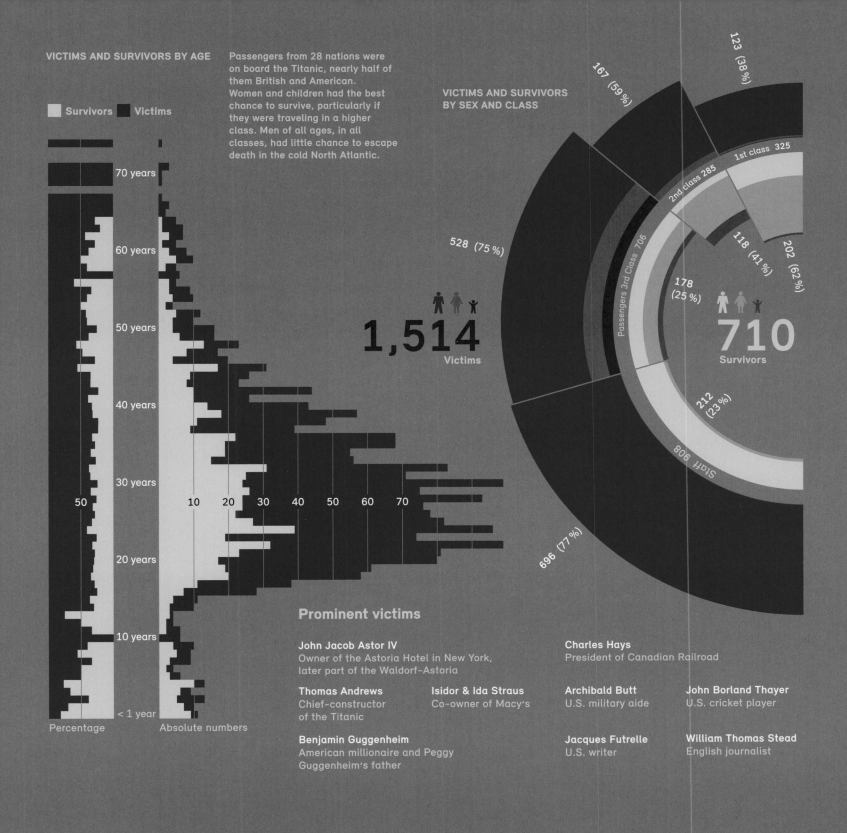
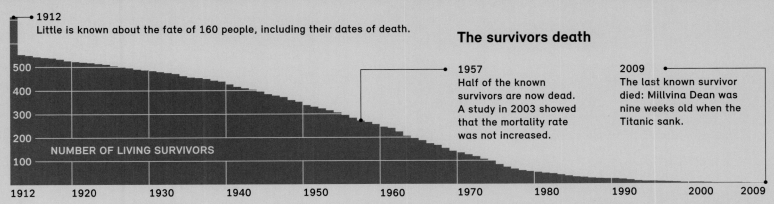

Titanic

THE DAYS THE EARTH STOOD STILL

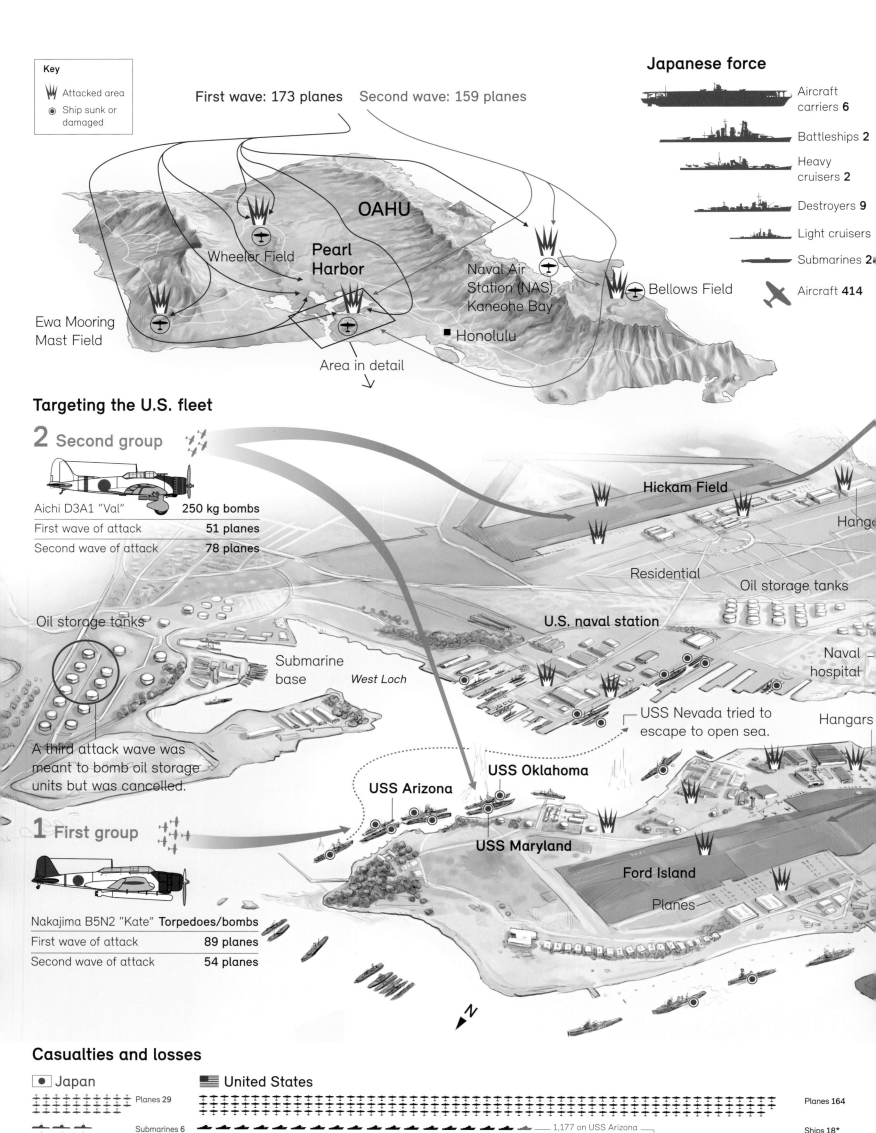

Route of task force

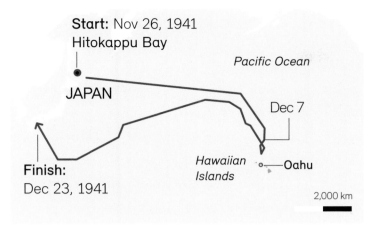

Start: Nov 26, 1941
Hitokappu Bay

Pacific Ocean

JAPAN

Dec 7

Finish:
Dec 23, 1941

Hawaiian Islands — Oahu

2,000 km

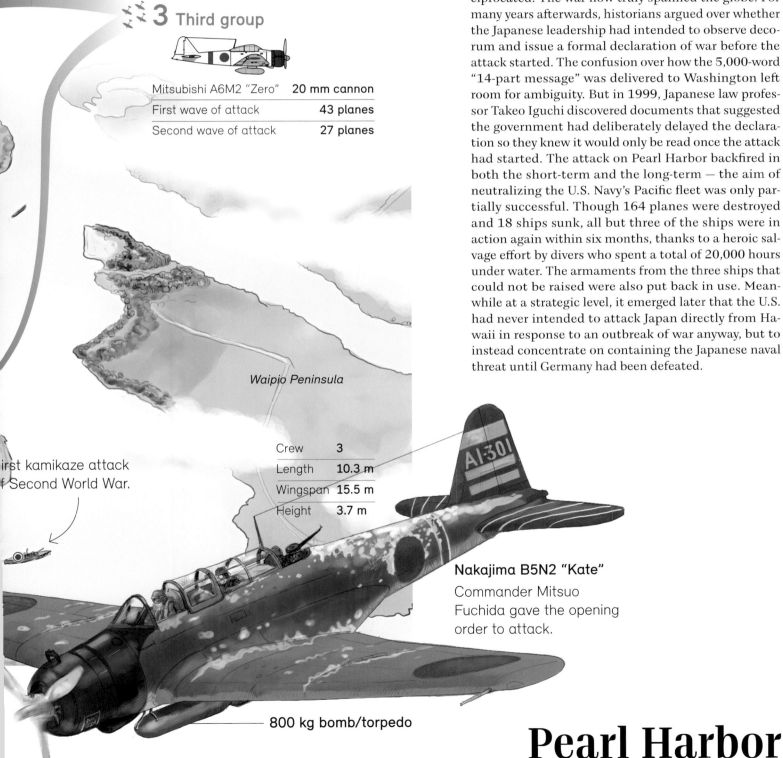

3 Third group

Mitsubishi A6M2 "Zero"	20 mm cannon
First wave of attack	43 planes
Second wave of attack	27 planes

Waipio Peninsula

irst kamikaze attack
f Second World War.

Crew	3
Length	10.3 m
Wingspan	15.5 m
Height	3.7 m

Nakajima B5N2 "Kate"
Commander Mitsuo Fuchida gave the opening order to attack.

800 kg bomb/torpedo

Waking the Giant

How Japan's devastating preemptive strike on the United States Navy won the battle, but lost the war.

The attack on Pearl Harbor on the morning of December 7, 1941 was as psychologically devastating as the Kennedy assassination and 9/11 — a profound shock to U.S. prestige and invincibility. But more significantly, for the Second World War it wiped out the argument for non-interventionism overnight. On December 8, even as the search for survivors was still under way, the U.S. declared war on Japan. By December 11, fresh American military operations led Germany and Italy to declare war on the U.S., whose government duly reciprocated. The war now truly spanned the globe. For many years afterwards, historians argued over whether the Japanese leadership had intended to observe decorum and issue a formal declaration of war before the attack started. The confusion over how the 5,000-word "14-part message" was delivered to Washington left room for ambiguity. But in 1999, Japanese law professor Takeo Iguchi discovered documents that suggested the government had deliberately delayed the declaration so they knew it would only be read once the attack had started. The attack on Pearl Harbor backfired in both the short-term and the long-term — the aim of neutralizing the U.S. Navy's Pacific fleet was only partially successful. Though 164 planes were destroyed and 18 ships sunk, all but three of the ships were in action again within six months, thanks to a heroic salvage effort by divers who spent a total of 20,000 hours under water. The armaments from the three ships that could not be raised were also put back in use. Meanwhile at a strategic level, it emerged later that the U.S. had never intended to attack Japan directly from Hawaii in response to an outbreak of war anyway, but to instead concentrate on containing the Japanese naval threat until Germany had been defeated.

Pearl Harbor

THE DAYS THE EARTH STOOD STILL

0:15–3:00
Paratroopers land in the hinterland of the coast.

3:15
The beaches are bombarded, the first troops land on a small island just off "Utah Beach."

6:30
H-Hour on "Utah" and "Omaha Beach."

7:30–7:45
H-Hour on the beaches "Gold," "Sword," and "Juno."

12:03
British troops meet up with paratroopers at the Orne bridges.

13:00
The 4th U.S. Infantry Division meet up with the 101st Airborne at Pouppeville.

16:00
Tanks advance inland from "Omaha Beach."

24:00
Five bridgeheads on the shore are fortified, the liberation of Europe has begun.

Liberating Europe

Why June 6, 1944 was to become known as "The Longest Day" in the Second World War.

They met at "Piccadilly Circus" at night — but not the bustling London intersection. The Allied fleet, some 5,000 ships, converged on a secret spot with this codename midway across the narrow English Channel that separated the British Isles from Nazi-occupied France. They were about to embark on the biggest and most complex landing operation in military history —

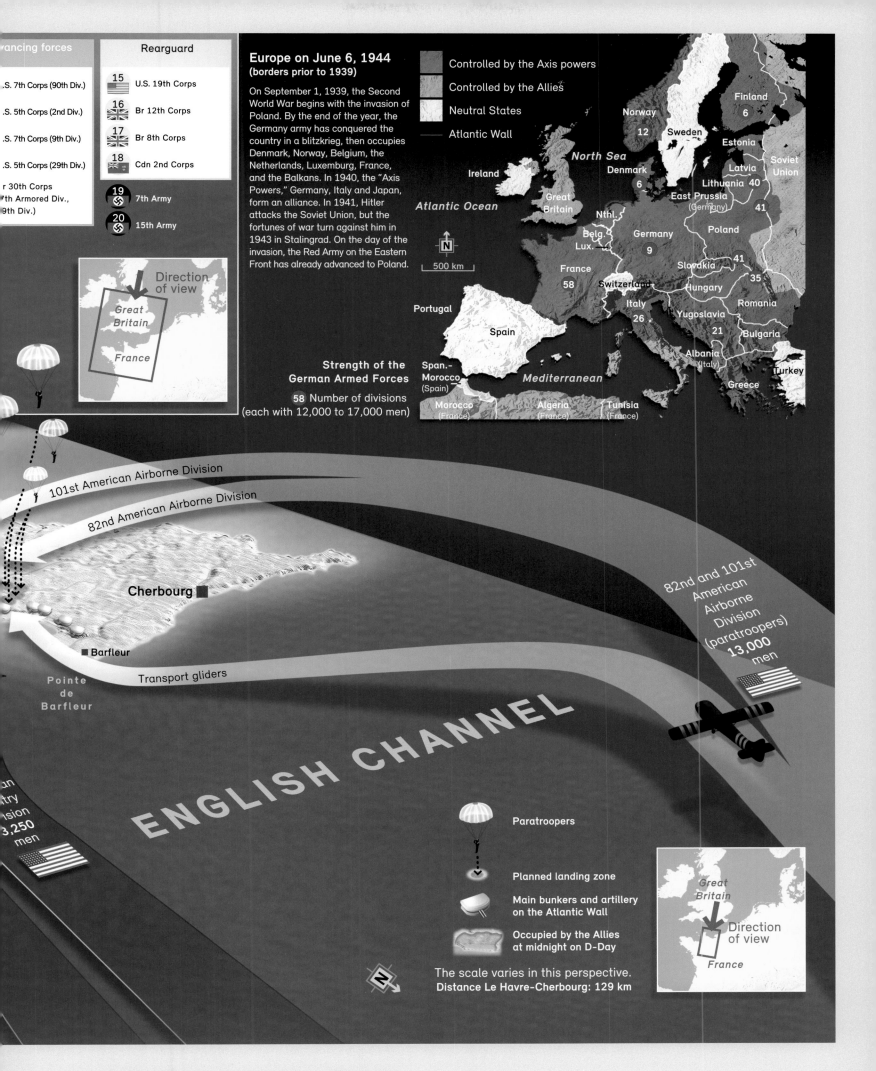

to put 160,000 soldiers on enemy territory, and begin "Operation Overlord" — the liberation of Europe. Around 12,000 British, Canadian, and American soldiers were to lose their lives on D-Day, along with between 4,000 and 9,000 German troops. But it was the beginning of the end of the most terrible war in human history.

D-Day

THE DAYS THE EARTH STOOD STILL

Candy Bombing for Freedom

The Berlin Airlift: how the West defied a Soviet Blockade by an act of logistical genius.

The Cold War had barely started when one of its greatest crises struck. It was June 1948 — the cataclysm of the Second World War was still a fresh memory when, in swift reaction to the introduction of the new Deutschmark currency in the western sectors of Berlin, the Soviet Union capitalized on the geographical advantage it had in controlling all the surrounding territory. Communist forces soon circled the city, cut off its road and rail access, and shut down power stations supplying electricity to the West. The U.S., Britain, and France faced a stark choice: give up the German capital and abandon its people to their fate under a Soviet dictatorship — or embark on the risky, uncertain strategy of supplying the city by air. What happened over the next 14 months became part of Berlin's mythology, and sealed an enduring bond between the German capital and the USA. Literally everything had to be flown in — coal to heat homes, industrial materials to feed factories, and of course food and clothing for the two million inhabitants. Planes landed at three different airports, were unloaded in less than 30 minutes — and then often carried people out to safety. Since then, no American president comes to Berlin without paying tribute to the special friendship forged by the U.S. Air Force's "Candy Bombers," who dropped sweets for the children as they were coming in to land — and older West Berliners won't hear a word said against the "Amis."

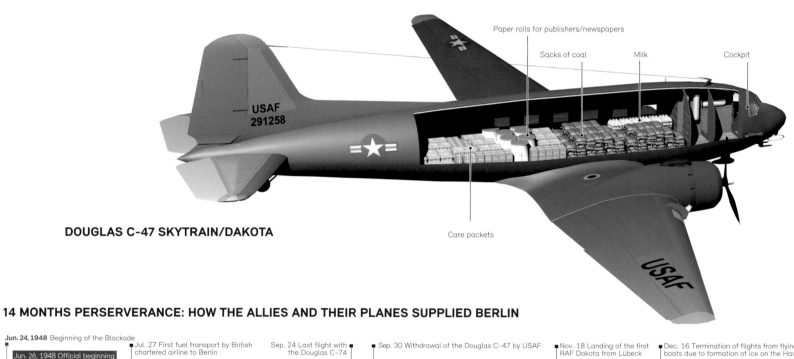

DOUGLAS C-47 SKYTRAIN/DAKOTA

14 MONTHS PERSEVERANCE: HOW THE ALLIES AND THEIR PLANES SUPPLIED BERLIN

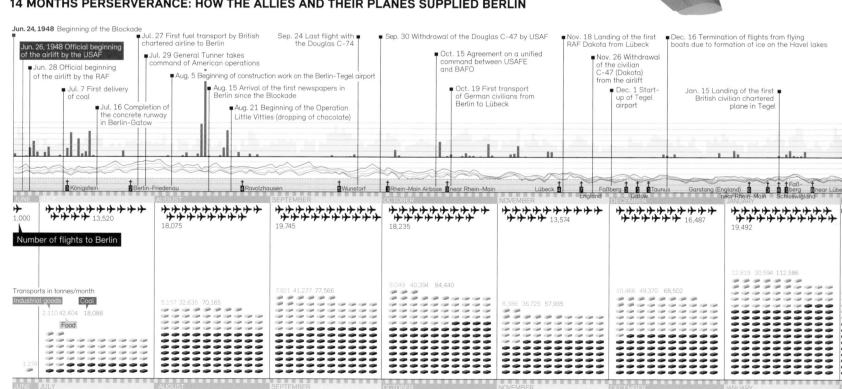

THE TYPES OF AIRCRAFT USED

The Western powers employed a total of 18 different types of aircraft during the airlift, flying a total of 279,114 flights. There were also a variety of types of civil aircraft from chartered and private airlines who put their aircraft at the service of the airlift.

WORKHORSES
The C-47s available at the beginning were gradually replaced by the more efficient C-54s. For the most part, these were used for the transport of coal. In addition, the British also used some Avro Yorks as well as the brand-new Hastings.

TANKER AIRCRAFT
The tankers flew mainly from Schleswig to Gatow. A broad and varied range of aircraft was used. Indeed, many of the tanker aircraft in use were converted Second World War bombers.

*approximate data. **number unknown
L/W/H: Length/Width/Height

FLYING BOATS
Taking off from Hamburg-Finkenwerder on the Elbe, the flying boats then landed on the Wannsee in Berlin. For three months, this route was used mainly to transport salt.

SPECIAL TRANSPORTERS
The C-82 had a hinged door at its stern and so was very suitable for the transport of vehicles and construction machinery. The Globemasters and the Stratofreighters were used mainly to transport supplies between Germany and the USA. They mostly transported complete replacement engines for the C-54 fleet.

OTHER TYPES OF AIRCRAFT IN USE
Some other civilian cargo aircraft were also used. The ones listed here are well-known, including planes from the French Air Force. Four Junkers Ju 52s each flew three times from Wunstorf to Berlin and back. Another C-47 from the French Air Force took off from Baden-Baden to Berlin — though only to supply their own troops.

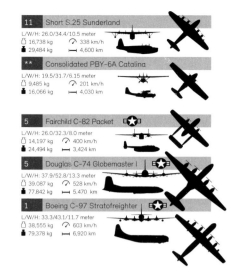

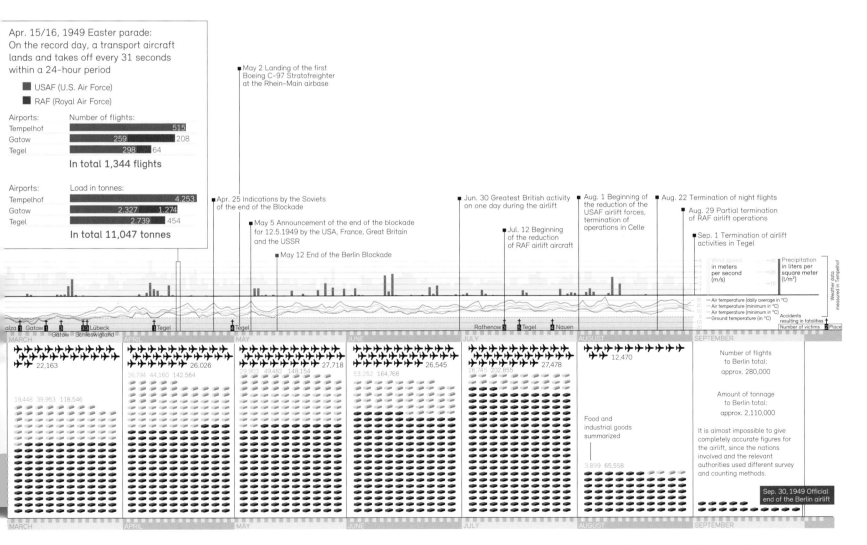

Berlin Airlift
THE DAYS THE EARTH STOOD STILL

GERMANY 1948/1949

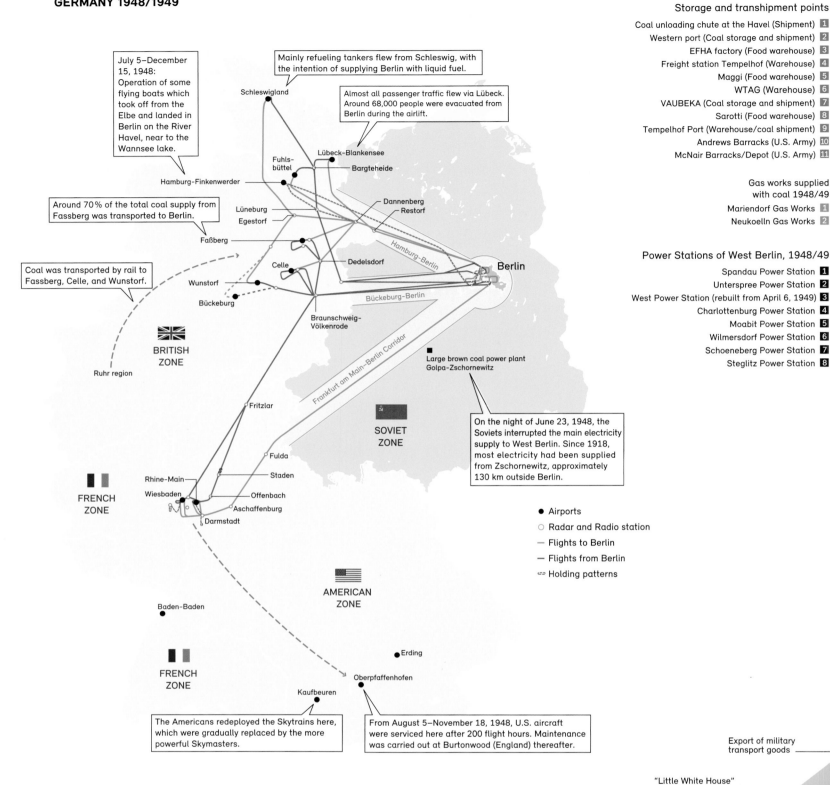

TEMPELHOF AIRPORT

The aircraft, which landed every minute, would line up on the apron of Tempelhof Airport. Within 15 minutes, the cargo of an aircraft would be loaded onto trucks by the waiting task forces (1). The cargo would then be reported to the headquarters (2) and the counting station (3) before it would be sent to one of the two ramps. Coal sacks would be sent to the coal ramp (4) where they would be emptied and transported in wagons. A steam engine would carry the coal to the VAUBEKA transhipment point, 4.5 kilometers away on the Teltow Canal. This journey would take a full hour. The other general cargo and food would be sent to the municipal ramp (5), where they would be sorted and then brought to the numerous storage and transhipment points using civilian trucks (see the map of Berlin on the right). The airport was closed for good on October 30, 2008.

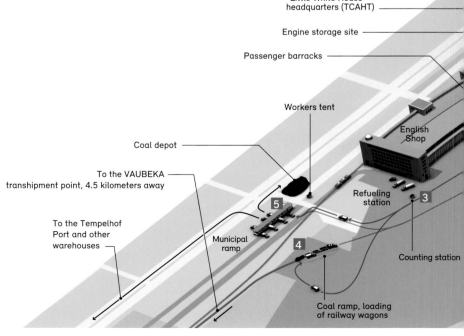

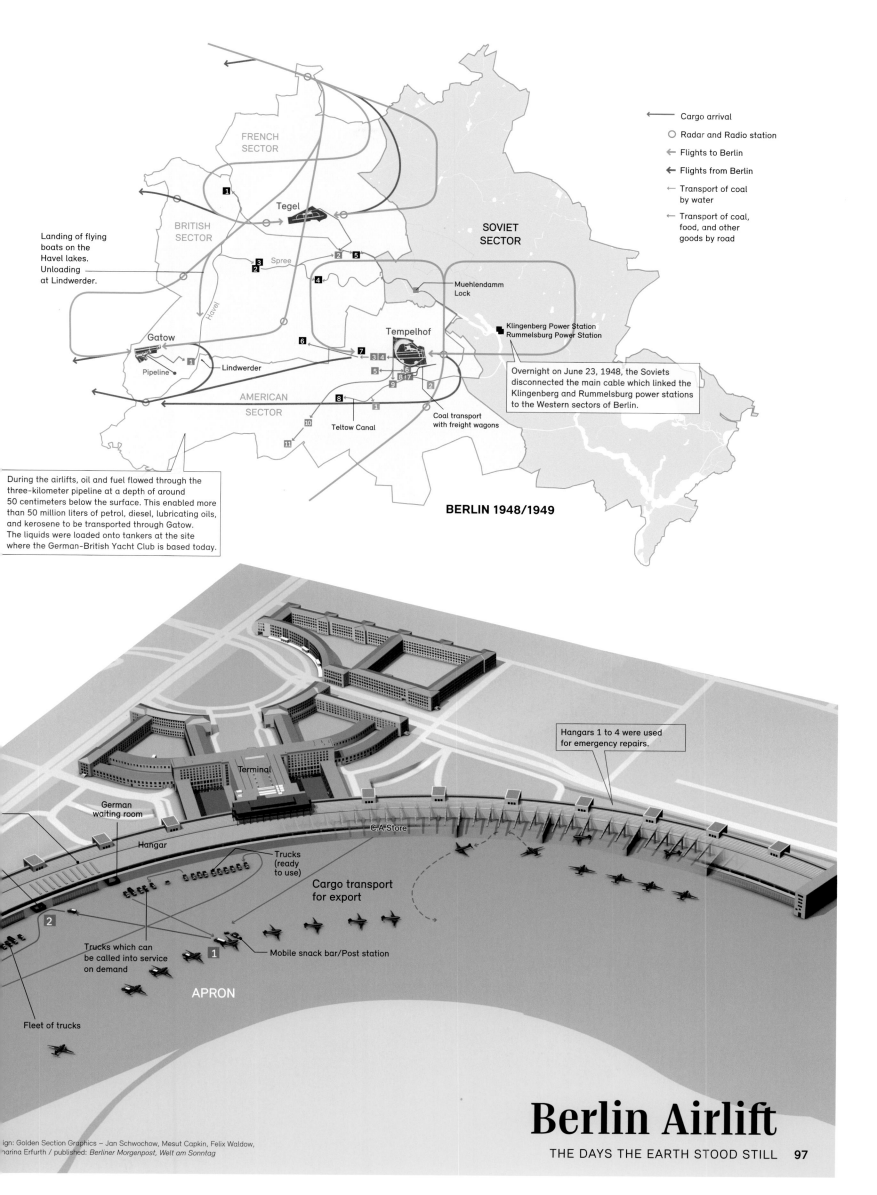

Touching the Ceiling

On May 29, 1953, two men finally realized the dreams of a legion of mountain climbers.

British mountain climber George Leigh Mallory was touring the United States in 1923, raising cash for his attempt on the summit of Mount Everest, when he answered probably the stupidest question a reporter could ask a mountain climber — "Why do you want to climb Mount Everest?" "Because it's there," said Mallory, and those three words became probably the pithiest thing anyone needs to say about human nature and ambition. As if he needed to, the mountaineer later explained what he meant: "Everest is the highest mountain in the world, and no man has reached its summit. Its existence is a challenge. The answer is instinctive, a part, I suppose, of man's desire to conquer the universe." A year later, Mallory was missing. He and his climbing partner Andrew "Sandy" Irvine disappeared from view less than a thousand feet from the summit of the mountain, and though his remains were finally found in 1999, no one knows for sure if Mallory reached the top before he died. But that does not diminish the achievement of New Zealander Edmund Hillary and Nepalese Indian Tenzing Norgay less than 30 years later. This is how they did it.

Effects of altitude on human body

The heroes of the final assault

Edmund Percival Hillary
1919–2008
New Zealand
34 years old

Tenzing Norgay
1914–1986
Nepal
39 years old

The expedition
More than 400 people carried 7.5 tonnes of baggage

- 13 British team
- 34 Sherpas
- 362 Porters
- Many yaks

- Wool mittens, two pairs
- Wooden and steel ice axe
- The team experimented with two oxygen systems — one with closed circuit and the other open; Hillary and Tenzing used the latter
- 20kg
- Woolen suit as insulating layer
- Insulated leggings and boots
- Crampons tied to boots
- Norgay and Hillary climbed roped together
- 10 meters
- Near the summit, Hillary had to fix Norgay's mask, as it was covered with ice.
- Ice axe with the flags of Nepal, UK, India, and UN.

Profile of the track

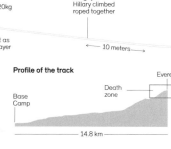

Base Camp — Death zone — Everest — 14.8 km

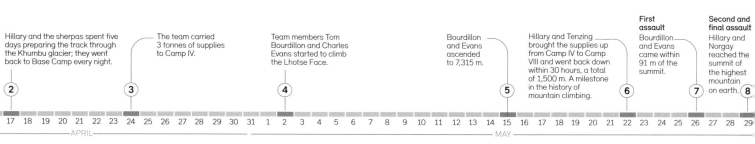

1. The team set up Base Camp after more than a month of hard marching from Kathmandu.
2. Hillary and the sherpas spent five days preparing the track through the Khumbu glacier; they went back to Base Camp every night.
3. The team carried 3 tonnes of supplies to Camp IV.
4. Team members Tom Bourdillon and Charles Evans started to climb the Lhotse Face.
5. Bourdillon and Evans ascended to 7,315 m.
6. Hillary and Tenzing brought the supplies up from Camp IV to Camp VIII and went back down within 30 hours, a total of 1,500 m. A milestone in the history of mountain climbing.
7. **First assault** — Bourdillon and Evans came within 91 m of the summit.
8. **Second and final assault** — Hillary and Norgay reached the summit of the highest mountain on earth.

APRIL — MAY

Design: South China Morning Post — Adolfo Arranz/SCMP
Published: *South China Morning Post*

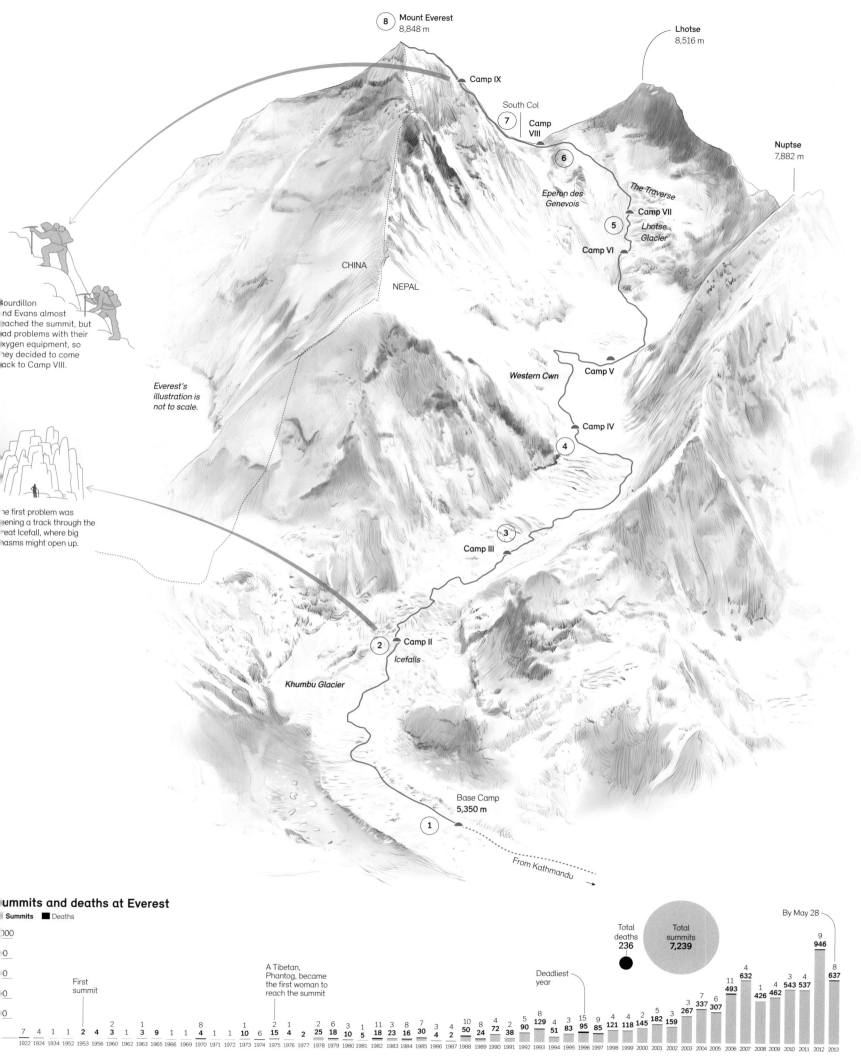

Mount Everest
THE DAYS THE EARTH STOOD STILL

Going MAD

When the world teetered on the edge of "Mutual Assured Destruction."

In October 1962, the grizzled Soviet Premier Nikita Khrushchev decided to test the U.S. President John F. Kennedy, barely 18 months in office. The young upstart, he decided, could be pushed a little further. Following a secret agreement with Cuban leader Fidel Castro, the Soviet Union began building medium-range nuclear weapons bases on the Caribbean island, just over 160 kilometers from the U.S. coast, and within range of all of America's major cities. After all, it was only fair Cold War maneuvering, seeing that the U.S. already had similar missiles in Turkey and Italy that were capable of destroying Moscow, Leningrad, Kiev, and Minsk. But when U.S. spy planes discovered the bases, it sparked a crisis that took the world closer to a full-scale nuclear war than it had ever come before. Instead of attacking Cuba directly — which he'd tried disastrously once before at the Bay of Pigs — Kennedy opted to blockade the island to prevent the Soviet ships from delivering the rockets. There were several flashpoints in the ensuing stand-off, with U.S. planes shot down and Soviet ships being fired on as they attempted to run the blockade. But negotiations continued — the Soviets initially eyeing West Berlin as a grand prize — and were finally resolved when Khrushchev promised to withdraw the missiles if the U.S. agreed to dismantle the Turkish and Italian missiles, which Kennedy did — though only in secret.

October 1962

TUE 16 — Kennedy learns of the missiles on Cuba and calls a group of advisers together: The Executive Committee (ExComm). The reaction from the USA is discussed in secret sessions. These result in the following options:
★ Naval blockade of Cuba
★ Air attacks on the missiles
★ Invasion of Cuba

WED 17 — The first U.S. army units are mobilized. Air reconnaissance discovers further missiles on Cuba.

THUR 18 — State visit from the Soviet Foreign Minister Gromyko to Washington. Kennedy does not reveal that the missiles have been discovered on Cuba. Gromyko maintains that the Soviet Union does not have offensive weapons against the USA stationed there.

FRI 19 — In order to avoid the public becoming aware of the situation, Kennedy leaves Washington on an election trip which has been planned for quite some time.

SAT 20 — Kennedy returns to Washington under the pretext of a minor illness. The ExComm decides on a **naval blockade of Cuba**. However, air attacks and an invasion of Cuba are also being prepared.

SUN 21 — An Air Force general informs Kennedy that he cannot guarantee one hundred percent destruction of the missiles in the case of an air attack in Cuba.

MON 22 — U.S. military forces worldwide are put on a higher state of alert. The newspapers announce a speech by the President. Khrushchev fears the discovery of the missiles and an attack by the USA. In a speech broadcast on radio, Kennedy informs the public, demands the missiles be withdrawn, and announces the naval blockade of Cuba ("quarantine") for the 24th of October. In a letter to Khrushchev, the President repeats his demands and warns against "a war nobody can win."

TUE 23 — Khrushchev denounces the blockade as "outright banditry" and describes the missiles on Cuba as defense weapons. Secretly, he has the majority of cargo ships on course to Cuba ordered to turn back. The USA nuclear bombers and intercontinental missiles are, for the first and only time, put on the second highest state of alert (DEFCON 2).

WED 24 — The blockade comes into effect. Khrushchev warns the USA of Soviet submarines, but declares his readiness to participate in a summit meeting, if no force is exercised against Soviet ships.

THUR 25 — The UN secretary-general Sithu U Thant appeals to both sides to take a few weeks to reflect on the situation. The Soviet UN Ambassador claims to know nothing about medium-range missiles on Cuba and is embarrassed by the U.S. Ambassador with photographic evidence.

FRI 26 — The Soviet Ambassador in Washington offers to withdraw the missiles, if the USA officially declares that it will not use military force against Cuba in the future.

SAT 27 — "Black Saturday": A U-2 reconnaissance aircraft is shot down over Cuba by a Soviet anti-aircraft defense missile and the U.S. navy forces the Soviet submarine "B-59" to surface. The commander of the submarine believes himself to be at war and is about to launch a nuclear torpedo. The ExComm is confronted with the demand from Moscow that, in return for the withdrawal of the Cuban missiles, the American Jupiter missiles are also to be withdrawn from Europe. U.S. Attorney General Robert Kennedy meets with the Soviet Ambassador in Washington.

SUN 28 — Radio Moscow announces the withdrawal of the missiles from Cuba in return for a guarantee from the USA that it will not invade Cuba. Informally, the Soviet Union still expects the withdrawal of U.S. medium-range missiles from Europe in the future. In November, the first missiles leave Cuba and the blockade is lifted.

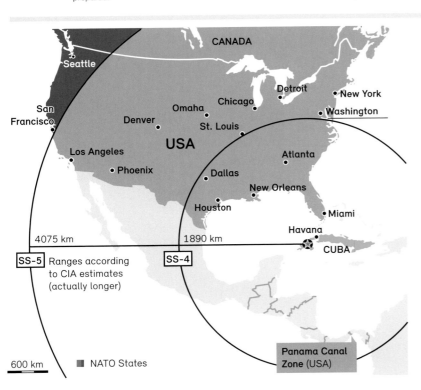

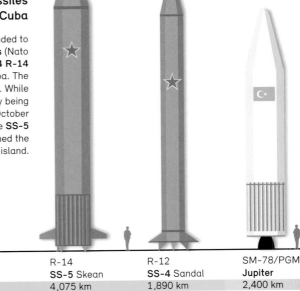

Medium-range missiles for Cuba

The Soviet Union intended to station **36 R-12 missiles** (Nato description **SS-4**) and **24 R-14 missiles (SS-5)** on Cuba. The operation was top secret. While the **SS-4s** were already being made ready for use in October 1962, only a few of the **SS-5** nuclear warheads reached the Caribbean island.

The **nuclear warheads** for these missiles have at least ten times the explosive power of the Hiroshima bomb.

	R-14 SS-5 Skean	R-12 SS-4 Sandal	SM-78/PGM Jupiter
Range	4,075 km	1,890 km	2,400 km
Length	24 m	22 m	18 m
Diameter	2.4 m	1.7 m	2.6 m
Start weight	86 t	42 t	49 t

Design: dpa-infografik – Andreas Brühl / published: *Hellweger Anzeiger*

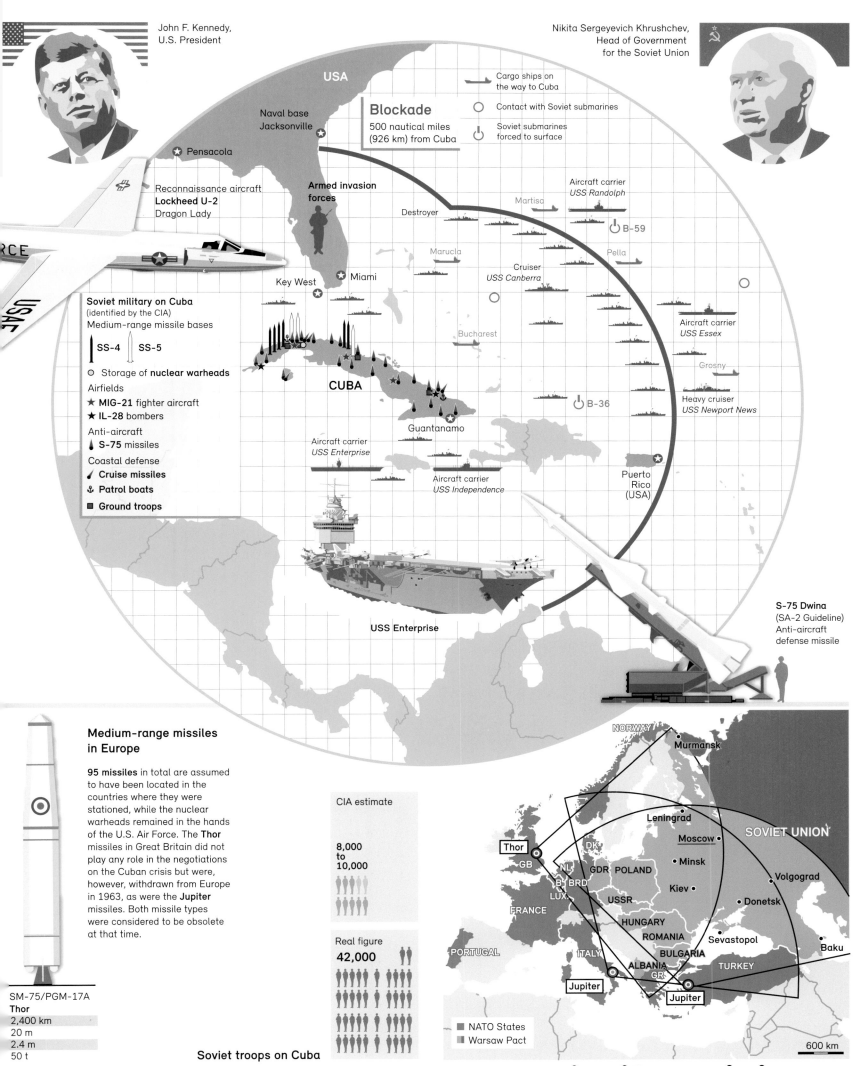

Cuban Missile Crisis

THE DAYS THE EARTH STOOD STILL

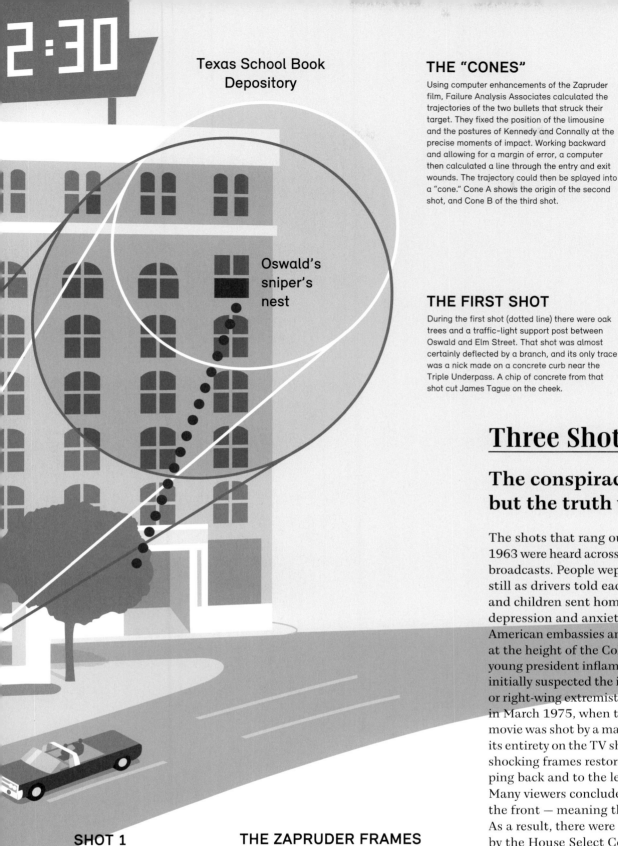

THE "CONES"

Using computer enhancements of the Zapruder film, Failure Analysis Associates calculated the trajectories of the two bullets that struck their target. They fixed the position of the limousine and the postures of Kennedy and Connally at the precise moments of impact. Working backward and allowing for a margin of error, a computer then calculated a line through the entry and exit wounds. The trajectory could then be splayed into a "cone." Cone A shows the origin of the second shot, and Cone B of the third shot.

THE FIRST SHOT

During the first shot (dotted line) there were oak trees and a traffic-light support post between Oswald and Elm Street. That shot was almost certainly deflected by a branch, and its only trace was a nick made on a concrete curb near the Triple Underpass. A chip of concrete from that shot cut James Tague on the cheek.

Three Shots in Dallas

The conspiracy was always seductive, but the truth was probably simpler.

The shots that rang out from Dealey Plaza on November 22, 1963 were heard across the world. TV stations interrupted their broadcasts. People wept in the streets. Traffic came to a standstill as drivers told each other the news. Schools were closed and children sent home. There were reports of higher rates of depression and anxiety, while switchboards were jammed at American embassies and consulates around the world. Coming at the height of the Cold War, the assassination of the popular young president inflamed the general paranoia, as many people initially suspected the involvement of communist governments or right-wing extremists. But the real conspiracy theories began in March 1975, when the Zapruder film (because this amateur movie was shot by a man called Zapruder) was first broadcast in its entirety on the TV show Good Night America. With the most shocking frames restored, it showed the president's head snapping back and to the left when the fatal third shot struck him. Many viewers concluded that the bullet must have come from the front — meaning that at least two shooters were involved. As a result, there were immediate calls for a new investigation by the House Select Committee, which eventually decided in 1979 that there had "probably" been a conspiracy, but failed to point to any particular groups that may have been involved, and expressly ruled out the involvement of the CIA, the Soviet Union, and the mafia. The conspiracy theory has since swollen even further in the public consciousness. On the publication of the initial Warren Report in September 1964, only 31 percent of Americans believed that Lee Harvey Oswald had not acted alone, but various polls in the past decade show that as many as 75 percent of people believe there was a conspiracy to kill the president. But as this graphic shows, reconstructions have found that all the shots that killed John F. Kennedy were most likely to have come from the sixth floor of the Book Depository building, and that there is no reason to believe that anyone other than the Marxist activist Oswald was involved.

THE ZAPRUDER FRAMES

The moments of impact of Oswald's second and third shots can be established by analyzing frames of the Zapruder film. The second shot (Cone A) hit both the President and Governor Connally just as their limousine emerged into Zapruder's view from behind a freeway sign. Careful analysis points to the impact of Oswald's second shot at frames 223–224. The third shot (Cone B), in full view of Zapruder, hit Kennedy in the back of his head at frame 313.

SHOT 1

0 SECONDS

FRAMES
160 to 166

JFK

THE DAYS THE EARTH STOOD STILL

THE GUN

Oswald's Italian Second World War Mannlicher-Carcano was purchased from a mail-order house in Chicago. The rifle cost $12.78, the 4× telescopic sight $7.17. In the Marines, Oswald was proficient with an M-1 rifle at distances of up to 180 meters — without the benefit of a telescopic sight. He had practiced to become equally effective with the Mannlicher-Carcano. The sling, adapted from the belt of a Navy pistol holster, provided additional steadiness. A brown bag, 8 centimeters longer than the disassembled rifle, was found in the sniper's nest.

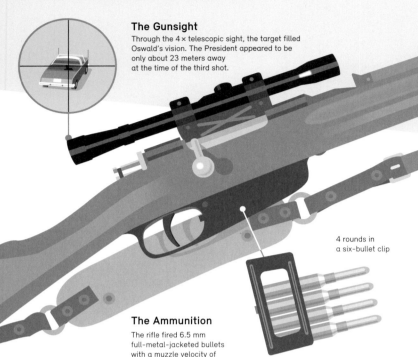

The Gunsight
Through the 4× telescopic sight, the target filled Oswald's vision. The President appeared to be only about 23 meters away at the time of the third shot.

The Ammunition
The rifle fired 6.5 mm full-metal-jacketed bullets with a muzzle velocity of over 2,000 feet per second.

4 rounds in a six-bullet clip

MARKSMANSHIP

Did Oswald have time to fire three shots? Enhancements of the Zapruder film lead to the answer. His first shot missed. He had at least 3 seconds to reload, aim, and fire the second shot, which hit both Kennedy and Connally. He then had another 5 seconds — ample time — for the third shot, which killed the President.

Bolt Action
The bolt action can easily be executed in a fraction of a second.

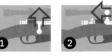

1. Push bolt up …
2. Pull back (to eject case and position next cartridge) …
3. Push forward …
4. Push down (to lock bolt).

THE SINGLE BULLET

Oswald's second shot, the first to strike, is the most contentious. It is variously called the "magic" or "pristine" bullet by conspiracy theorists, who contend that no single bullet could have so seriously wounded both men. The bullet needed no magic and was not pristine. Its trajectory, based on the Failure Analysis computations and the Zapruder film, is reconstructed here.

View from above
The trajectory, plotted in accordance with the exact postures of both men, was not significantly altered until the bullet was slightly deflected by Connally's rib. ▶

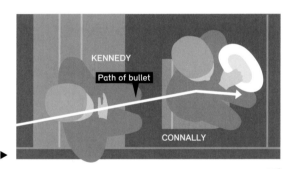

KENNEDY — Path of bullet — CONNALLY

BULLET SPEED
518–549 meters per second

FRAMES 223, 224

KENNEDY
- Entry wound in the back 6.5 mm in diameter.
- Bullet grazed tip of a vertebra in the neck, slightly splintering the bone.
- Exit wound in throat.
- Cavity momentarily caused by bullet's passage.
- Bullet tumbling.

CONNALLY
- Entry wound in right shoulder was 3 cm long — the exact length of the bullet — indicating the bullet was tumbling end over end.

457–488 meters per second

- Exit wound below the right nipple was large — nearly 5 cm in diameter — and ragged; the bullet was still tumbling.
- Traverses chest and shatters fifth right rib.

Stetson hat.

274 meters per second

- Entry wound at top of right wrist was ragged and irregular. The bullet, now traveling backward, fractured the radius bone.

When the bullet came to rest in Connally's left thigh, having lost more than 80 percent of its velocity, it was just able to penetrate skin.

122 meters per second

FRAME 230

By frame 226 the President began to show a neurological reflex — known as the Thorburn position — to spinal injury. His arms jerked up to a fixed position, hands nearly at his chin, elbows pushed out.

104

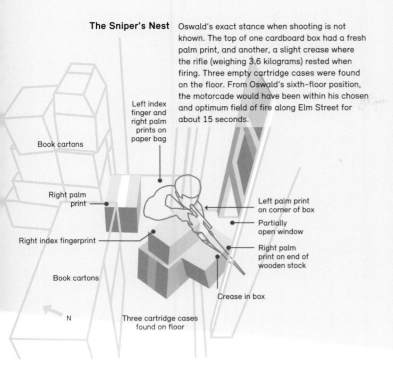

The Sniper's Nest Oswald's exact stance when shooting is not known. The top of one cardboard box had a fresh palm print, and another, a slight crease where the rifle (weighing 3.6 kilograms) rested when firing. Three empty cartridge cases were found on the floor. From Oswald's sixth-floor position, the motorcade would have been within his chosen and optimum field of fire along Elm Street for about 15 seconds.

Lee Harvey Oswald, a U.S.-trained assassin.

Many of the conspiracy theories around Lee Harvey Oswald's involvement in Kennedy's assassination center on the rifle he used, and whether he could have fired off three shots within the time the Zapruder film showed. But Oswald was a skilled marksman trained in the U.S. Marines who knew where to position himself to get the easiest shot at the President, and the bolt-action rifle he owned could be reloaded in fractions of a second. Oswald's past has been endlessly scrutinized in an attempt to find evidence for or against a conspiracy. His lonely, withdrawn childhood, his training in the Marines, his defection to the Soviet Union, where skeptical authorities thwarted his ambitions to study in Moscow by sending him to work at an electronics factory in Minsk. That work left him unhappy enough to return to a surprisingly forgiving United States, where he unsuccessfully attempted to kill the outspoken anti-communist and racist General Edwin Walker. He was also involved in handing out pro-Cuban literature in New Orleans. On the afternoon of November 22, 1963, he was initially arrested in a movie theater after acting suspiciously and then pulling a revolver on the arresting officers. He was first charged with the murder of Dallas patrolman J. D. Tippit, and then with the Kennedy assassination. He denied both charges, told the press that he was a "patsy," before being abruptly killed two days later by nightclub operator Jack Ruby while being transferred to a county jail. His grave bears only his surname.

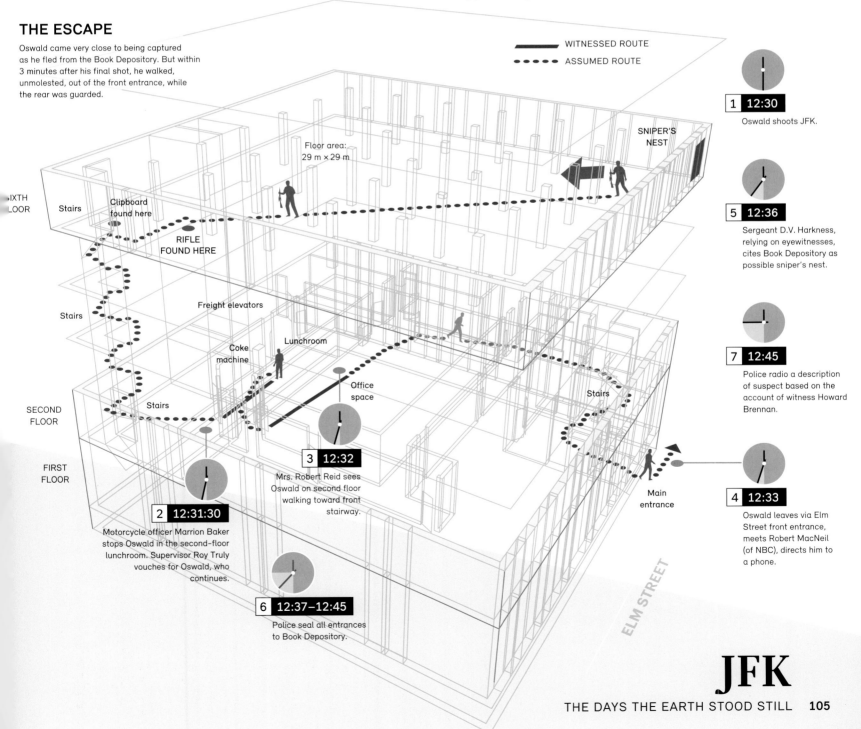

THE ESCAPE

Oswald came very close to being captured as he fled from the Book Depository. But within 3 minutes after his final shot, he walked, unmolested, out of the front entrance, while the rear was guarded.

1 **12:30** Oswald shoots JFK.

2 **12:31:30** Motorcycle officer Marrion Baker stops Oswald in the second-floor lunchroom. Supervisor Roy Truly vouches for Oswald, who continues.

3 **12:32** Mrs. Robert Reid sees Oswald on second floor walking toward front stairway.

4 **12:33** Oswald leaves via Elm Street front entrance, meets Robert MacNeil (of NBC), directs him to a phone.

5 **12:36** Sergeant D.V. Harkness, relying on eyewitnesses, cites Book Depository as possible sniper's nest.

6 **12:37–12:45** Police seal all entrances to Book Depository.

7 **12:45** Police radio a description of suspect based on the account of witness Howard Brennan.

JFK

THE DAYS THE EARTH STOOD STILL

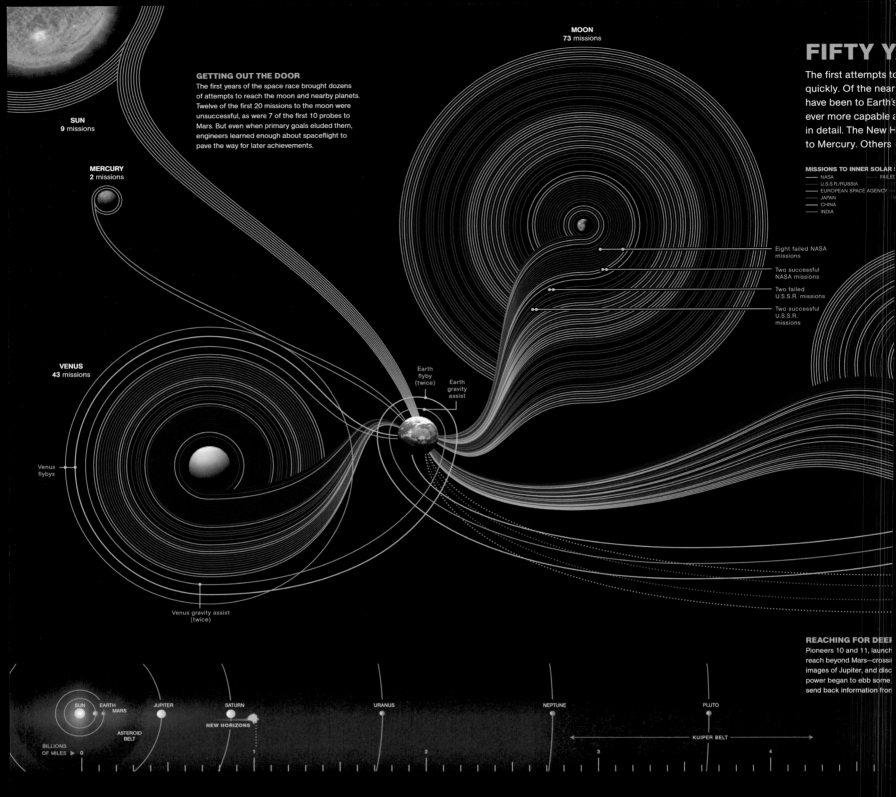

Far Out, Man

The history of space exploration so far. A visualization of all the places we've been.

As far as space exploration goes, we are still hopelessly shy homebodies. We're like Emily Dickinson, forever pottering around a draughty house and waiting for visitors. The only place we've visited personally is our moon. And even there we were just tourists — seeing the sights, going for a stroll, collecting a few souvenirs, flying home. And we aren't showing much interest in going back either. We have of course sent a few exploratory robots out — to Mars, for one thing. There is even a plan to establish a colony on the red planet by 2023, which received 78,000 applications for a one-way, the-rest-of your-life-ticket within two weeks. Meanwhile, our yearning to make contact is expressed most sentimentally with Voyager 1, the one human-made object that has traveled farther than any other. Since its launch in 1977, it has journeyed over 22 billion kilometers among the planets, and is due to reach interstellar space in the next couple of years, still carrying its message in a bottle — the famous golden record with recordings of Mozart, babies crying, whales singing, and waves crashing on a shore. If the aliens ever find it, they'll think we're a pretty slushy bunch.

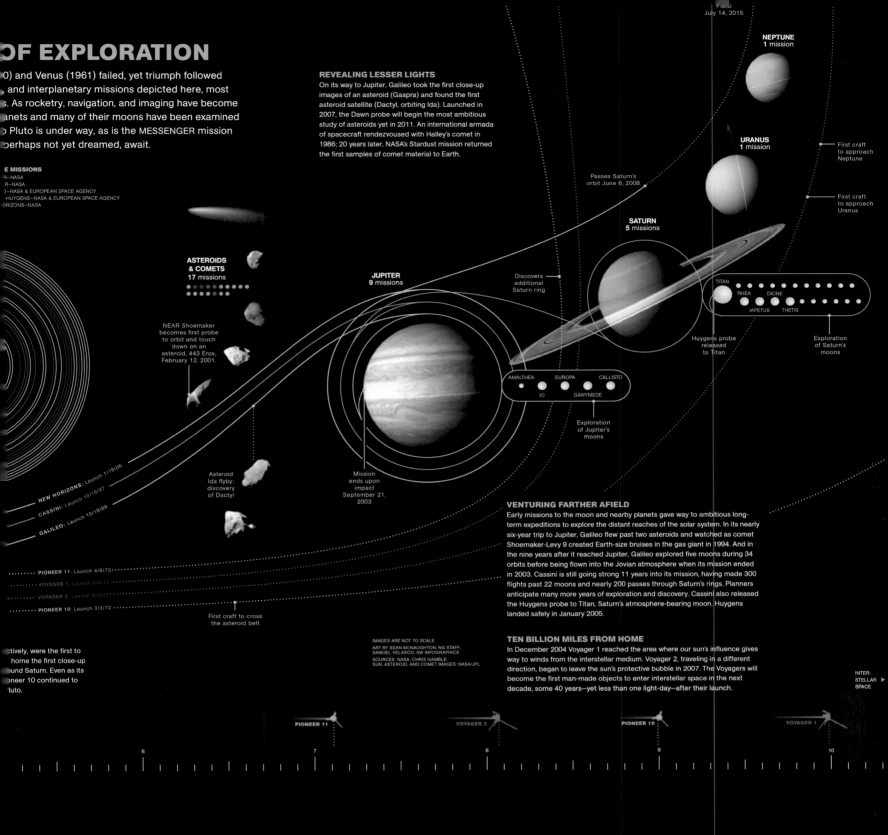

The Space Age

The rise and fall of the world's first reusable spacecraft.

On January 28, 1986, one of NASA's four original Space Shuttles exploded shortly after launch, killing all seven astronauts. The Challenger tragedy was a serious blow not only to the U.S. ego and its space exploration ambitions, but to the dreams of every child. Together with their parents, children witnessed Neil Armstrong hopping on the moon and heard his voice so succinctly say: "One small step for man, one giant leap for mankind." Those kids had their imaginations fired up by the spectacular space battle that closed the 1979 James Bond blockbuster *Moonraker*. The real Space Shuttle was actually yet to celebrate its maiden launch when Roger Moore piloted the screen version (armed with a laser cannon), but we already knew that this was how we wanted space travel to be. We didn't want our astronauts to be shot mindlessly into space on the top of an unwieldy, disposable rocket. We wanted them to fly up on a winged spacecraft shaped like a killer whale, and then glide back down gracefully, having explored the universe and taken humanity one step closer to its secrets. The individual Shuttle names — Discovery, Challenger, Endeavour, Atlantis — were clearly meant to provoke a deep primeval desire to go further and find new worlds. Unfortunately, of course, the Shuttle was incapable of flying beyond low Earth orbit. And it couldn't get anywhere at all without the help of enormous disposable boosters, or, if being transported, piggybacking on a Boeing 747. The noble Space Shuttle was not retired until 2011, having served some 15 years longer than it was originally designed to do. At the moment, U.S. astronauts going to the International Space Station have to hitch a ride on Russian Soyuz rockets. We are still waiting for a successor to take us further.

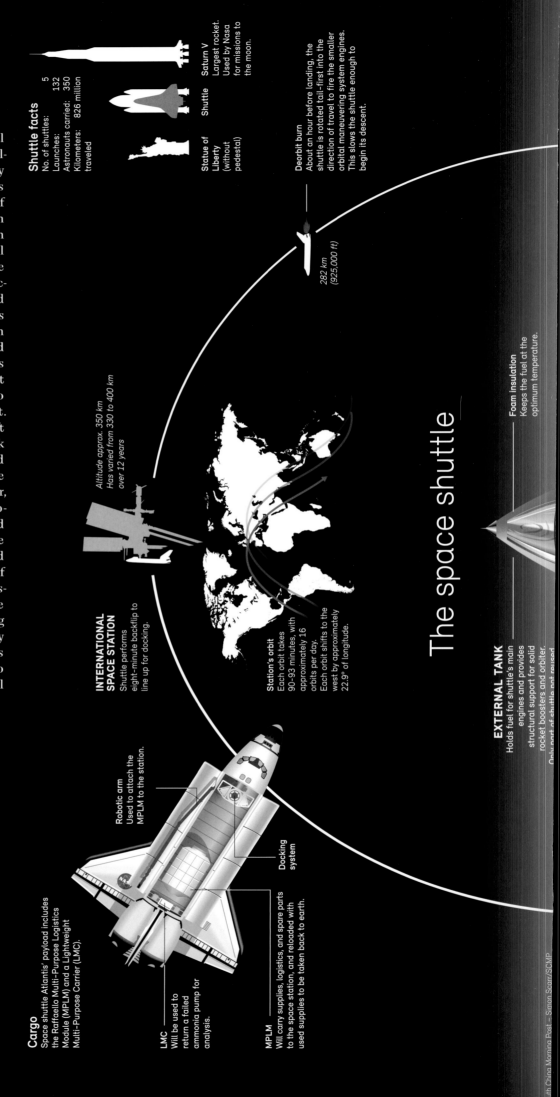

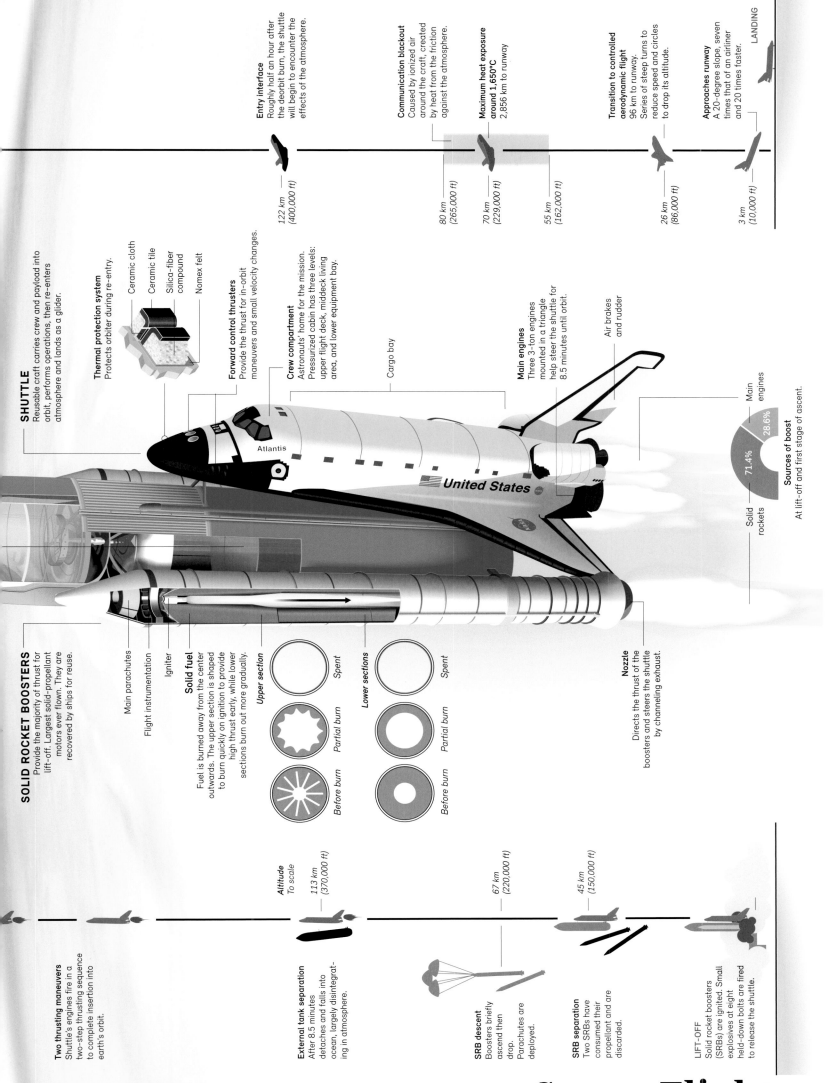

Space Flight
THE DAYS THE EARTH STOOD STILL

Attack on the World Trade Center

The images of the aircraft being flown into the World Trade Center on September 11, 2001, are unforgettable and are locked in the memory of our society forever. In the four attacks, a total of 2996 people lost their lives, including 19 terrorists. This and the following pages give an insight into the knowledge that has been gained over the last years about these events.

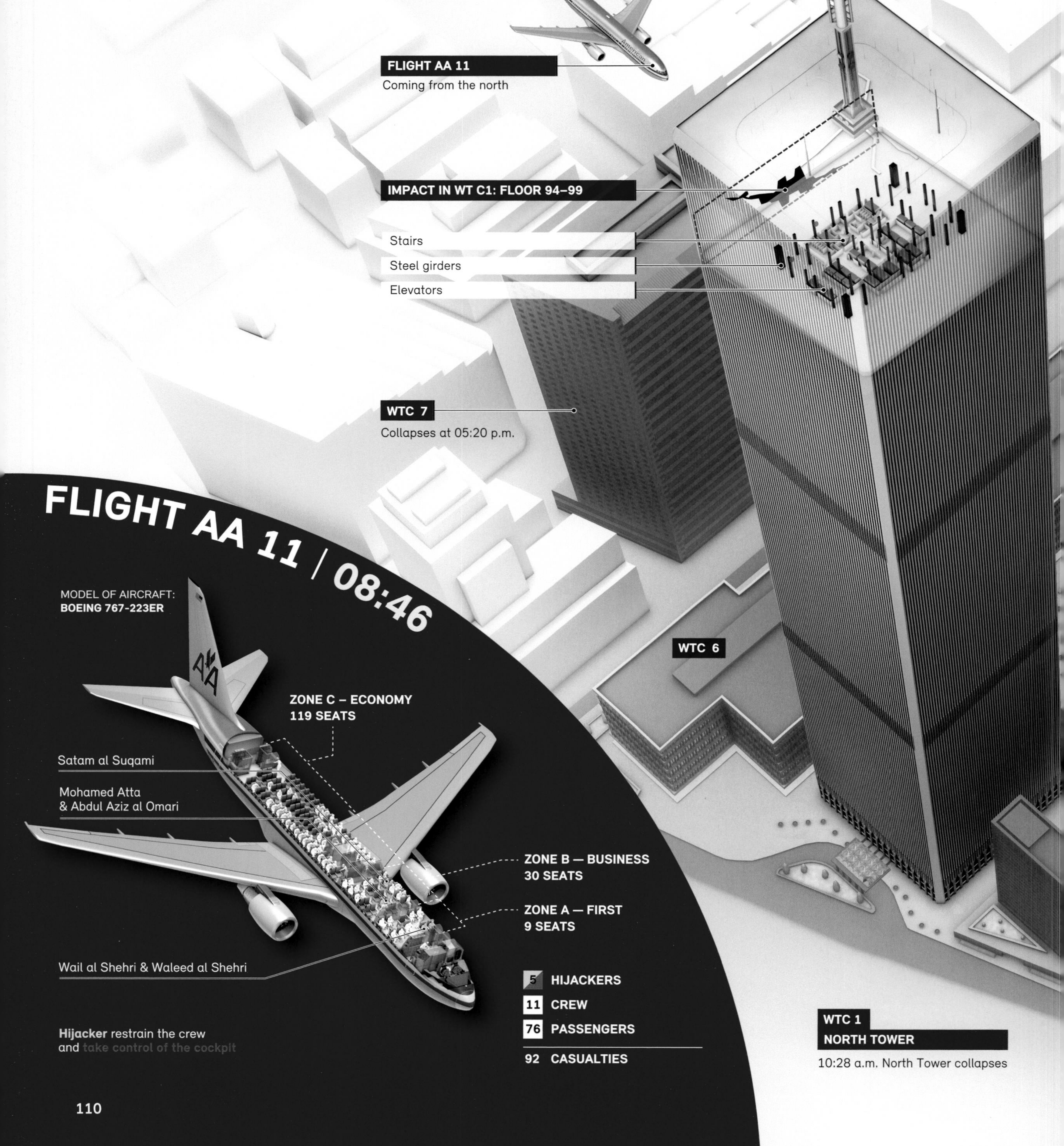

FLIGHT AA 11
Coming from the north

IMPACT IN WT C1: FLOOR 94–99

Stairs
Steel girders
Elevators

WTC 7
Collapses at 05:20 p.m.

WTC 6

WTC 1
NORTH TOWER
10:28 a.m. North Tower collapses

FLIGHT AA 11 / 08:46

MODEL OF AIRCRAFT:
BOEING 767-223ER

Satam al Suqami

Mohamed Atta
& Abdul Aziz al Omari

Wail al Shehri & Waleed al Shehri

ZONE C — ECONOMY
119 SEATS

ZONE B — BUSINESS
30 SEATS

ZONE A — FIRST
9 SEATS

5	HIJACKERS
11	CREW
76	PASSENGERS
92	CASUALTIES

Hijacker restrain the crew and take control of the cockpit

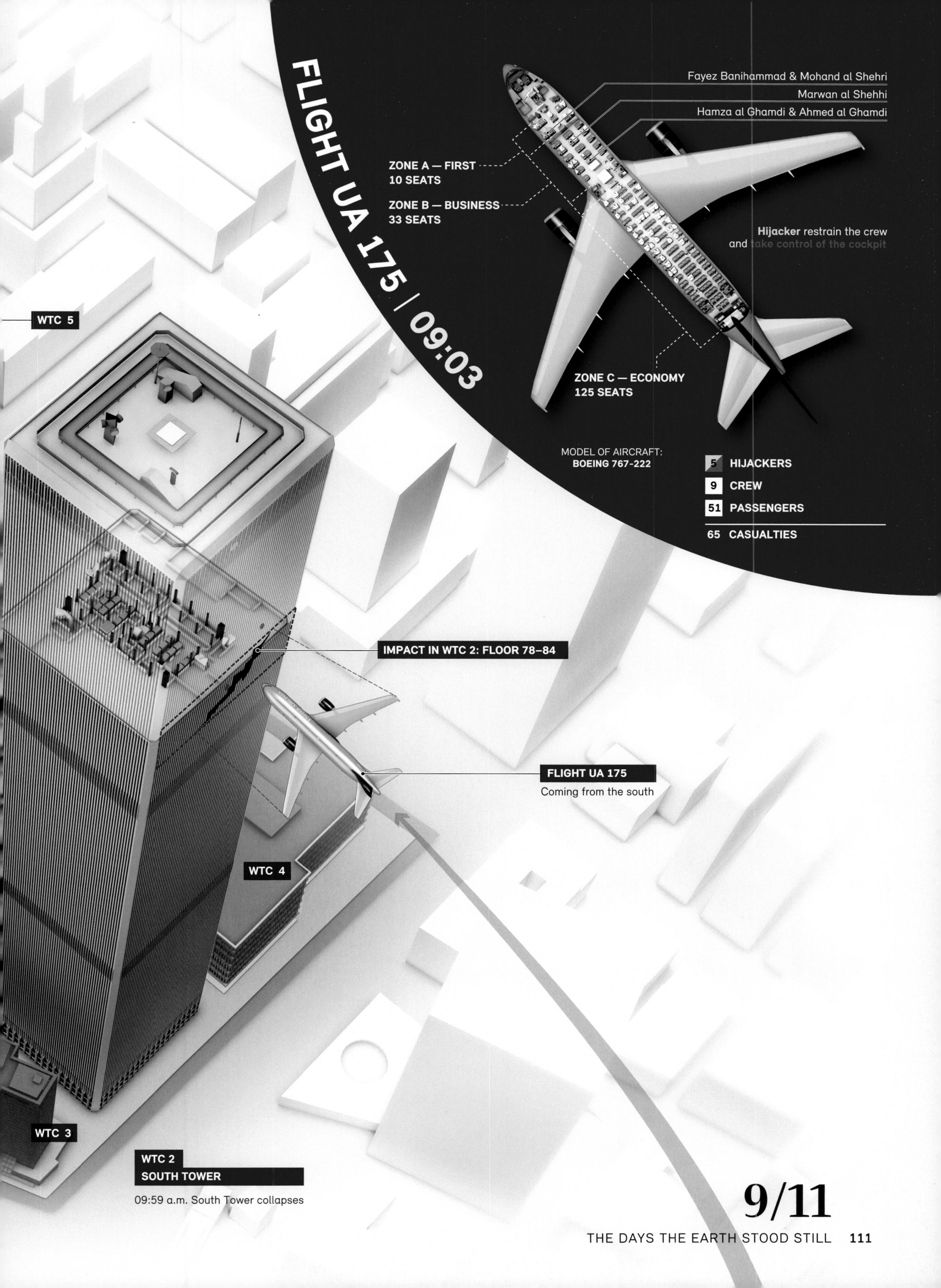

FLIGHT UA 175 | 09:03

Fayez Banihammad & Mohand al Shehri
Marwan al Shehhi
Hamza al Ghamdi & Ahmed al Ghamdi

ZONE A — FIRST
10 SEATS

ZONE B — BUSINESS
33 SEATS

Hijacker restrain the crew and take control of the cockpit

ZONE C — ECONOMY
125 SEATS

MODEL OF AIRCRAFT:
BOEING 767-222

5	HIJACKERS
9	CREW
51	PASSENGERS
65	CASUALTIES

WTC 5

IMPACT IN WTC 2: FLOOR 78–84

FLIGHT UA 175
Coming from the south

WTC 4

WTC 3

**WTC 2
SOUTH TOWER**
09:59 a.m. South Tower collapses

9/11

THE DAYS THE EARTH STOOD STILL

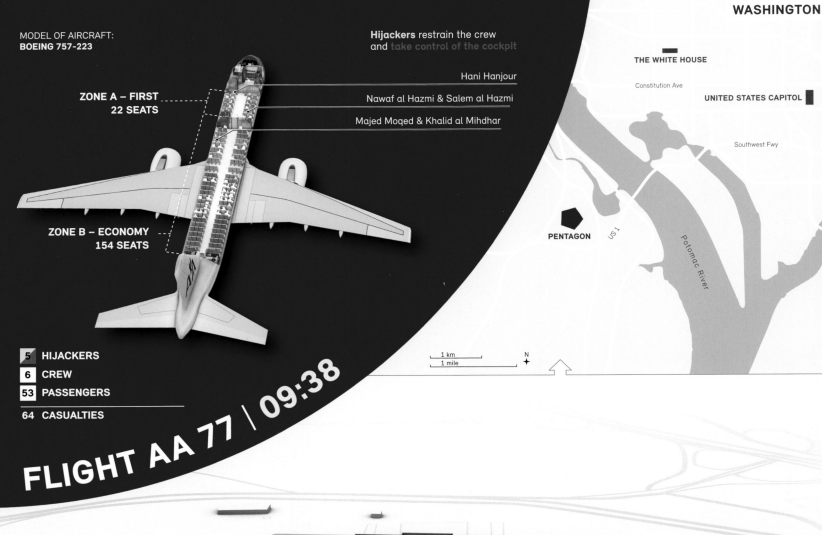
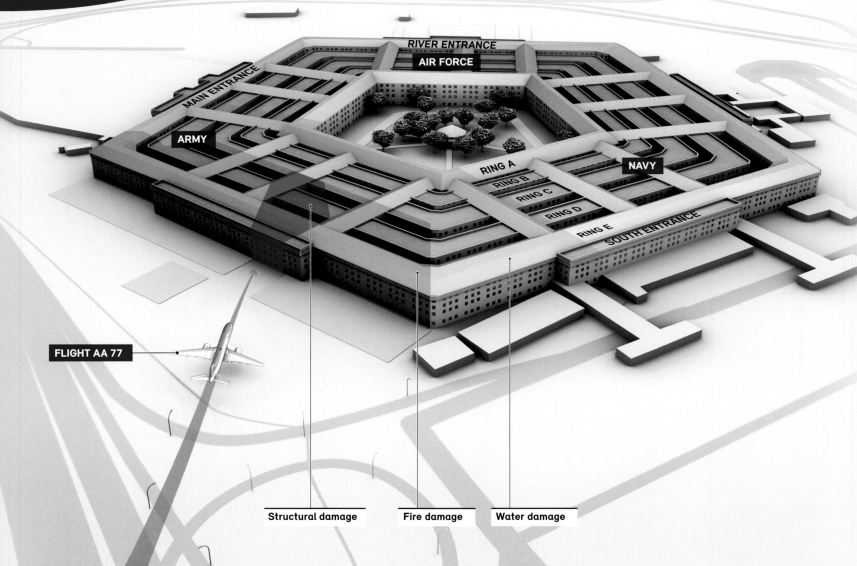

Third Target: the Pentagon. And the White House?

There were 125 people who died on the attacks on the Pentagon. Not everything is known about what happened on flight United 93. However it is quite certain that the crew and some passengers had found out what had happened in New York. Aware of the fact they would likely not survive the flight, they invaded the cockpit to try to regain control of the plane, or crash it away from urban areas. They succeeded with that second goal. None of the 44 passengers survived.

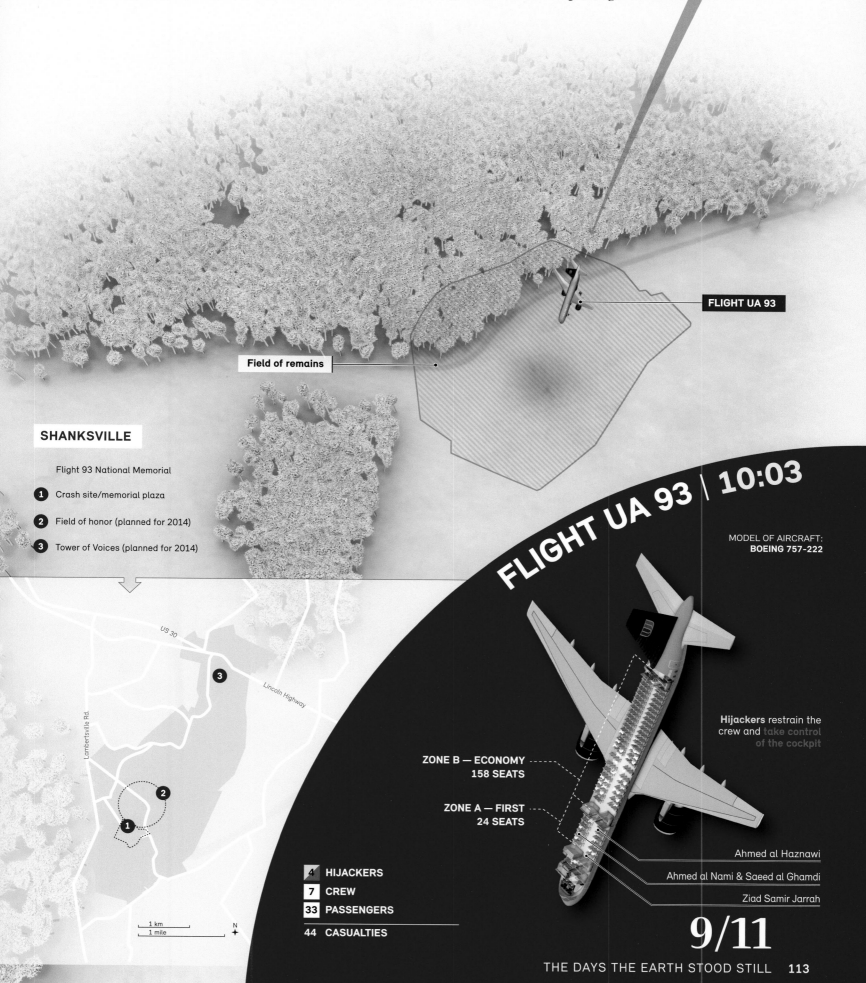

FLIGHT UA 93

Field of remains

SHANKSVILLE

Flight 93 National Memorial

1. Crash site/memorial plaza
2. Field of honor (planned for 2014)
3. Tower of Voices (planned for 2014)

FLIGHT UA 93 | 10:03

MODEL OF AIRCRAFT: **BOEING 757-222**

Hijackers restrain the crew and take control of the cockpit

ZONE B — ECONOMY
158 SEATS

ZONE A — FIRST
24 SEATS

Ahmed al Haznawi
Ahmed al Nami & Saeed al Ghamdi
Ziad Samir Jarrah

4	HIJACKERS
7	CREW
33	PASSENGERS
44	CASUALTIES

9/11
THE DAYS THE EARTH STOOD STILL

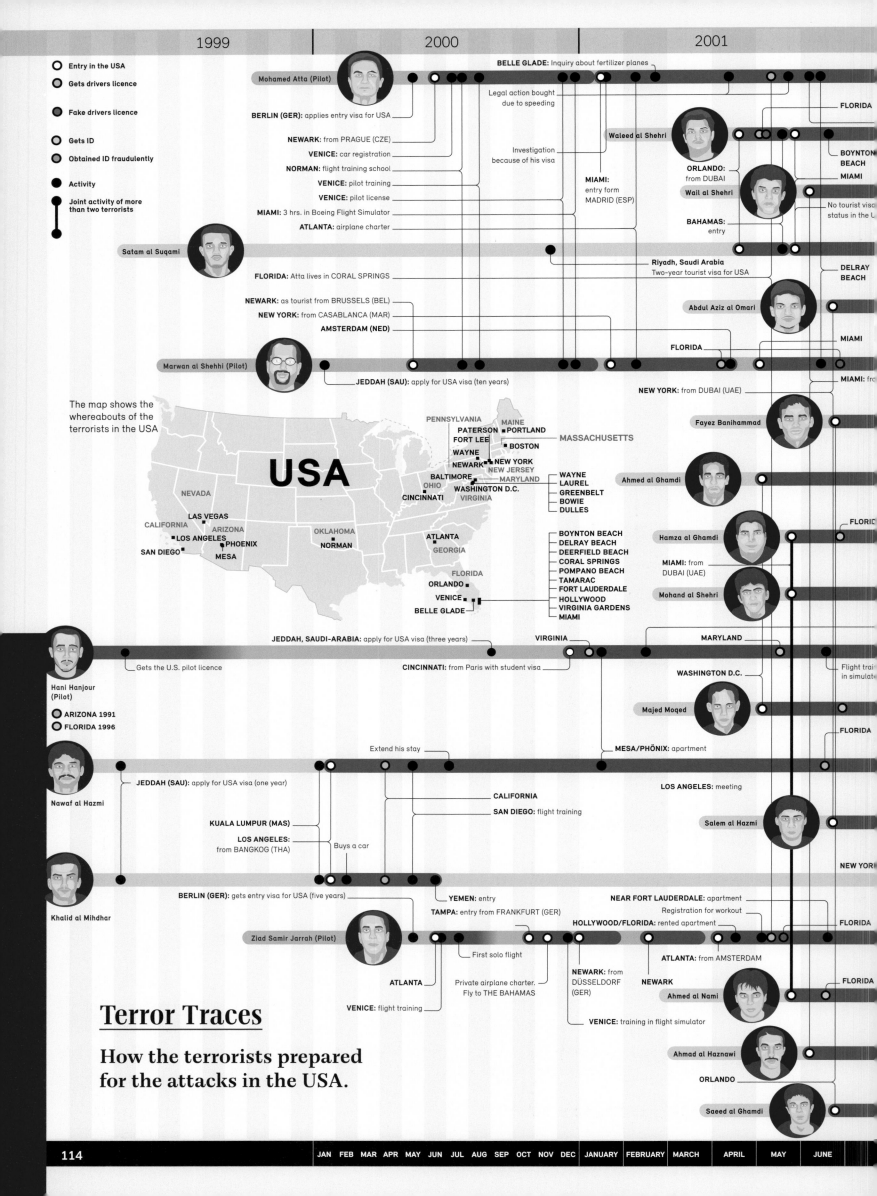

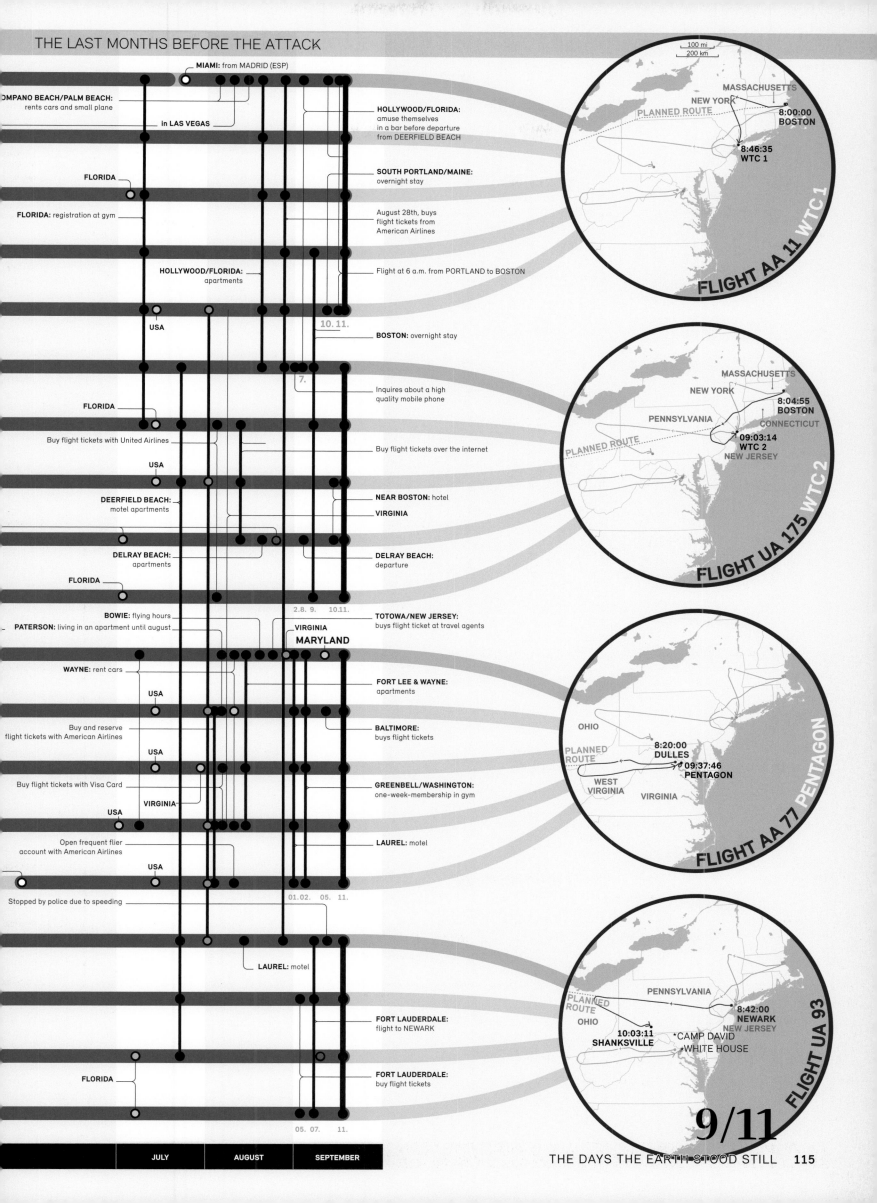

The Price of War

Counting the cost — both human and financial — of 9/11.

Over a decade later, the terrorist attacks of September 11, 2001 remain the single most decisive event of the twenty-first century so far. Its significance for global power relations has barely diminished, even if U.S. foreign policy has changed tack a little since. On the other hand, while there was much talk at the time of the psychological blow to U.S. self-confidence, that seems to be one of the more transient outcomes. Much more permanent was the damage caused to people's lives — both to military servicemen and women in the U.S., and civilian populations abroad. In Iraq and Afghanistan, swift initial victories borne by the patriotic euphoria gave way to

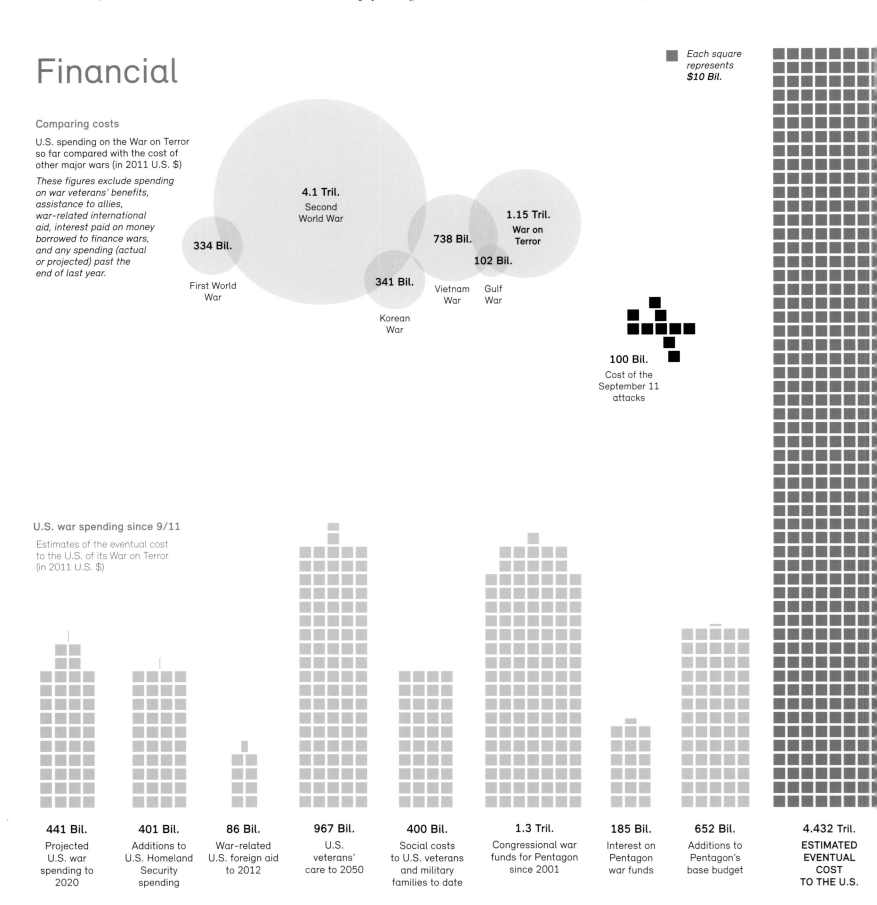

long, agonizing, mismanaged reconstruction periods that turned bloodier and more chaotic as they dragged on. Even the economic benefits — the lucrative Iraqi oil contracts often mentioned as the "real" reason for the "War on Terror" — have largely gone elsewhere, especially to China. Ten years after 9/11, U.S. and NATO forces — facing increasingly war-fatigued populations — were looking for a way to extract themselves as painlessly as possible. This graphic attempts to count up the damage done on that Tuesday morning in the fall of 2001.

Human

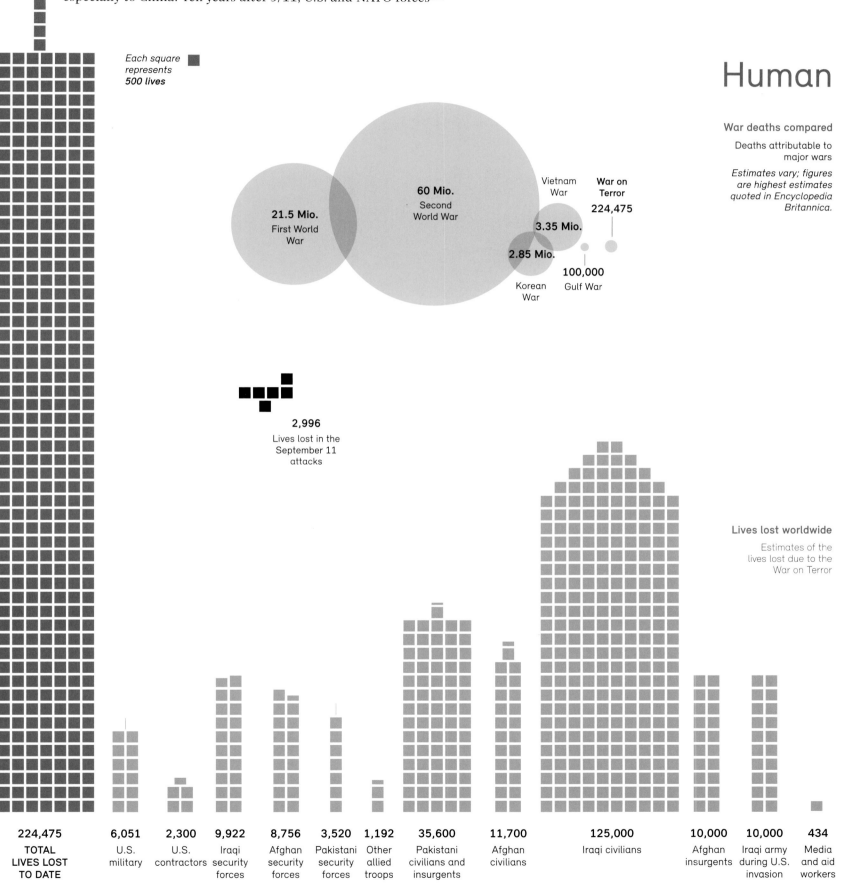

Each square represents **500 lives**

War deaths compared

Deaths attributable to major wars

Estimates vary; figures are highest estimates quoted in Encyclopedia Britannica.

- 21.5 Mio. First World War
- 60 Mio. Second World War
- 2.85 Mio. Korean War
- 3.35 Mio. Vietnam War
- 100,000 Gulf War
- 224,475 War on Terror

2,996 Lives lost in the September 11 attacks

Lives lost worldwide

Estimates of the lives lost due to the War on Terror

| 224,475 TOTAL LIVES LOST TO DATE | 6,051 U.S. military | 2,300 U.S. contractors | 9,922 Iraqi security forces | 8,756 Afghan security forces | 3,520 Pakistani security forces | 1,192 Other allied troops | 35,600 Pakistani civilians and insurgents | 11,700 Afghan civilians | 125,000 Iraqi civilians | 10,000 Afghan insurgents | 10,000 Iraqi army during U.S. invasion | 434 Media and aid workers |

9/11

THE DAYS THE EARTH STOOD STILL

Meltdown in Japan

An earthquake, a tsunami, and a nuclear disaster.

On March 11, 2011, at 2:46 p.m. local time, the Tohokou earthquake struck 70 kilometers off the coast of Japan with such force that the entire country shifted two meters, and the Earth's axis wobbled. As well as from the devastation, which cost around 16,000 lives, the resulting tsunami knocked out the cooling system at a coastal nuclear power station dating from the 1970s. This graphic shows what happened next in the worst nuclear catastrophe since 1986, when the fallout from Chernobyl, Ukraine released a radioactive cloud over Central Europe and desolated an entire city. Nuclear power stations work by firing neutrons into uranium atoms, and then controlling the resulting chain reaction. The reactions heat

REACTOR MODEL

The three stricken reactors at the Fukushima I (Daiichi) Nuclear Power Plant were all boiling water reactors (BWR) developed by General Electric, Toshiba, and Hitachi, and owned by TEPCO, the Japanese energy giant that operates most of the nuclear power stations in the country. This reactor is based on an older U.S. model from the 1960s, shown in the schematic on the right.

The fuel elements, regulated by control rods and circulating pumps, heat the water inside the reactor pressure vessel (RPV). The resulting steam drives the turbines, which in turn powers a generator. The superheated steam is then condensed inside a condenser by a separate coolant system. The cooled feed water is warmed in a pre-heater before being led back into the RPV.

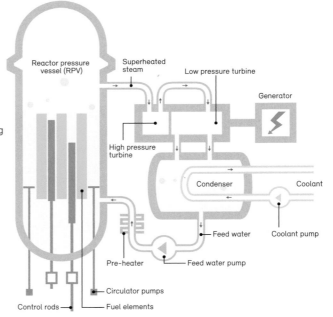

FUKUSHIMA DAIICHI NUCLEAR POWER PLANT

UNITS 1, 2, and 3 As the earthquake struck, the reactors in units 1, 2, and 3 were in operation, while those in units 4, 5, and 6 had been shut down for routine inspection and maintenance. Reactors 1, 2, and 3 "scrammed," or went into automatic shutdown, and emergency generators immediately started up to power coolant systems and electronics. But these were swamped by the ensuing 15-meter tsunami, which caused severe structural damage to the buildings. As a result, all three reactor cores largely melted within three days of the tsunami. There were also explosions inside all three buildings, caused by the ignition of leaked hydrogen gas. After two weeks, the three units could be stabilized with water addition, but no proper cooling system could be installed for the decay heat from the fuel. As yet, there have been no reported deaths or cases of radiation sickness as a result of the accident, but over 100,000 people had to be evacuated.

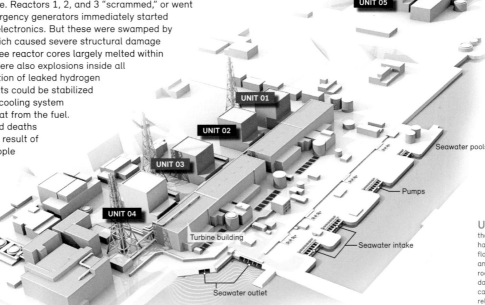

UNIT 4 Though it was not in operation at the time of the quake, unit 4's 548 fuel rods had been moved to a fuel pool on an upper floor of the building. This began to overheat, and four days later a hydrogen explosion in the rooftop area, which incidentally further damaged unit 3, tore through the building, causing more concern that radiation could be released.

- Hydrogren gas explosion
- Fires under control
- Under control

	REACTOR UNIT 01	REACTOR UNIT 02	REACTOR UNIT 03	REACTOR UNIT 04
Status	Active	Active	Active	Inactive
Reactor type	BWR	BWR	BWR	BWR
Fuel	URANIUM	URANIUM	PLUTONIUM	PLUTONIUM
Manufacturer	General Electric	General Electric/Toshiba	Toshiba	HITACHI
Operator	Tokya Electric Power Co. (TEPCO)	Tokya Electric Power Co. (TEPCO)	Tokya Electric Power Co. (TEPCO)	Tokya Electric Power C (TEPCO)
Grid capacity	439 MWe	760 MWe	760 MWe	760 MWe
Total capacity	460 MWe	784 MWe	784 MWe	784 MWe
Heat capacity	1,380 MWt	2,381 MWt	2,381 MWt	2,381 MWt
Completed	October 10, 1970	October 5, 1973	June 9, 1974	January 28, 1978
Connected to grid	November 17, 1970	December 24, 1973	October 26, 1974	February 24, 1978
Capacity	2,637,414 MWh	4,903,293 MWh	4,937,601 MWh	5,462,108 MWh
Total output	76,892,000 MWh	142,199,000 MWh	149,809,000 MWh	148,263,000 MWh
Shutdown	March 26, 2011	2014	2016	2018

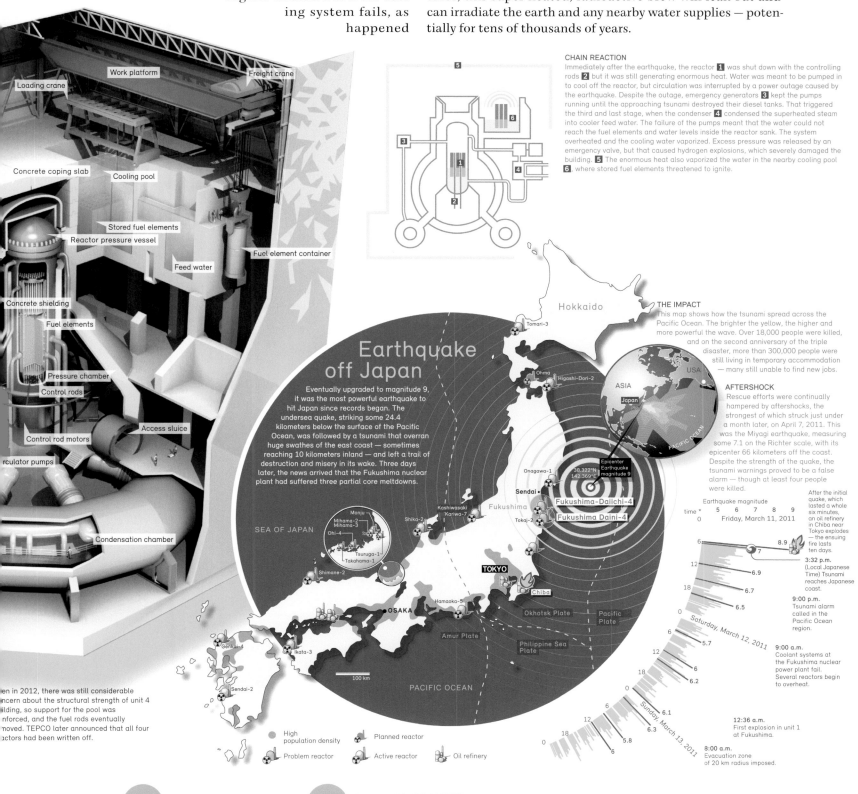

water which turns to steam to drive turbines to create electricity. Under normal operation, the inside of a reactor reaches temperatures of around 270 degrees Celsius. But if the cooling system fails, as happened in Fukushima, the uranium fuel rods heat up to 1,200 degrees. Then they melt and create a soup of metal and uranium that pools at the bottom of the reactor. If the reactor's casing ruptures, this super-heated, radioactive brew will leak out and can irradiate the earth and any nearby water supplies — potentially for tens of thousands of years.

CHAIN REACTION

Immediately after the earthquake, the reactor **1** was shut down with the controlling rods **2** but it was still generating enormous heat. Water was meant to be pumped in to cool off the reactor, but circulation was interrupted by a power outage caused by the earthquake. Despite the outage, emergency generators **3** kept the pumps running until the approaching tsunami destroyed their diesel tanks. That triggered the third and last stage, when the condenser **4** condensed the superheated steam into cooler feed water. The failure of the pumps meant that the water could not reach the fuel elements and water levels inside the reactor sank. The system overheated and the cooling water vaporized. Excess pressure was released by an emergency valve, but that caused hydrogen explosions, which severely damaged the building. **5** The enormous heat also vaporized the water in the nearby cooling pool **6**, where stored fuel elements threatened to ignite.

THE IMPACT

This map shows how the tsunami spread across the Pacific Ocean. The brighter the yellow, the higher and more powerful the wave. Over 18,000 people were killed, and on the second anniversary of the triple disaster, more than 300,000 people were still living in temporary accommodation — many still unable to find new jobs.

AFTERSHOCK

Rescue efforts were continually hampered by aftershocks, the strongest of which struck just under a month later, on April 7, 2011. This was the Miyagi earthquake, measuring some 7.1 on the Richter scale, with its epicenter 66 kilometers off the coast. Despite the strength of the quake, the tsunami warnings proved to be a false alarm — though at least four people were killed.

Fukushima

THE DAYS THE EARTH STOOD STILL

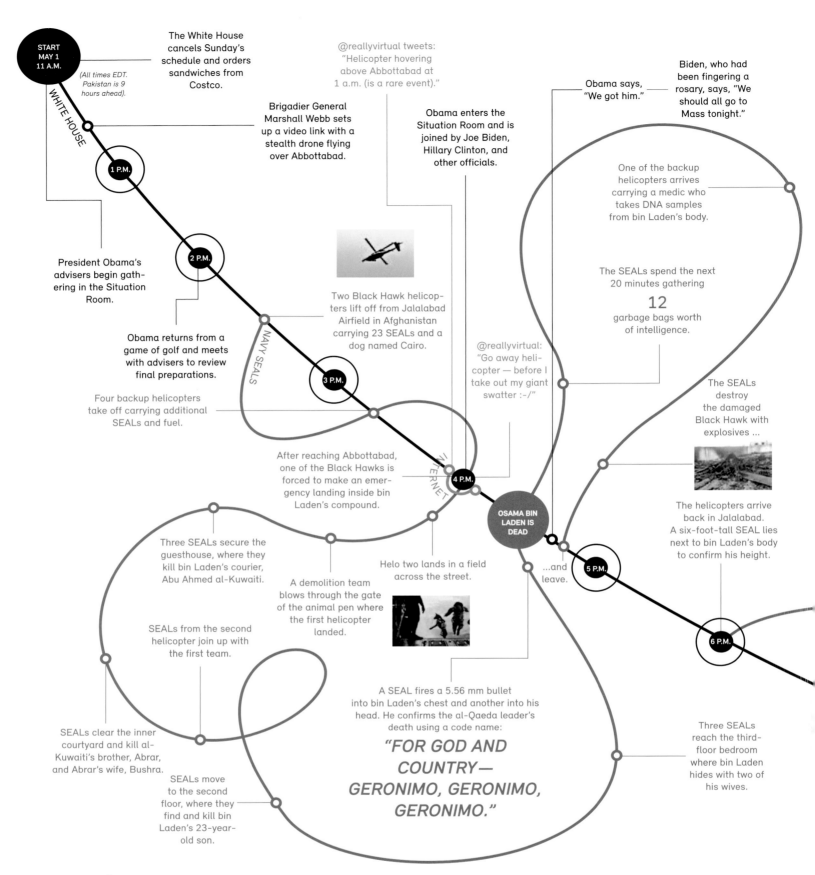

Geronimo, EKIA

The day that U.S. Navy Seals killed Osama bin Laden — in real life and on the Internet.

On May 2, 2011, at around 1 a.m. somewhere in Abbottabad, Pakistan, a thirty-something-year-old software consultant was being bothered by a helicopter hovering over his hometown. As many computer geeks would have done in his position, he decided this was an event odd enough to tweet about, which he did under his online moniker @reallyvirtual. When the helicopter noise was still there a few minutes later, he tweeted about it again, warning the troublesome aircraft that he would "take out my giant swatter." It was a good thing he didn't try it. Though he knew nothing about it, it turned out that Sohaib Athar was live-blogging the bloody conclusion of the CIA's decade-long search for the world's most infamous terrorist, which was taking place 2.5 kilometers from his home. May 1, 2011 — as it still was in the U.S. — turned out to be the day that Osama bin Laden, spiritual head of al-Qaeda, the Islamist group responsible for the September 11 attacks, was killed in his own home by a troop of U.S. Navy Seals sent there for that purpose. Seven and a half hours later, President Barack Obama announced the successful raid on television. Geronimo's status, to use Osama's codename, had changed from perennial threat to EKIA — Enemy Killed In Action.

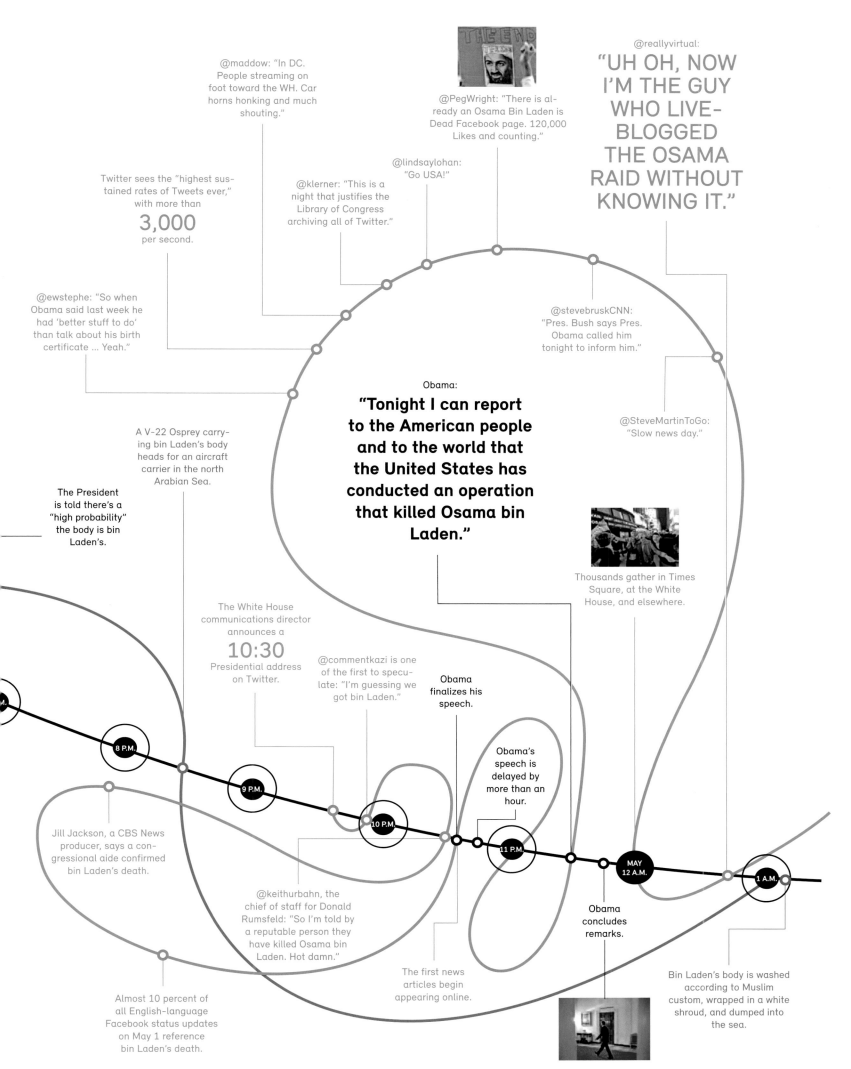

Killing Osama bin Laden

THE DAYS THE EARTH STOOD STILL

THE GOOD LIFE

What are you having for breakfast? It's a simple question, but your answer will say a lot about where you're from and how you start your day.

These small, seemingly insignificant decisions shape who and where we are. When we travel, it's the tiny details more than the major differences in which we find ourselves fascinated.

"If we could travel in time, we would without a doubt feel the same unease. Chocolate tastes different in India. TV was different in 1970. Speed limits are unfamiliar in Italy. Beer labels are different in Nigeria. Sports were something else in third-century Persia. These don't seem very important or very surprising statements, but they are in fact hugely significant. Everything in this chapter is a form of cultural expression, and has been chosen to help you reflect on what exactly that means. Briefly, culture is what we choose to do with our time, and how we decide to spend the fruits of our work. It's what we consume, and how our consumption patterns shift and change. It's how we demonstrate who we are, and the subtle ways in which we communicate with those around us.

Culture is the expression of arbitrary rules agreed upon to represent a collective experience. We are a people who eat this, listen to that. We wear this (this season at least), we go on holiday there, we spend our money on that. And thanks to these choices in food, sports, pets, art, music, movies, and TV, millions of people around the world can share in our expression. These images help describe some of those choices in a visually informative manner.

We can learn a lot about ourselves by understanding the cultural choices we didn't make, and those that our parents and grandparents did (that we would never be caught dead choosing). It's true that we don't often think about the origins of what we consume — who picked this banana, where did it come from, how did it get to our store, why is it so popular, and how can something that has traveled so far cost so little? Once you start to think about these questions, it's difficult to stop. So next time you're chewing on your breakfast, look at what's on your spoon, and remember: none of this is "normal" or "natural." Pretty much everything you see and hear has been constructed by the culture that surrounds it, as a form of expression designed to be a part of a wider context of "here" and "now." It is the water we swim in.

So good morning to you, dear reader, and enjoy every culturally appropriated morning mouthful. Just keep your eyes open as you chew for those insignificant things in your house that seem obvious to you, but yet bizarre or unusual for someone from somewhere else. Here's a clue: it's everything.

Filling Up Your Life

What do we actually do with our lives?

"Life is what happens to you while you're busy making other plans," John Lennon once sang. Perceptive words, as usual from the great man, but even the visionary Beatle could not have predicted the sheer gorge of prevarication that opened up in the twenty-first century. Or, more accurately, the numerous gorges. Obviously the Internet is the major one, the Grand Canyon of distraction, but modern life in general has basically become a field of time-sinks. Picking a path through your day so that you actually achieve something has become a mission that takes supreme logistical planning executed with military precision and discipline. What are you doing reading this, anyway? And the problem is universal — prevarication is the curse of every culture in the world. Set forth here is an average Russian's life, broken down into terrifying numbers.

Watching TV
07:01:00
years months weeks

Work
06:10:02
years months weeks

Personal hygiene
04:10:02
years months weeks

124

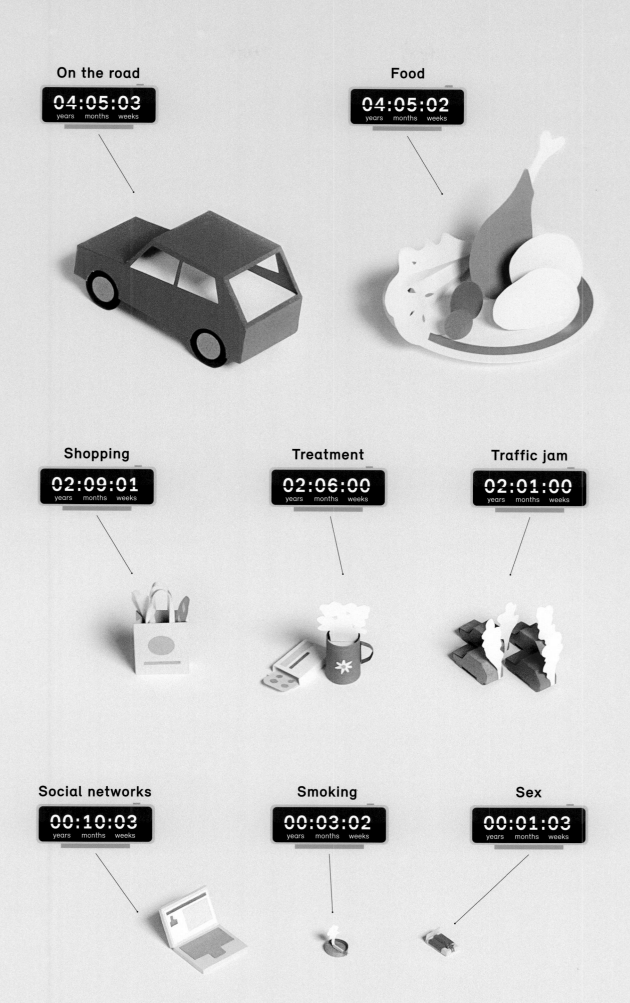

Family Time

THE GOOD LIFE

Where Does All the Money Go?

Income breakdown Italian-style.

It's the same story at the start of every month. You've completed the paperwork, turned the grinder, mined the coalface, and finally you're allowed to get the cash to indulge your pleasures. And then it all melts away as the necessities close in: food, shelter, clothing, health. Before you know it you're back down to zero. But no, look! Not quite — almost zero. You've got a couple of hundred bucks left over. What now? Add it to your meager mound of savings? Or blow it on whatever your guilty pleasure is? The answer is always the same. Hell, you deserve it. Consumer spending has diversified massively in the past 100 years or so. In the olden days, almost all of the household budget went on basics — food and housing. Those are still the two biggest expenses that most families face, but other expenditures — especially telecommunications and media — have now also become essential. Doing without them risks being degraded to a second-class citizen. This graphic breaks down the expenditures of an average family in the developed world — in this case, an Italian one. Interestingly, according to one statistical analysis by researchers Eurostat, Italy's consumer spending is influenced by the fact that small independent shops — as opposed to supermarkets — play a bigger part in Italian culture than elsewhere. One international clothing retailer's survey from 2012 found that independent retailers and street markets still account for 34 percent of Italian clothes shopping. And while we're on the subject of clothes, you might be happy to hear that the EU financial crisis did not dent the Italian love of fashion — according to the same survey, Italian spending on clothes barely dropped despite rising unemployment. Though they did learn to seek out more bargains.

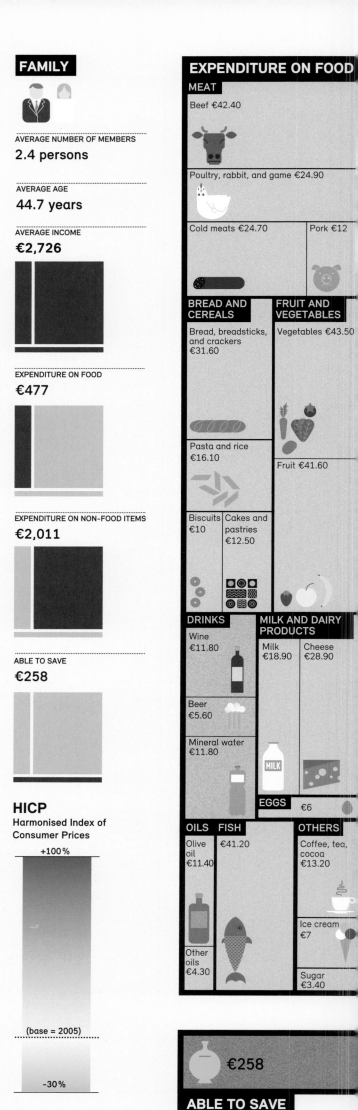

Design: Samuel Granados with Gianluca Seta / published: *La Lettura – Corriere della S*

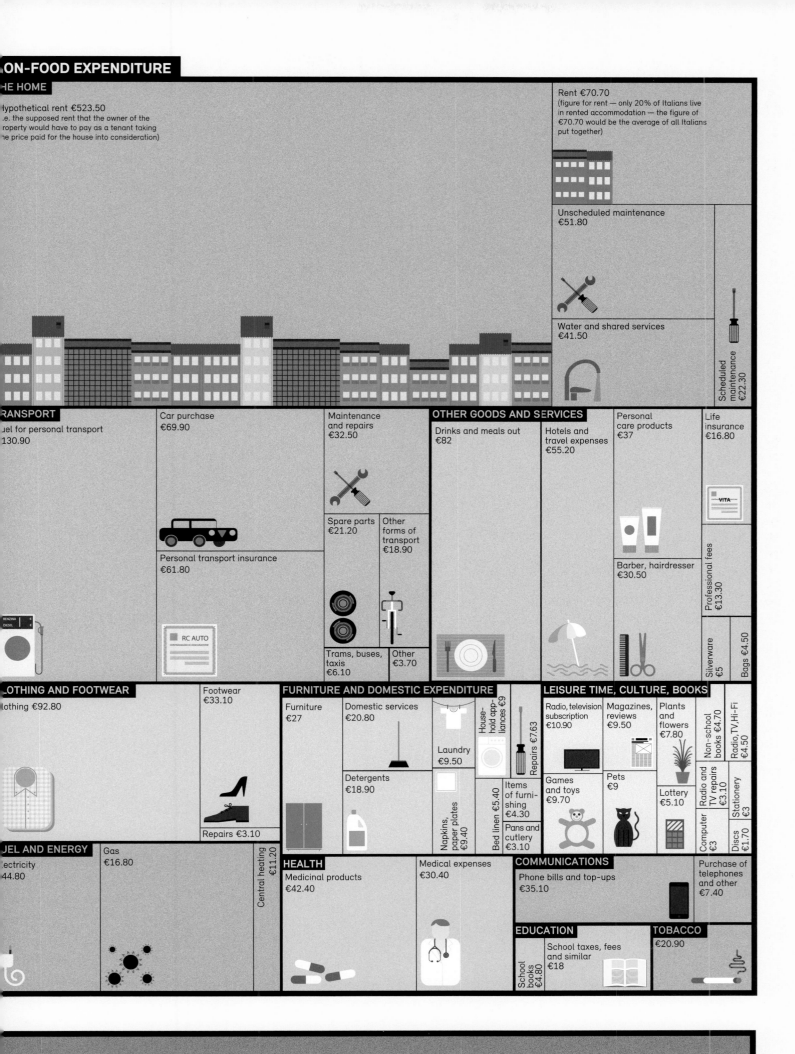

Family Spendings

Custom-made Breakfast
The most important meal of the day — papercraft-style.

You're on a romantic weekend in Rome. Your lover is still asleep, deep in dreamless repose amid the quilts and pillows. You've decided to surprise them with breakfast in bed. What do you serve? A) White sausage, mustard, and wheat beer? B) A soup of slow-cooked fava beans? C) Sausage, bacon, fried potatoes, and coffee? Or D) A Mediterranean-style light pastry and an espresso? This graphic constitutes a handy guide that serves up some of the world's various breakfast customs. They're made out of paper, lest you think that a perfectly good Bavarian sausage has gone to waste.

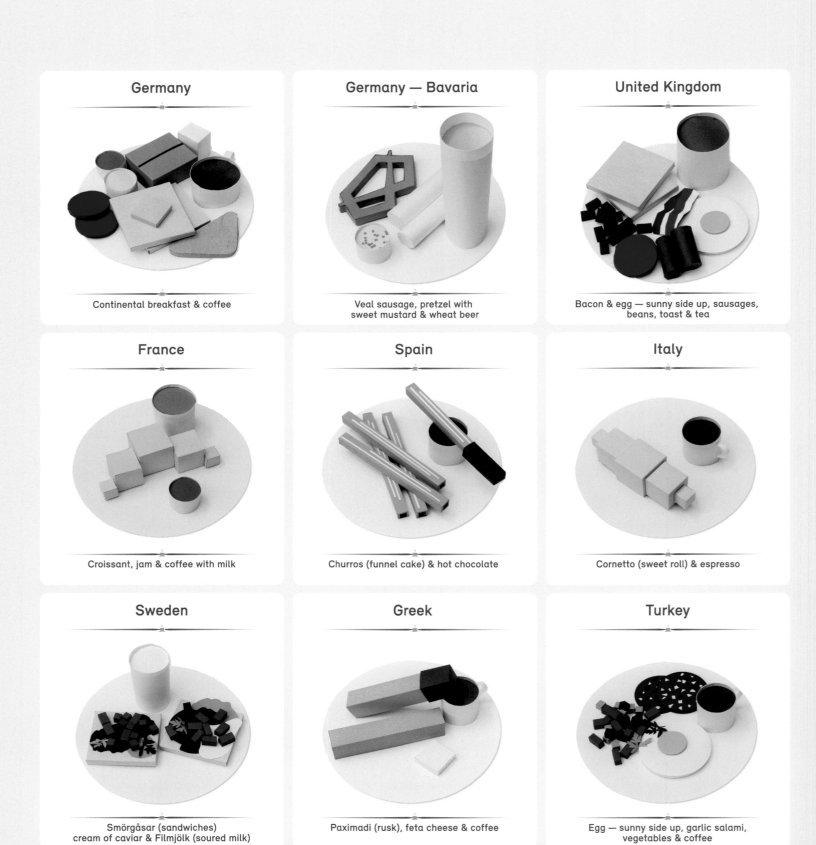

Russia

Kasha (oatmeal) & coffee

Canada

Pancakes with syrup & coffee

USA

Egg — sunny side up, sausages, bacon, roasted potatoes & coffee

Costa Rica

Egg — sunny side up, rice with beans, baked bananas & coffee

Guatemala

Egg — sunny side up, beans, tortillas, baked bananas & coffee

Mexico

Huevos Motuleños

Tortillas, egg — sunny side up, beans, baked bananas & coffee

United Arab Emirates

Hummus (chickpeas cream), ragout, pita bread & tea

India

Dual Chapati (pita bread), lentils, rice & spicy tea

China

Jiaozi (filled samosa) & hot soya milk

Japan

bento box for kids

Finger food from fish, meat & vegetables

Egypt

Ful Medames (soup of slow-cooked fava beans) & coffee

Tanzania

Mandazi (fried dough balls) & Chai Ya rangi (black tea)

Breakfast

THE GOOD LIFE

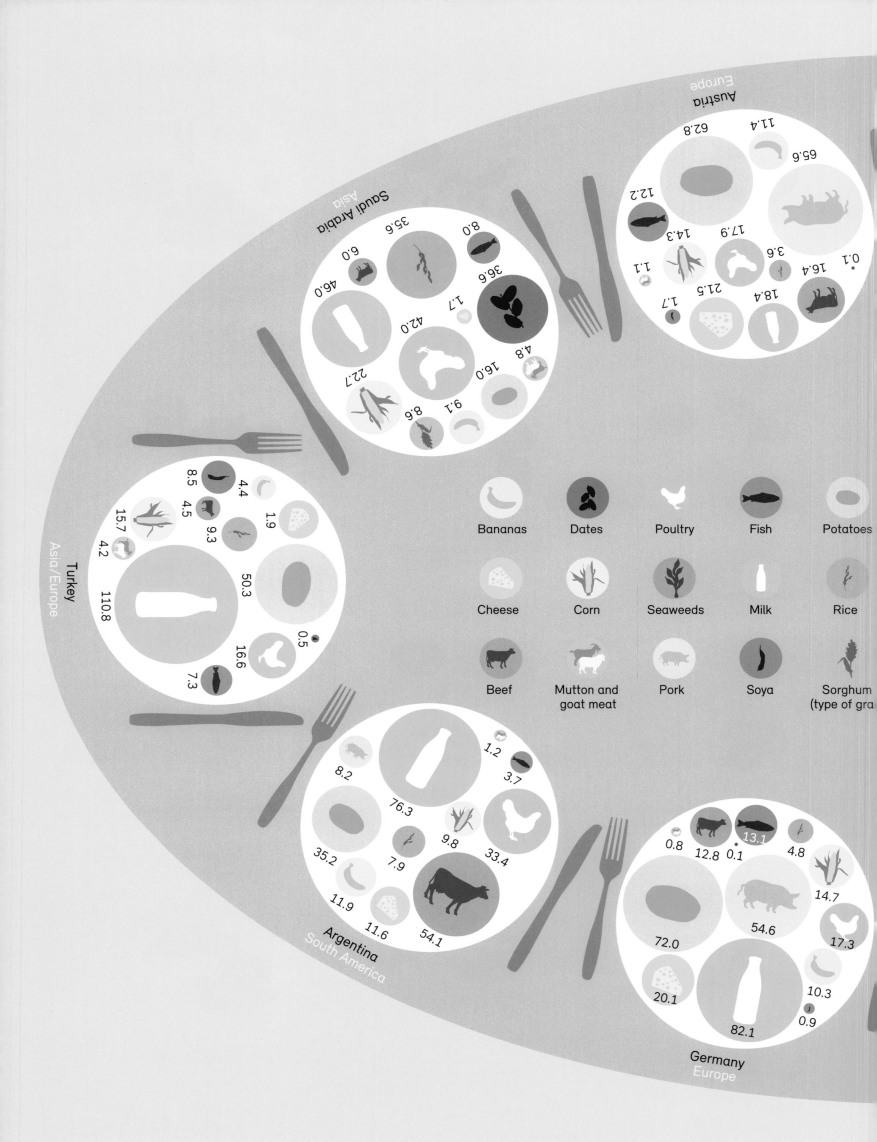

The Curse of Being Delicious

How we are hunting down our favorite fish, the tuna.

Those little cans in the supermarket look so innocent, don't they? And why do they make them so small when the animals they come from are sometimes true giants of the open waters — growing up to 4.5 meters and weighing as much as a full grown cow, much bigger, as the graphic below shows, than the average scuba diver? Nearly everyone loves tuna. It's the

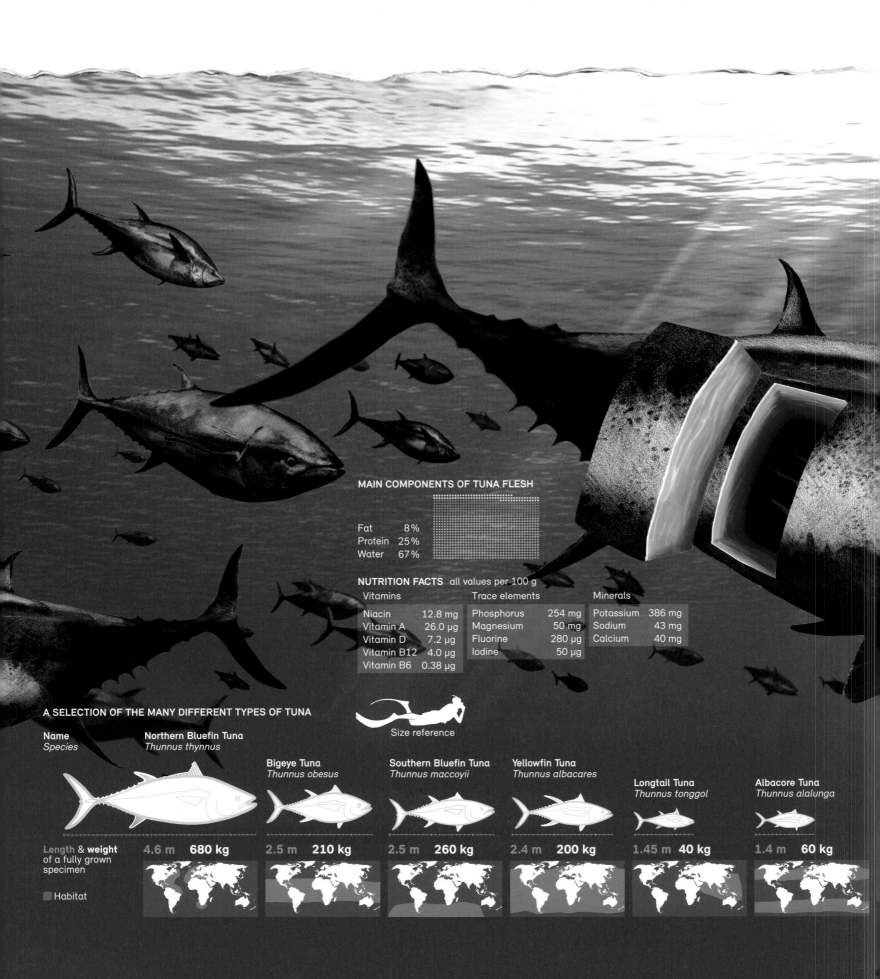

MAIN COMPONENTS OF TUNA FLESH

Fat 8%
Protein 25%
Water 67%

NUTRITION FACTS all values per 100 g

Vitamins		Trace elements		Minerals	
Niacin	12.8 mg	Phosphorus	254 mg	Potassium	386 mg
Vitamin A	26.0 µg	Magnesium	50 mg	Sodium	43 mg
Vitamin D	7.2 µg	Fluorine	280 µg	Calcium	40 mg
Vitamin B12	4.0 µg	Iodine	50 µg		
Vitamin B6	0.38 mg				

A SELECTION OF THE MANY DIFFERENT TYPES OF TUNA

Size reference

Name / Species	Length & weight of a fully grown specimen
Northern Bluefin Tuna — *Thunnus thynnus*	4.6 m — 680 kg
Bigeye Tuna — *Thunnus obesus*	2.5 m — 210 kg
Southern Bluefin Tuna — *Thunnus maccoyii*	2.5 m — 260 kg
Yellowfin Tuna — *Thunnus albacares*	2.4 m — 200 kg
Longtail Tuna — *Thunnus tonggol*	1.45 m — 40 kg
Albacore Tuna — *Thunnus alalunga*	1.4 m — 60 kg

Habitat

one healthy food that actually tastes nice. Even your cat loves it. The downside is of course that we're hunting them — like the North American buffalo in the nineteenth century — to the point that we are threatening entire ecosystems in our oceans. Wars have been fought over that yummy grayish-pink meat, and conservationists warned early in 2013 that the numbers of bluefin — highly valued in the sushi world — had dropped by over 96 percent since fishing began. Illustrating this drastic scarcity, one Pacific bluefin sold for $1.78 million at an auction in Tokyo in January 2013.

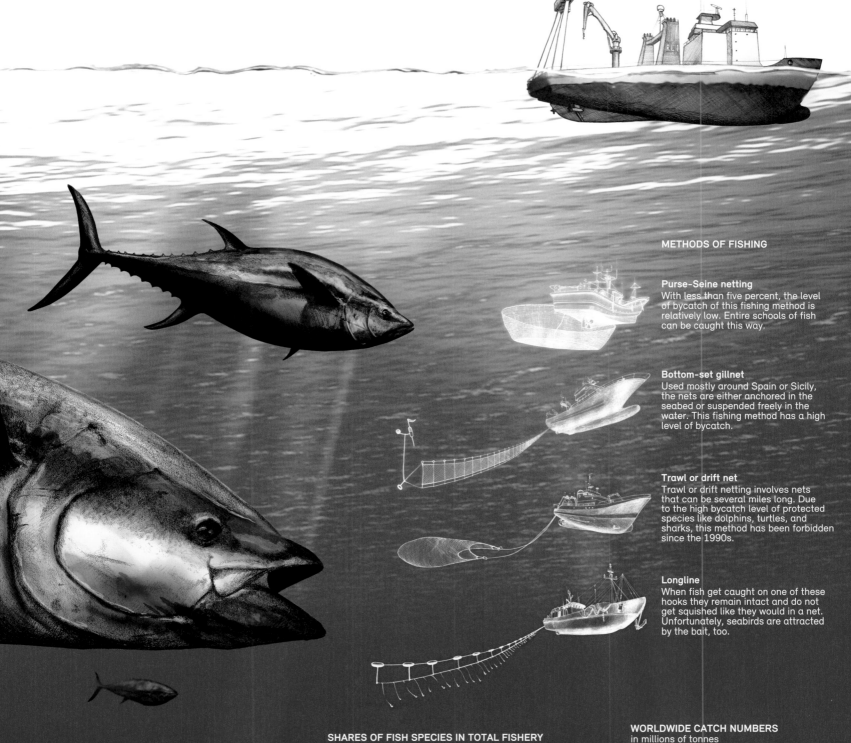

METHODS OF FISHING

Purse-Seine netting
With less than five percent, the level of bycatch of this fishing method is relatively low. Entire schools of fish can be caught this way.

Bottom-set gillnet
Used mostly around Spain or Sicily, the nets are either anchored in the seabed or suspended freely in the water. This fishing method has a high level of bycatch.

Trawl or drift net
Trawl or drift netting involves nets that can be several miles long. Due to the high bycatch level of protected species like dolphins, turtles, and sharks, this method has been forbidden since the 1990s.

Longline
When fish get caught on one of these hooks they remain intact and do not get squished like they would in a net. Unfortunately, seabirds are attracted by the bait, too.

MONTHLY IMPORT OF TUNA BY CONTINENT

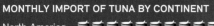

SHARES OF FISH SPECIES IN TOTAL FISHERY

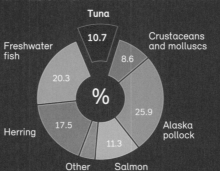

Tuna 10.7
Crustaceans and molluscs 8.6
Alaska pollock 25.9
Salmon 11.3
Other 17.5
Herring 20.3
Freshwater fish

WORLDWIDE CATCH NUMBERS
in millions of tonnes

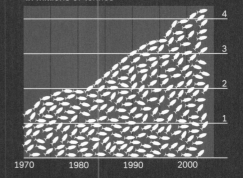

Tuna

THE GOOD LIFE

Fruit flies

In today's world, fruit has traveled thousands of miles to get to your bowl.

German shoppers, like most in the developed world, have no need for seasons. When it comes to our food, they are no longer a necessary condition of everyday life. All these fruits and vegetables are dependent on some time of year or another, in some region of the world or other. This graphic shows that not only are 50 percent of vegetables and 80 percent of fruit imported, but whether we like it or not, bananas have often traveled vast distances to get to our dinner tables. Indeed, anyone who has tasted a banana in, say, South America, then eaten one in Europe, knows that a transatlantic trip in a freezer container on an ocean liner often has a detrimental effect on the quality of the produce.

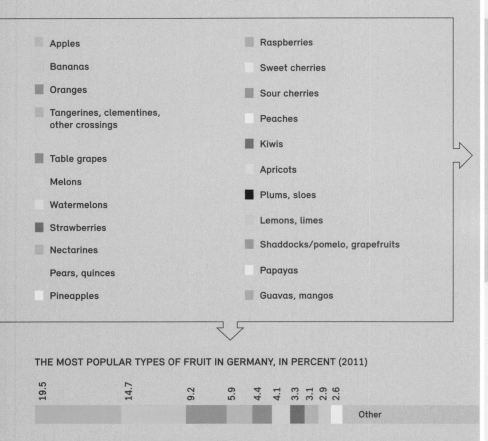

- Apples
- Bananas
- Oranges
- Tangerines, clementines, other crossings
- Table grapes
- Melons
- Watermelons
- Strawberries
- Nectarines
- Pears, quinces
- Pineapples
- Raspberries
- Sweet cherries
- Sour cherries
- Peaches
- Kiwis
- Apricots
- Plums, sloes
- Lemons, limes
- Shaddocks/pomelo, grapefruits
- Papayas
- Guavas, mangos

THE MOST POPULAR TYPES OF FRUIT IN GERMANY, IN PERCENT (2011)

19.5 | 14.7 | 9.2 | 5.9 | 4.4 | 4.1 | 3.3 | 3.1 | 2.9 | 2.6 | Other

In the countries surrounding the Mediterranean, an average of 700 grams of fruit and vegetables per person are eaten per day. In Germany, however, people are satisfied with 250 grams. Compared to the EU-27 countries, only Estonia, Finland, Ireland, and Lithuania fall behind Germany in the consumption of fruit and vegetables.

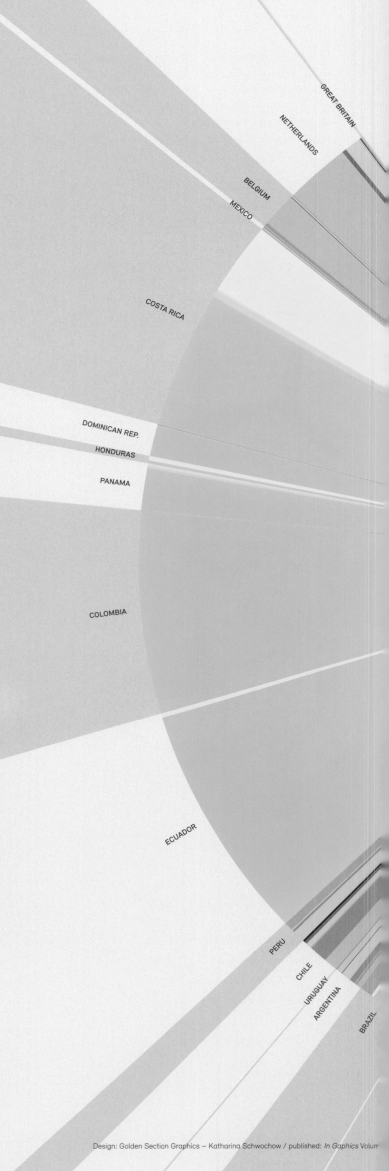

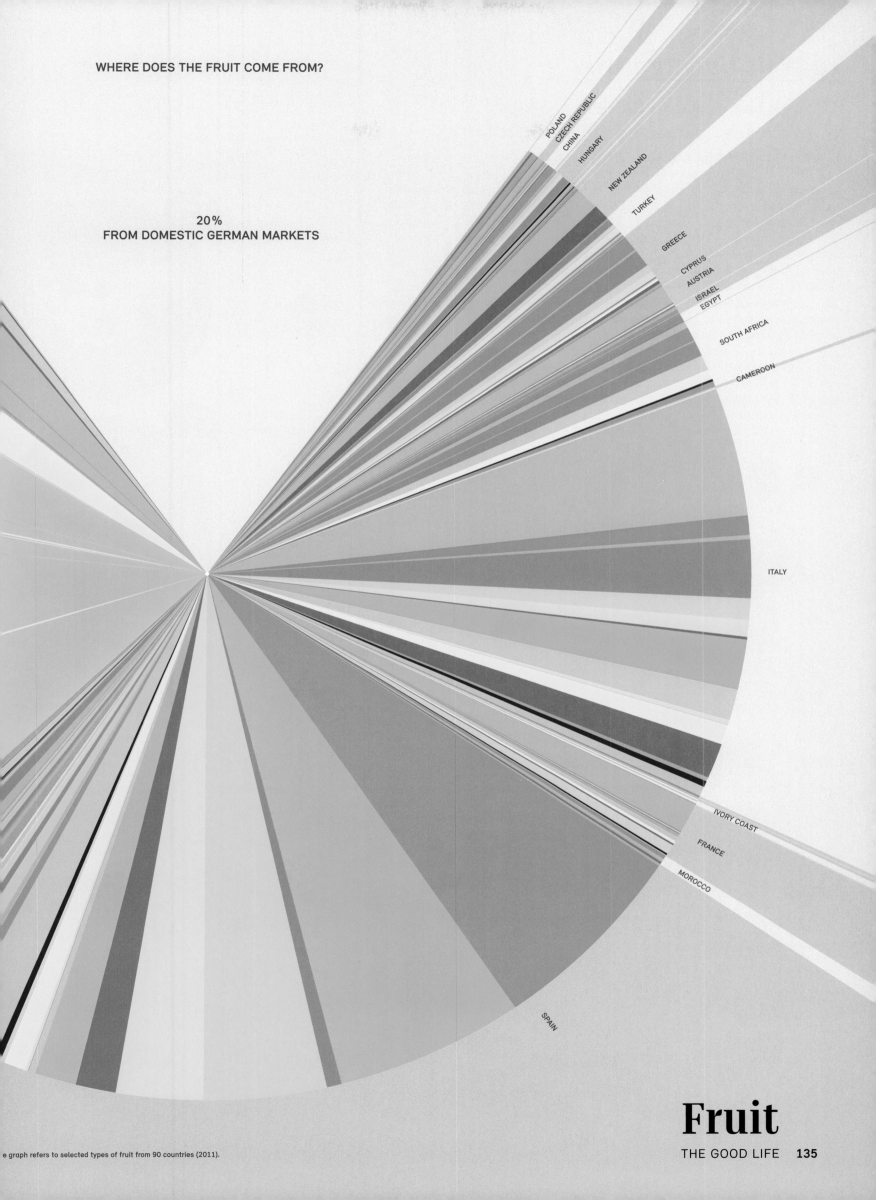

Organic Undergrowth

Though the demand for organic food is increasing, the actual area that farming takes up is still surprisingly small.

Some say it's a fad for the middle classes, but the trend is seemingly unstoppable. Organic stores used to be small, strange-smelling shops run by unrepentant hippies on the corners of high streets in the more enlightened college towns. Now they're major supermarket and restaurant chains that stock everything from pesticide-free olives to bleach-free cleaning products. The global market for organic produce has more than tripled in the first decade of the twenty-first century. Accordingly, the land given over to organic farming is slowly expanding, with Italy, of all places, growing its wheat, fruit, and olives in a greener way than anywhere else. But as this graphic shows, the demand for produce farmed in a "sustainable" way, to use the appropriate buzzword, is still almost exclusively limited to North America and Europe.

ONLY FRUIT AND VEGETABLES OF YOUR COUNTRIES

In 2007 worldwide **32.2 million hectares** were allocated to organic farming: **0.8%** of the total cultivated area (in the 141 countries included in the Ifoam report). The largest area (20 million hectares) is occupied by pastures, as is clear from the diagram on the intended use:

1. TEMPORARY
Arable land dedicated over the years to several organic products
Total 4.746,656 ha

2. PERMANENT
Lands under fixed cultivation, such as cocoa, coffee, and rubber
Total 1,877,380 ha

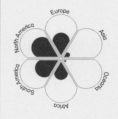

3. PASTURES
For animals reared naturally
Total 20,009,411 ha

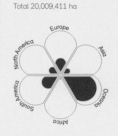

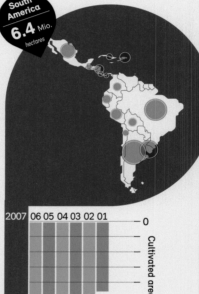

North America 2.2 Mio. hectares

South America 6.4 Mio. hectares

THE WEST WANTS CONTROLLED ORIGIN VEGETABLES

The global market for organic products is growing every year by approx. **5 billion dollars**. The demand is concentrated in North America and Europe: the two areas together generate **97%** of total turnover worldwide.

North America 43% · Europe 54% · Asia · South America · Africa · Oceania

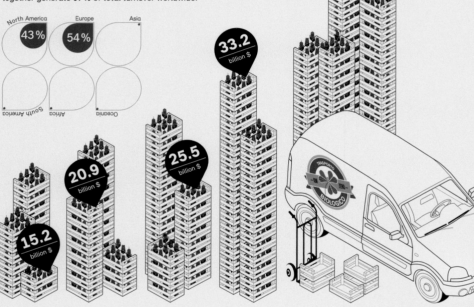

1999: 15.2 billion $
2001: 20.9 billion $
2003: 25.5 billion $
2005: 33.2 billion $
2007: 46.1 billion $

WHERE TO GROW FREE FROM PESTICIDES

Hectares of land (arable land plus pastures) dedicated to organic: absolute number and percentage of the total (2007 data)

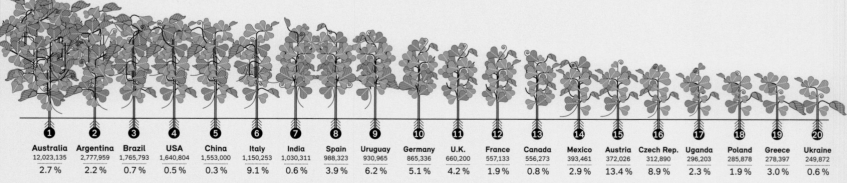

#	Country	Hectares	%
1	Australia	12,023,135	2.7%
2	Argentina	2,777,959	2.2%
3	Brazil	1,765,793	0.7%
4	USA	1,640,804	0.5%
5	China	1,553,000	0.3%
6	Italy	1,150,253	9.1%
7	India	1,030,311	0.6%
8	Spain	988,323	3.9%
9	Uruguay	930,965	6.2%
10	Germany	865,336	5.1%
11	U.K.	660,200	4.2%
12	France	557,133	1.9%
13	Canada	556,273	0.8%
14	Mexico	393,461	2.9%
15	Austria	372,026	13.4%
16	Czech Rep.	312,890	8.9%
17	Uganda	296,203	2.3%
18	Poland	285,878	1.9%
19	Greece	278,397	3.0%
20	Ukraine	249,872	0.6%

Organic farming

For a farm, it means producing with particular attention to protection of the environment, without using synthetic chemicals and using natural resources in a sustainable manner.

Natural enemies of parasites

Birds and ladybirds feed on plant parasites. These are animals that find refuge in hedges and ditches. Man can exploit their presence to his benefit and use them to fight insects harmful to agriculture.

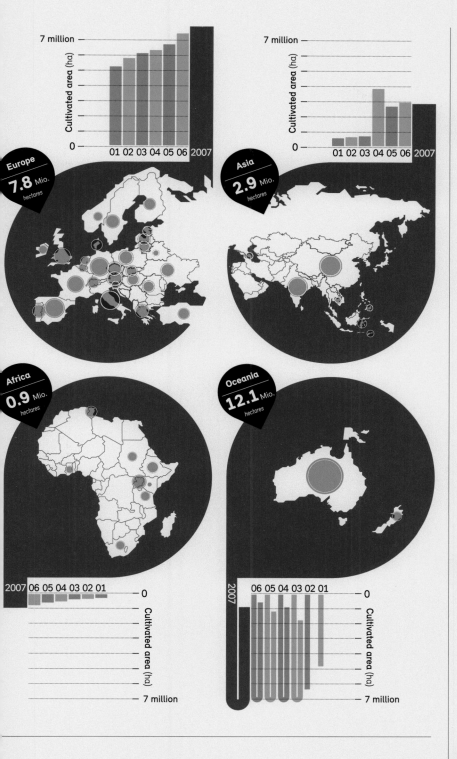

WHO IS ON THE TOP STEP OF THE PODIUM?

According to that emerging from the geography of the areas allocated to organic farming, divided by product, Italy is the country with the greatest extension of organically grown wheat, cereals, citrus fruits, grapes, and olives.

Wheat (hectares of organic cultivation)
1. ITALY 143,598
2. UNITED STATES 115,601
3. CANADA 79,278
4. GERMANY 73,000
5. UKRAINE 50,423
6. FRANCE 34,364

Cereals
1. ITALY 241,430
2. UNITED STATES 228,109
3. GERMANY 181,000
4. CANADA 154,152
5. SPAIN 116,864
6. UKRAINE 105,477

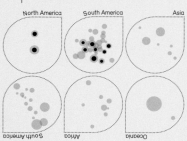
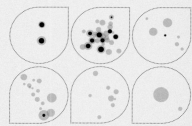

Citrus fruits
1. ITALY 22,062
2. CUBA 4,195
3. UNITED STATES 4,107
4. GHANA 3,760
5. MEXICO 3,201
6. SPAIN 3,165

Coconuts
1. PHILIPPINES 14,106
2. MEXICO 8,031
3. DOMINICAN REP. 3,025
4. CUBA 1,056
5. EL SALVADOR 1,024
6. IVORY COAST 875

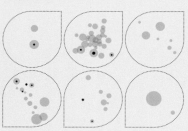
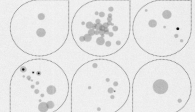

Cocoa
1. DOMINICAN REP. 79,401
2. ECUADOR 22,308
3. MEXICO 16,366
4. PERU 14,407
5. PANAMA 4,850
6. TANZANIA 4,316

Coffee
1. MEXICO 239,763
2. ETHIOPIA 108,560
3. PERU 72,174
4. TANZANIA 23,867
5. EAST TIMOR 21,325
6. UGANDA 17,721

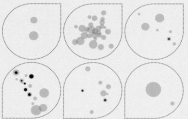
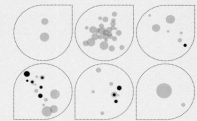

Grapes
1. ITALY 36,684
2. FRANCE 22,509
3. SPAIN 17,189
4. UNITED STATES 9,177
5. TURKEY 5,706
6. GREECE 4,554

Olives
1. ITALY 109,992
2. SPAIN 94,251
3. TUNISIA 89,324
4. GREECE 52,553
5. TURKEY 26,372
6. PORTUGAL 18,409

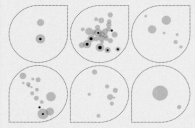
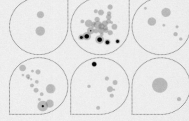

	21 Sweden	22 Portugal	23 Latvia	24 Tunisia	25 Finland	26 Denmark	27 Ethiopia	28 Romania	29 Peru	30 Turkey
	248,104	233,475	173,463	154,793	148,760	145,393	140,305	131,401	124,714	124,263
	8.0%	6.4%	9.8%	1.6%	6.5%	5.5%	0.4%	0.9%	0.6%	0.6%

Organic livestock raising
In organic livestock raising, the animals are taken to open grassland. It is important that their nutrition and care takes place with natural products.

Organic fertilization
Use of organic substances promotes the activity of soil organisms and the formation of humus. In this way, the soil becomes fertile and plants flourish.

Ecological equipment
Using agricultural machines with moderation, the invasiveness of the mechanical action on the soil will be minimized and microbiological activity will not be compromised.

Organic Farming

THE GOOD LIFE   137

Addicted to Chocolate

One inconspicuous South American bean — and the single most seductive food on the planet.

Possibly, 3,000 years ago, some Aztec parents threatened their unruly children with a sip of xocolātl, or "bitter water," if they didn't behave — one dose of that spicy cocoa-bean beverage and those naughty kids would soon be back in line. Whether or not this venerable Nahuatl word is indeed the etymological origin of the word chocolate (which came to English via Spanish) remains a linguistic debate, but one thing's for sure: the Spanish conquistadors changed the game when they came up with the idea of adding sugar to the concoction sometime after their first contact with the Theobroma cacao tree in 1519. After that minor modification, European demand skyrocketed, and cocoa-bean plantations sprang up in the Caribbean, West Africa, and plenty of other suitable regions within 20 degrees of the equator. Since then, chocolatiers have become veritable artists, blending concoctions from several nations, in search of a unique formula. Today, West African countries grow 70 percent of all cocoa beans (though they only make up one percent of world consumption.) Meanwhile the Swiss, at the other end of the scale, remain hopelessly hooked.

Key

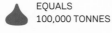 Top 10 Cocoa-Consuming Countries

 Top 10 Cocoa-Producing Countries

▼ EQUALS 100,000 TONNES

⇉ TRADE LINES TO THE UNITED STATES

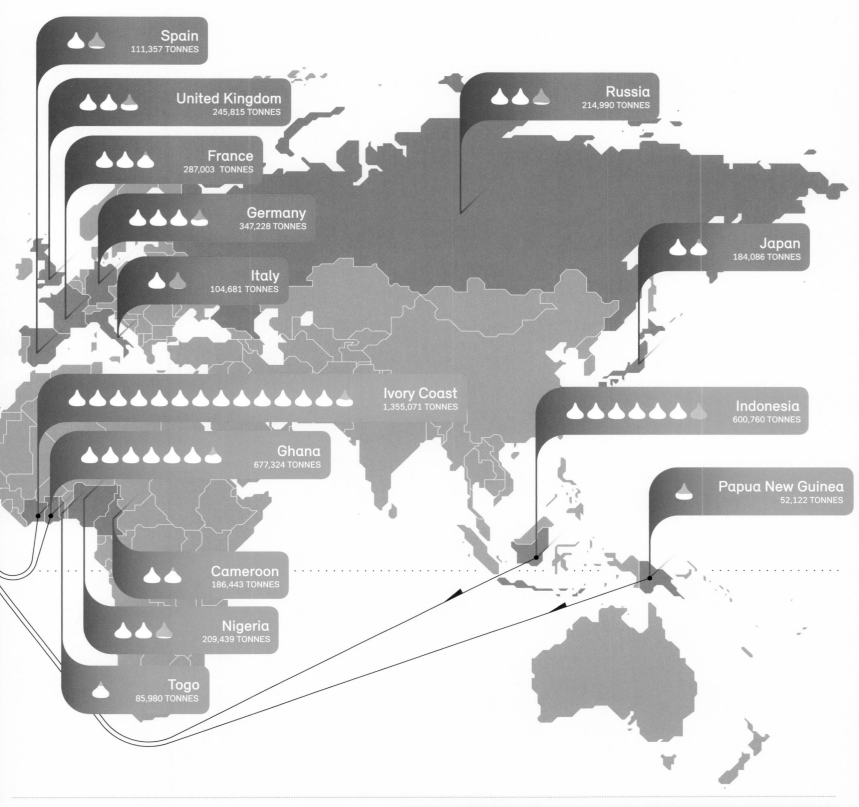

Top 12 Chocolate Countries Per Capita 2006

Which country boasts the biggest chocoholics? Here are the world's top chocolate consumers — dominated first by northern Europe, and second by predominantly Christian countries. Possible conclusion? Christmas, Easter, and cold weather equals shameless chocolate abuse.

SWITZERLAND	10.07 KG
UNITED KINGDOM	9.97 KG
GERMANY	9.61 KG
BELGIUM	8.94 KG
NORWAY	8.75 KG
AUSTRIA	7.26 KG
IRELAND	7.17 KG
DENMARK	6.89 KG
FINLAND	6.62 KG
SWEDEN	6.44 KG
AUSTRALIA	5.72 KG
UNITED STATES	5.44 KG

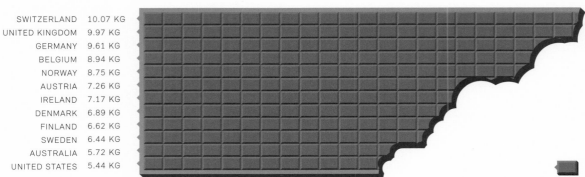

Chocolate

THE GOOD LIFE

Liquid Bread

The nectar of the gods.

Some historians tell us that one night, about 7,000 years ago, on a lush patch of land that is in Iran, a few people gathered and took their first sip of a strange brown home brew they had just made. Maybe they clinked their clay cups together and wished each other good health too. A few thousand years later, their odd experiment in fermenting cereal crops has become possibly the most popular drink ever, evolving into literally hundreds of thousands of variations in almost every culture in the world — even the Islamic Republic of Iran allows the production and sale of non-alcoholic varieties, possibly as a nod to their

SOME A LOT, SOME LITTLE. THE DRINKERS EXCHANGE

The Czechs drink like sponges, the Irish, Germans, and Australians hold their own, while in Britain overdoing the "Lager" or "Stout" could quickly ruin you. Consumption, costs, taxes, and prohibitions: an X-ray of the beer market on the five continents.

PUBLIC PLACES AND ALCOHOL

ALCOHOL ADVERTISING. HARD TIMES FOR ADVERTISING IN THE MEDIA

From the absolute — and absolutist — "Niet" in Iran, Egypt, and Algeria to, "anything goes" for advertising in Brazil, China, and South Africa.

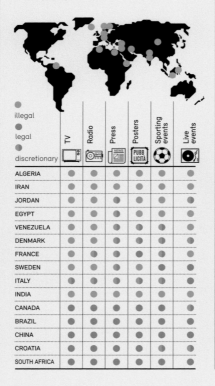

FROTH, FRAGRANCE, CREAM: CHOOSE THE RIGHT GLASS

The bible associates "Lager" beers with the mug or tulip glass and "dark" beers with goblets. It is true, but it's not enough. In addition to the shape, one must consider the type of glass and the thickness.

Tulip glass — *330 ml*
The flared mouth prevents excessive frothing and enhances the sense of smell. Suitable for Belgian abbey beers.

Mug — *400 ml*
The "classic." The German glass version, the "Mass," is perfect for "Märzen." For "Pale ales," you need the British glass.

Conical column — *330 ml*
Glass of medium thickness, a wide mouth to control the froth. It is ideal for the lively and fragrant Danish beers.

Altglass — *425 ml*
Cylindrical, so as not to exalt or mortify the froth. The very thin glass makes it suitable for the amber "Alt" beers.

Weizenbecker — *600 ml*
Capacity of half liter. The flare used to control abundant froth of wheat beers, the "Weissbier."

ancient beer-swilling forebears. Not that the ancient Mesopotamians can necessarily take the credit. It may have been the ancient Egyptians, or the Sumerians, or indeed the Chinese. Or just nature, since any cereal crop containing sugar can actually ferment spontaneously if there is enough wild yeast in the air. One thing we do know is that the Czechs, of all people, remain the world's champion beer drinkers — with a slightly staggering per capita intake of close to 160 liters per year.

HOW MUCH COCA-COLA FOR A BEER

Relationship between the price of a beer and a Coca-Cola: in Australia for the price of a beer you can buy just over half of a 500 ml Coke. In Iran a "Lager" beer costs 22 times more.

1	AUSTRALIA 0.61		100	IRAN 22
2	MONGOLIA 0.66		99	ALGERIA 10.13
3	MALAYSIA 0.67		98	GUATEMALA 5.25
4	ROMANIA 0.67		97	BOLIVIA 5.07
5	HUNGARY 0.67		96	PERU 4.22
6	PARAGUAY 0.7		95	ITALY 3.88
7	BELARUS 0.78		94	LITHUANIA 3.75
8	COSTA RICA 0.8		93	SEYCHELLES 3.68
9	CROATIA 0.8		92	NORWAY 3.47
10	PORTUGAL 0.87		91	ARGENTINA 3.29
11	VENEZUELA 0.87		90	CANADA 3.02
12	GUINEA 0.88		89	UKRAINE 3
13	DENMARK 0.89		88	KAZAKHSTAN 3
14	BULGARIA 0.91		87	GUYANA 3
15	CZECH REP. 0.93		86	SOUTH KOREA 3
16	MALAWI 0.94		85	AZERBAIJAN 3
17	UNITED STATES 1		84	COLOMBIA 2.82
18	POLAND 1		83	THAILAND 2.78
19	SLOVAKIA 1		82	ISRAEL 2.78
20	CENTRAL AFRICAN REP. 1.02		81	NICARAGUA 2.7

BEER AND SOFT DRINKS: IN EUROPE HOPS WIN

Per capita consumption of beer (A), carbonated soft drinks (B), and non-carbonated soft drinks (C). 2008 data.

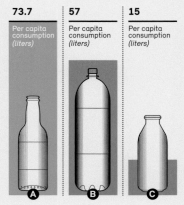

73.7 Per capita consumption (liters) — A
57 Per capita consumption (liters) — B
15 Per capita consumption (liters) — C

Closed glass — *330 ml*
Its tapered shape lifts the froth, preventing it from overflowing. Its thin glass enhances "Lager" beers.

Goblet — *250 ml*
Hemispherical shape, ideal to progressively lower the froth and enhance the fragrance.

Biconical column — *425 ml*
Wide at the center, tapered mouth. For Belgian "Pils" beers and for those who want the froth decapitated with a spatula.

Pint — *473 ml*
The inverted cone shape neutralizes the froth of "Bitter ales" and enhances the *Cream* of Stouts.

Beer
THE GOOD LIFE

The Guide to Good Taste

Sniff, sip, and slurp — how to tell a fine wine from a box wine.

Can you tell a Merlot from a Cabernet Sauvignon, a Claret from a Beaujolais? Can you distinguish the 400 separate aroma compounds thought to form complex flavors? Then you're in good company. A recent scientific study in the U.K. found that people could tell the cheapest rotgut from the finest French classics most of the time, but were hopelessly lost when it came to discriminating among the many varieties in between. Not that the experts faired any better. One long-term study, carried out by Californian winery owner Robert Hodgson, rocked the wine-tasting world in 2013. Testing several wines over several years, Hodgson found that even the top critics — whose finely-tuned palates often determined the price of a particular vintage — were embarrassingly inconsistent, rating the same wine differently on a different day. But, junk science or not, here's your guide to how to make it look professional. Just don't forget to spit out the sample.

Tasting, step by step

1 Serving

Wine tasting glass standard dimensions (vary within 2–3 mm)

Cup — 46 mm
Capacity around 215 ml
Radius 65 mm
Stem
Base — 65 mm
100 mm / 55 mm

Fill the glass to 1/4 or 1/3 of its volume

The wine bottle
Most common volume is 75 cl (750 ml), but sizes vary

Wine servings

Serving temperature

18°C to 15°C

Large red wine
Glass shape with a wider mouth offers more exposure to air, allowing wine to breathe

Wine/grape variety
Older red wines with complex aromas such as pinot noir, burgundy, Médoc, Rioja

14°C to 13°C

Small red wine
Narrower glass directs the bouquet to the nose

Light red wines such as Beaujolais, shiraz

14°C

Fortified wine
Smaller than a wine glass, but with a rounded bowl. Allows concentration of aromas on nose

Port, Madeira, mistelle, Marsala

14°C to 11°C

Sherry
Even smaller than a port glass, perfect to accentuate fruity aromas

Fino, manzanilla, pale cream, amontillado, dark cream

12°C to 11°C

Large white wine
A little smaller than a small red wine glass, but with a wide bowl

Chardonnay, white bordeaux, verdejo, burgundy

11°C to 10°C

Small white wine
Glass shape traps the aromas of the wine

Aromatic light whites: riesling, gewürztraminer

10°C

Rosé wine
Large opening directs wine to tip of the tongue, increasing sensitivity to sweetness

Young rosés: syrah, weissherbst, carignan, Cigales

8°C

Sweet wine
Rounded bowl allows the wine to be directed towards the back of the mouth

Sweet, unfortified wines: muscat, ice wine

9°C to 7°C

Flute
Tall narrow bowl keeps wine cold, allowing appreciation of the bubbles

Champagne, cava, vinho verde, Asti Spumante

Inside a grape

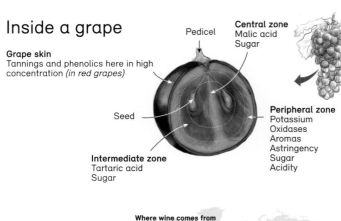

Grape skin
Tannings and phenolics here in high concentration (in red grapes)

Central zone
Malic acid
Sugar

Pedicel

Seed

Intermediate zone
Tartaric acid
Sugar

Peripheral zone
Potassium
Oxidases
Aromas
Astringency
Sugar
Acidity

Where wine comes from
N 50°
N 30°
S 30°
S 50°

The colors of wine

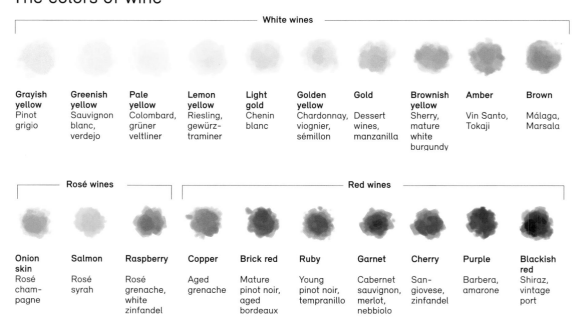

White wines

Grayish yellow	Greenish yellow	Pale yellow	Lemon yellow	Light gold	Golden yellow	Gold	Brownish yellow	Amber	Brown
Pinot grigio	Sauvignon blanc, verdejo	Colombard, grüner veltliner	Riesling, gewürztraminer	Chenin blanc	Chardonnay, viognier, sémillon	Dessert wines, manzanilla	Sherry, mature white burgundy	Vin Santo, Tokaji	Málaga, Marsala

Rosé wines | **Red wines**

Onion skin	Salmon	Raspberry	Copper	Brick red	Ruby	Garnet	Cherry	Purple	Blackish red
Rosé champagne	Rosé syrah	Rosé grenache, white zinfandel	Aged grenache	Mature pinot noir, aged bordeaux	Young pinot noir, tempranillo	Cabernet sauvignon, merlot, nebbiolo	Sangiovese, zinfandel	Barbera, amarone	Shiraz, vintage port

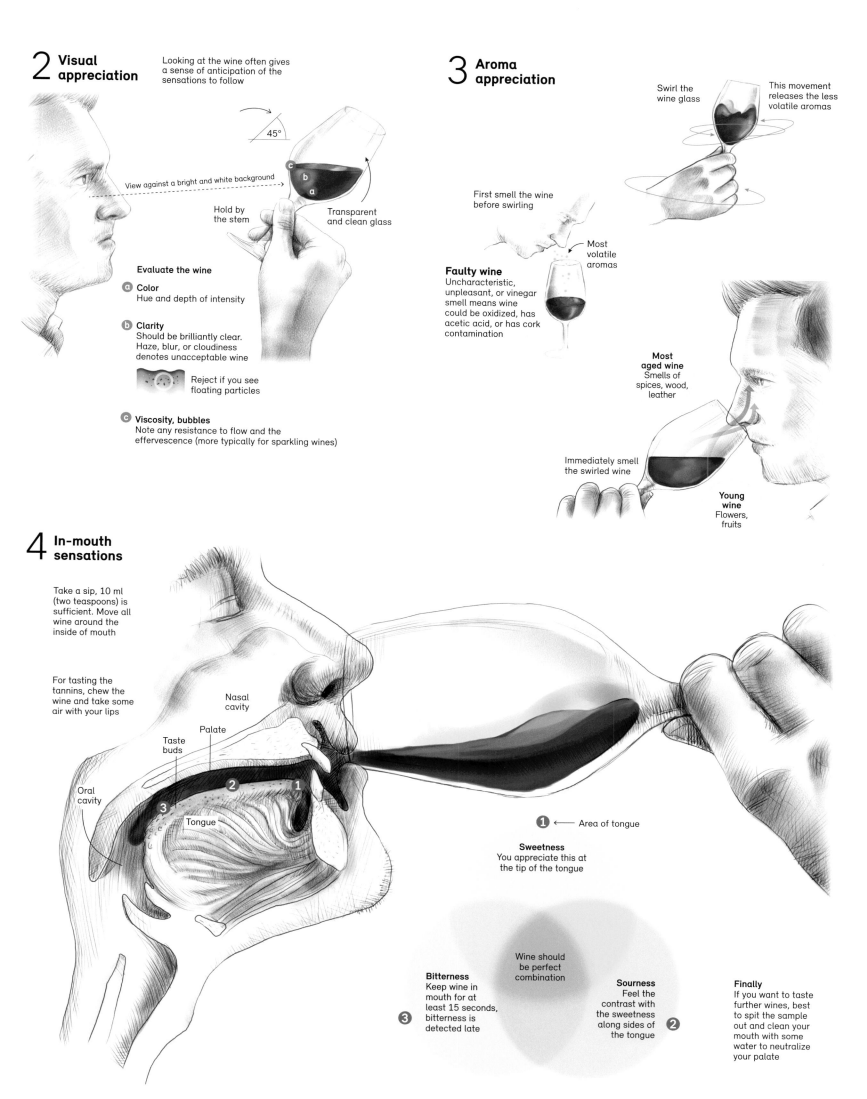

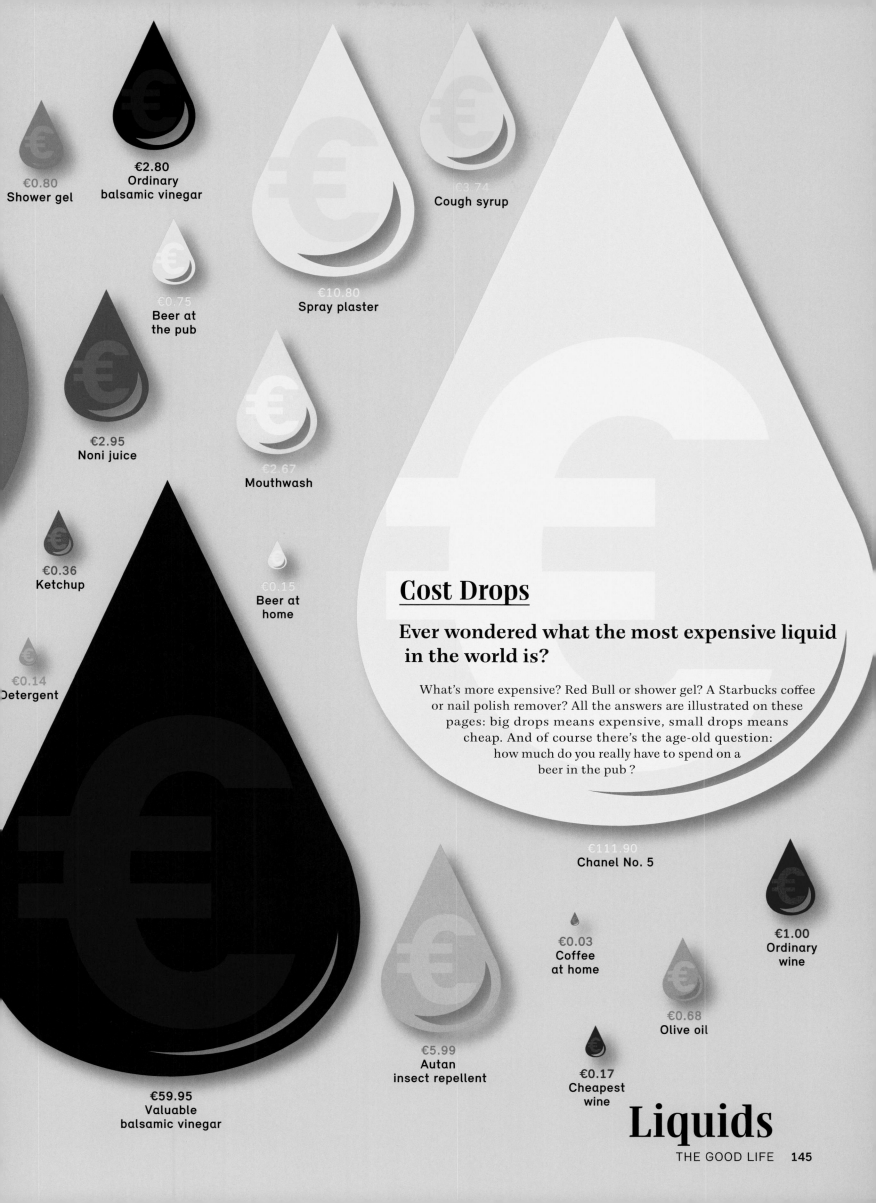

Cost Drops

Ever wondered what the most expensive liquid in the world is?

What's more expensive? Red Bull or shower gel? A Starbucks coffee or nail polish remover? All the answers are illustrated on these pages: big drops means expensive, small drops means cheap. And of course there's the age-old question: how much do you really have to spend on a beer in the pub?

€0.80 Shower gel
€2.80 Ordinary balsamic vinegar
€0.75 Beer at the pub
€10.80 Spray plaster
€3.74 Cough syrup
€2.95 Noni juice
€2.67 Mouthwash
€0.36 Ketchup
€0.15 Beer at home
€0.14 Detergent
€59.95 Valuable balsamic vinegar
€5.99 Autan insect repellent
€111.90 Chanel No. 5
€0.03 Coffee at home
€0.68 Olive oil
€0.17 Cheapest wine
€1.00 Ordinary wine

Liquids
THE GOOD LIFE

Circling the Planet

Sometime in the last 25 years, we all became globetrotters.

Maybe it was the rise of budget air travel, or perhaps it was the expansion of the consumer classes. Whatever happened, over the past generation or so, the population of the world has suddenly been granted the freedom to indulge its innate urge to seek out whatever is around the next corner, or over the next hill, or behind the next wall. Air travel has helped us to comprehend what is going on in other parts of the world, and tourism and air travel have become almost synonymous making the world a much smaller and more familiar place than it used to be. The side effects of flying include the emmission of carbon dioxide. Modern long distance planes consume about 4 liters, or roughly one gallon per passenger per 100 kilometers flying close to the speed of sound — which isn't as bad as you might expect — especially when compared to some SUVs.

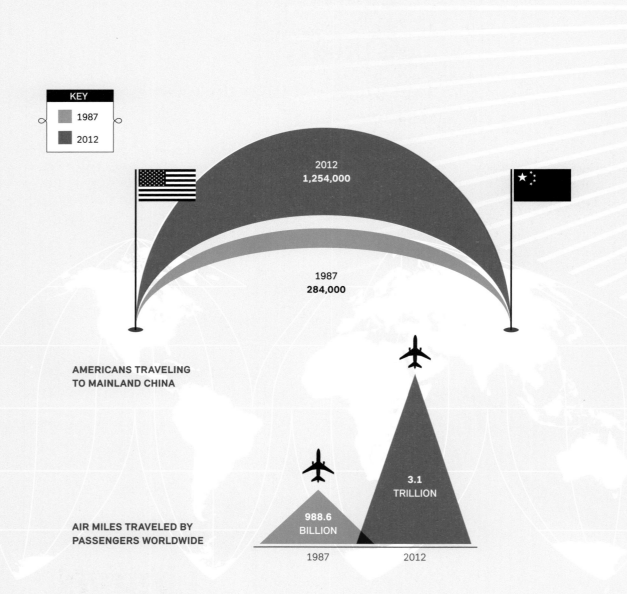

KEY
1987
2012

2012
1,254,000

1987
284,000

AMERICANS TRAVELING TO MAINLAND CHINA

AIR MILES TRAVELED BY PASSENGERS WORLDWIDE

988.6 BILLION — 1987
3.1 TRILLION — 2012

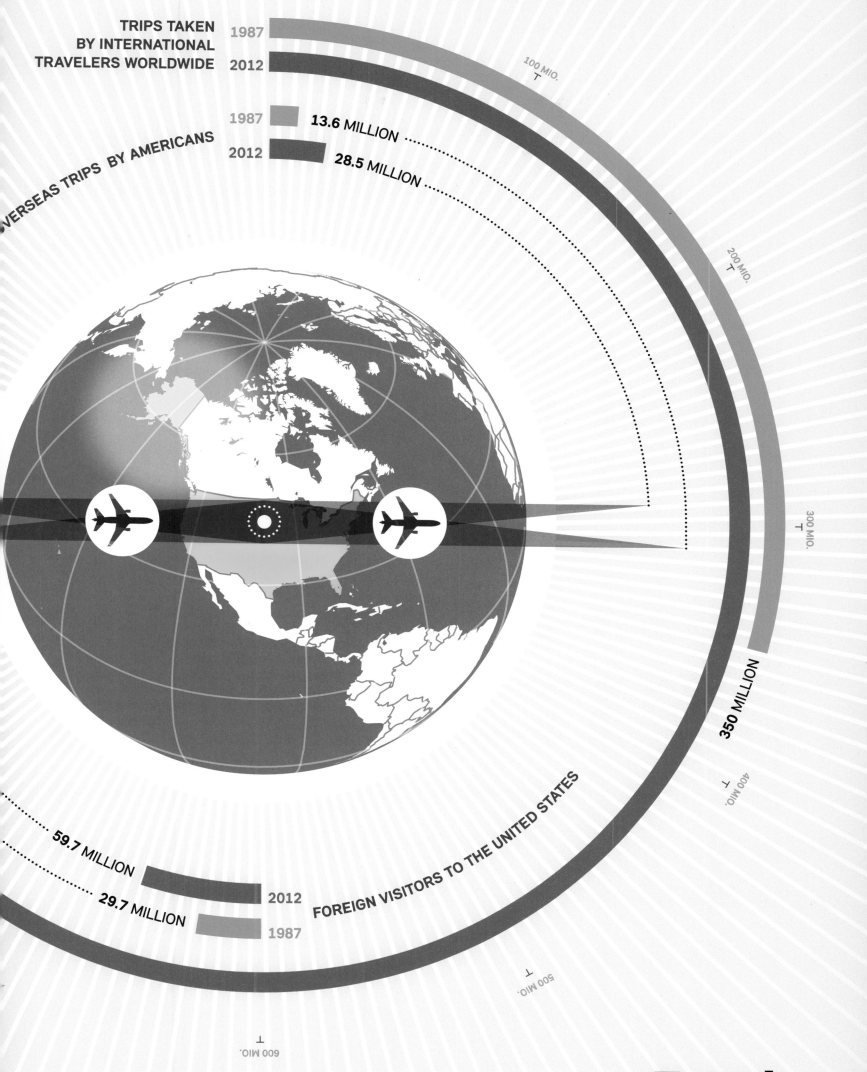

Travel
THE GOOD LIFE

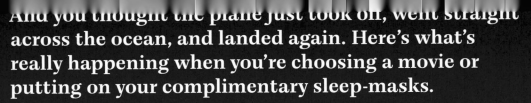

And you thought the plane just took off, went straight across the ocean, and landed again. Here's what's really happening when you're choosing a movie or putting on your complimentary sleep-masks.

Around 1,200 aircraft travel along the transatlantic superhighway every day — or, as it's officially known, the North Atlantic Organized Track System, which connects two of the world's busiest airports — New York's JFK to London's Heathrow. Like so many giant tubes of metal, the airplanes are slid into place in one of the designated channels at each end — which one depends on the weather conditions — and slide out at the other, after you've caught a few hours of cramped sleep.

128.7 km
96.6 km
96.6 km
128.7 km (10 min.)
600 meters
600 meters

FLIGHT LEVELS (METERS)
11,900
11,300
10,700
10,100
9,400
8,800

ORGANIZED TRACK SYSTEM

T
U
V
W
X

2 SAFETY ENVELOPE
Aircraft must keep minimum distances from one another in the track system, while maintaining constant altitude and speed.

1 GETTING IN LINE
Taking into account airlines' preferred routes, oceanic controllers at Gander, Newfoundland, organize aircraft approaching from different directions into position for the Atlantic crossing. This flight is entering the system on track V at 10,700 meters.

Concorde flew between 15,200 and 18,300 meters, far above the main traffic flow.

4 HALFWAY POINT
At 30°W, responsibility for the flight is transferred from Gander to Prestwick Oceanic Air Traffic Control in Scotland.

579.4 KILOMETERS

3 POSITION CHECK
Aircraft in oceanic airspace are out of radar contact for about four hours. Position reports are made by radio at every 10 degrees of longitude, and the information is used to update displays at the oceanic control centers.

Some flight levels are reserved for aircraft flying in the direction opposite the peak flow.

Aircraft crossing the main traffic flow (for example, Madrid to Los Angeles) are routed above or below the track system.

SHANWICK OCEANIC CONTROL AREA

UNITED KINGDOM
Prestwick
Shannon
IRELAND

ICELAND

GREENLAND

WESTBOUND (DAY)

30°W

GANDER OCEANIC CONTROL AREA

EASTBOUND (NIGHT)

EAST INTO THE NIGHT
As a result of passenger demand, time zone differences, and airport noise restrictions, North Atlantic air traffic has two peak flows: eastbound, leaving North America in the evening, and westbound, leaving Europe in the morning. Every 12 hours a new track system is prepared, to allow as many aircraft as possible to follow the most economical flight paths. Because of changing weather conditions, the track positions are rarely identical.

A B C D E
T U V W X

CANADA
Gander
NEWFOUNDLAND

JET STREAM
NORTH ATLANTIC OCEAN

Flight Paths

THE GOOD LIFE 149

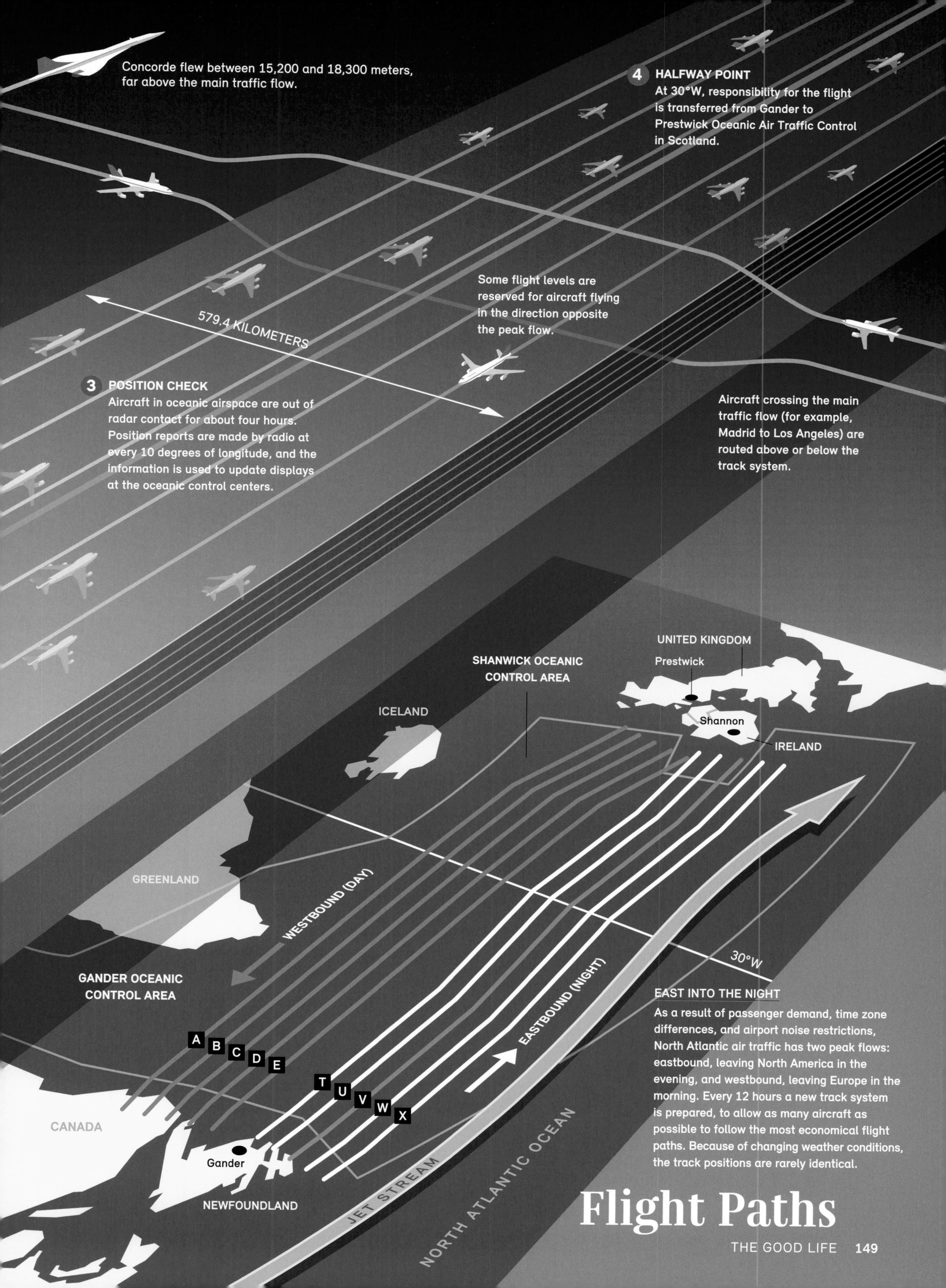

The Speed Need

Speed limits correlated in different countries around the world.

Germany's autobahns are still the only roads in the world where you can — in theory — put the pedal to the metal without fear of the law. The Germans cling to their highway freedom as if it were an article of the constitution (though 34.5 percent of German highways do have a limit of some kind). With the possible exception of the Beer Purity Law, no other piece of legislation has become so integral to national identity and no other German folklore — aside from perhaps the Oktoberfest — seems to make an equal impression on foreign visitors. Whether this lack of a general speed limit has something to do with the country's reputation as the world's favorite luxury car maker is up for debate. But then again, speed limits do not necessarily lower the number of road deaths, and Germany's autobahns are among the safest roads in the world.

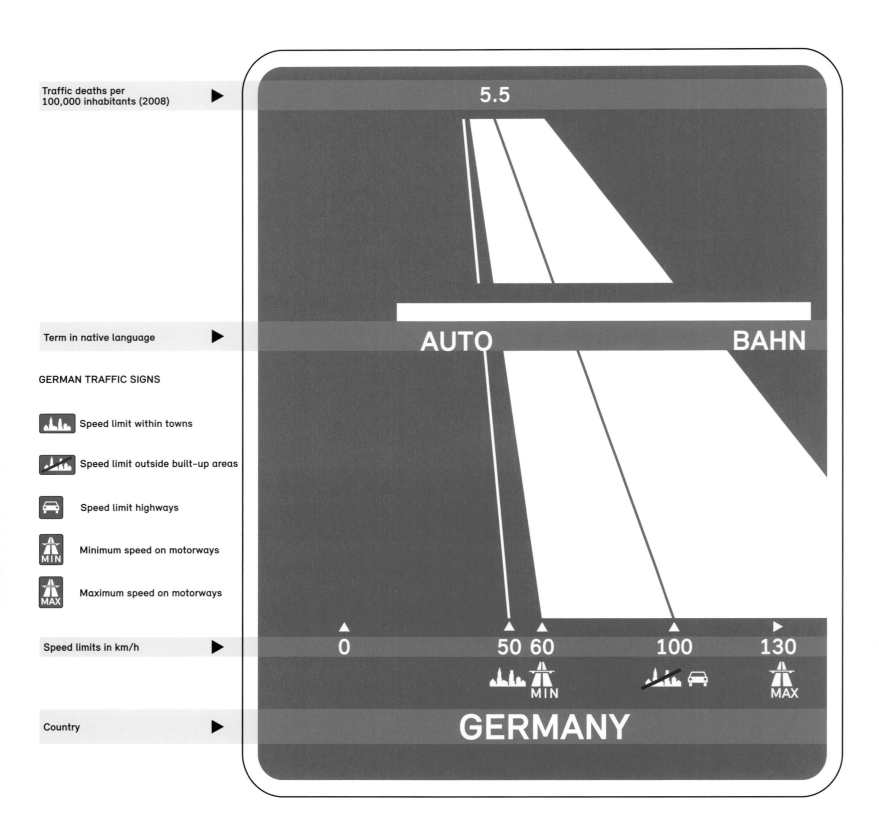

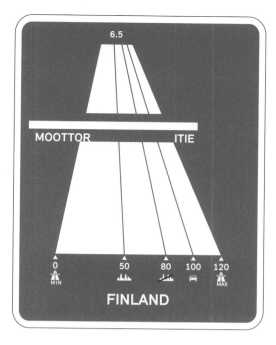
FINLAND

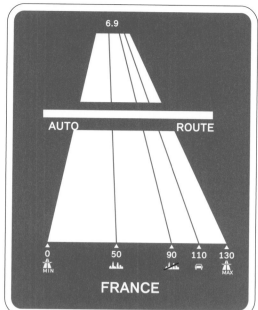
FRANCE

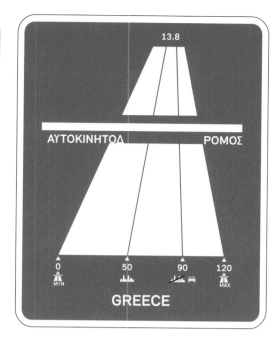
GREECE

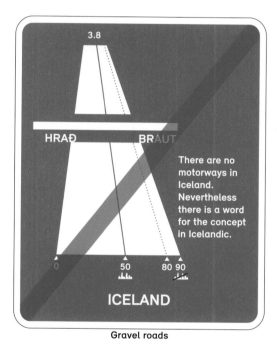
ICELAND

Gravel roads

There are no motorways in Iceland. Nevertheless there is a word for the concept in Icelandic.

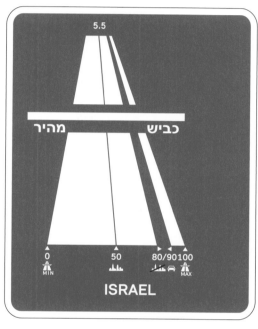
ISRAEL

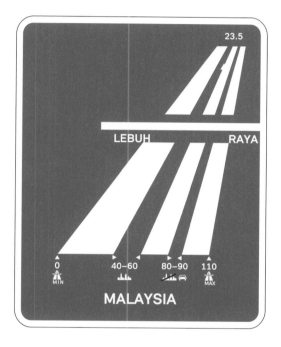
MALAYSIA

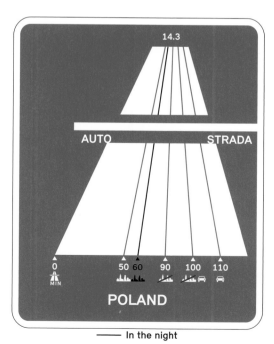
POLAND

— In the night

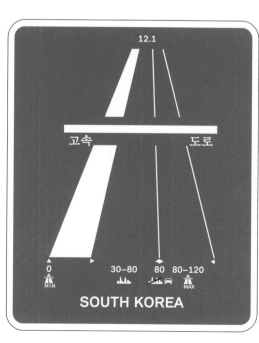
SOUTH KOREA

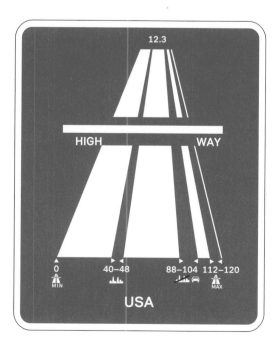
USA

Highways
THE GOOD LIFE

Jingle Sells

The American holiday season is a time for home, hearth, family, merry old men — and cold hard cash.

Every Westerner loves the holiday season — even the Scrooges might nibble a little gingerbread when they think no one is looking. But no one loves the yuletide more than retailers, especially in the United States, where businesses have come to count on the annual boom, when the music of the cash register becomes their Christmas carols. This retail boom has expanded to cover almost the whole of November and December, neatly bookended, you might say giftwrapped, by Cyber Monday — when everyone goes online for their last-minute shopping at the start of Thanksgiving week — and Christmas. It also helps to make some good news for Christmas — after a year of economic recession and downturn, the "retail is booming" story at the end of the evening news broadcast always helps put people in a cozy mood.

How much do Americans **consume** over the holidays?
(most recent dates available)

$9.3 billion
in jewelry-store sales during November and December, 2007.
$4.3 billion in sales for a typical two-month non-holiday period in 2007.

30–35 million
real Christmas trees are sold in the U.S. every year.

2007 HOLIDAY RETAIL SALES
$474.5 BILLION

IN 2007, THE AVERAGE CONSUMER PLANNED TO SPEND MORE THAN $800 ON HOLIDAY-RELATED SHOPPING AND MORE THAN $100 ON THEMSELVES

THE TYPICAL AMERICAN PLANNED TO SPEND:

$469.14 ON FAMILY MEMBERS
$94.69 ON CANDY AND FOOD
$90.13 ON FRIENDS
$49.76 ON DECORATIONS
$37.45 ON ACQUAINTANCES (E.G., TEACHERS, BABYSITTERS, CLERGY)
$32.21 ON GREETING CARDS & POSTAGE
$22.79 ON CO-WORKERS
$20.53 ON FLOWERS

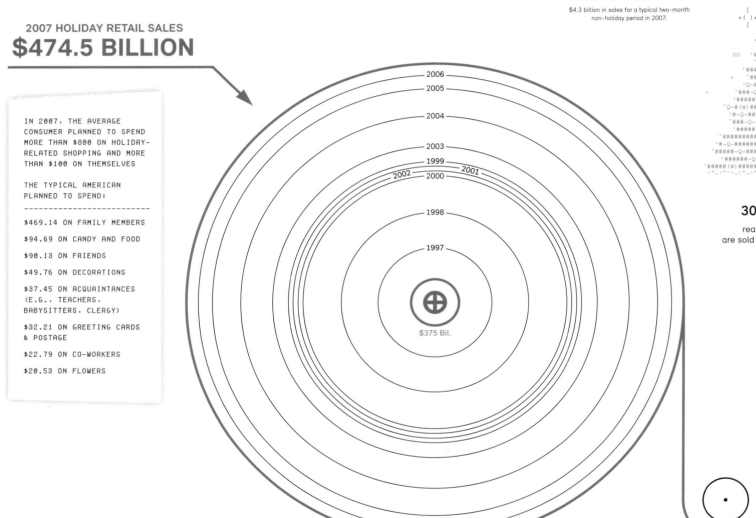

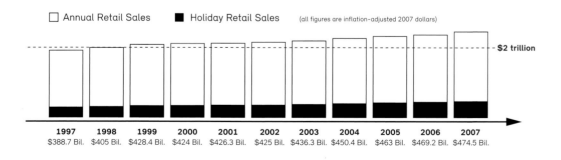

☐ Annual Retail Sales   ■ Holiday Retail Sales   (all figures are inflation-adjusted 2007 dollars)

1997	1998	1999	2000	2001	2002	2003	2004	2005	2006	2007
$388.7 Bil.	$405 Bil.	$428.4 Bil.	$424 Bil.	$426.3 Bil.	$425 Bil.	$436.3 Bil.	$450.4 Bil.	$463 Bil.	$469.2 Bil.	$474.5 Bil.

$2 trillion

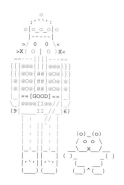

$5.8 billion

pent in stores on hobbies, toys, and games during November and December, 2007.

$2.3 billion in sales for a typical two-month non-holiday period in 2007.

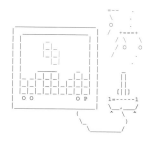

$19.8 billion

in computer and video game, console, and accessory sales during November and December, 2007.

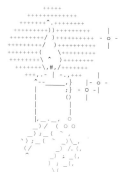

$700 billion

in candle sales during the 2007 holiday season.

That's 350 million pounds of wax.

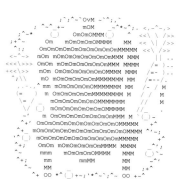

67 million

turkeys are eaten at Thanksgiving and Christmas.

The turkey industry produced an estimated 269.8 million turkeys in 2007.

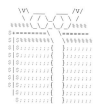

20 billion

cards, letters, and packages are delivered by the U.S. Postal Service between Thanksgiving and Christmas.

192 billion pieces of mail are delivered by the U.S. Postal Service during the rest of the year.

$26.3 billion

estimated worth of gift cards were purchased during the 2007 holiday season.

275 million

first class cards and letters were mailed on December 17, 2007, the busiest mailing day of the year.

131 million

pounds of eggnog were sold during the 2006 holiday season.

Thanksgiving | DECEMBER | Christmas | JANUARY

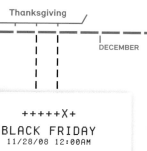

+++++X+
BLACK FRIDAY
11/28/08 12:00AM

FRIDAY FOLLOWING THANKSGIVING IS CALLED BLACK DAY BECAUSE HISTORICALLY IS THE DAY RETAILERS N TO TURN A PROFIT FOR YEAR. THE PHRASE HAS ITS GINS IN A BUSY 1975 SHOP SEASON THAT TOOK PLACE ITE AN ECONOMIC DOWN , CAUSING BUSINESSES' NCE SHEETS TO GO OUT OF RED AND INTO THE BLACK.

[YBER IVI0NDAY

A TERM COINED BY SHOP.ORG—THE NATIONAL RETAIL FEDERATION'S DIGITAL DIVISION—FOR THE MONDAY FOLLOWING THANKSGIVING, WHEN AMERICANS START MAKING GIFT PURCHASES ONLINE, ESPECIALLY FROM THE WORKPLACE. ACCORDING TO INTERNET INFORMATION PROVIDER COMSCORE, CYBER MONDAY, 2006, SAW $608 MILLION IN ONLINE RETAIL SPENDING, WHICH WAS THE HEAVIEST ONLINE SHOPPING DAY ON RECORD AT THE TIME.

ONLINE HOLIDAY SALES
(NOVEMBER 1-DECEMBER 27)

2007 $28 BILLION
2006 $24.2 BILLION

ONLINE SALES
(JANUARY-OCTOBER)

2007 $93.6 BILLION
2006 $79.7 BILLION

(ALL NUMBERS IN INFLATION-ADJUSTED 2007 DOLLARS)

SHOPPING WIND[]W

THE BUSIEST SHOPPING PERIOD OF THE YEAR TENDS TO FALL BETWEEN DECEMBER 17 AND DECEMBER 23.

IN 2007, ABOUT 14% OF SHOPPERS ARRIVED AT A RETAIL STORE BY 4 A.M. THAT SAME YEAR, 40% OF SHOPPERS STARTED MAKING HOLIDAY PURCHASES BEFORE HALLOWEEN.
IN 2006, RETAILERS ADDED 596,000 EMPLOYEES TO HANDLE THE HOLIDAY SHOPPING PERIOD.

THANKS FOR SHOPPING

DEBT

IN NOVEMBER AND DECEMBER OF 2007, CONSUMERS ACCUMULATED AN ESTIMATED $12.8 BILLION IN NEW DEBT, ACCORDING TO CARDTRAK, AN ORGANIZATION THAT TRACKS CREDIT CARD USE. THAT IS 16% OF THE TOTAL DEBT ACCUMULATED THAT YEAR.

ACCORDING TO A 2007 SURVEY, ONE THIRD OF CONSUMERS WERE STILL PAYING OFF DEBT FROM THE 2006 HOLIDAY SEASON.

THANK YOU

Christmas

Dedicated Followers of Fashion

From Savile Row to Marks and Spencer, no one loves fashion in the same way as the Brits.

At the bottom of every shopping bag of designer clothes, there's that nasty little slip of paper, a downer that punctures the euphoria of your shopping spree — the receipt. This graphic follows the money — more specifically, the money that Brits spend every year on their threads, innit. One thing this graphic makes clear: clothes shopping is a pursuit in itself. We buy clothes not only to adorn our bodies, but also to pass the time.

1 | Where do we buy clothes?

* Arcadia owns Dorothy Perkins, Evans, Top Shop, Miss Selfridge, Wallis, Burton, and Top Man

● In 2004, the top five clothing retail groups — Marks & Spencer, Next, Arcadia Group,* Matalan and Bhs — accounted for 44.8% of total sales.

Share of clothing market
(percentage of total sales, 2004)

- Clothing specialists (includes M&S & Debenhams): 66.1
- Department and variety stores: 12.4
- Home shopping (mail order, direct selling, and Internet): 8.8
- Grocers (Tesco, Asda, etc.): 5.4
- Sports shops: 4.3
- Textiles specialists: 2.0
- Other: 1.0

Most patronized clothing outlets
(percentage of respondents who said they had used outlet from April 2004 to April 2005, selected outlets)

%	Outlet
42%	Marks & Spencer
27%	Asda (George)
25%	Next (incl. Directory)
24%	Tesco
24%	Matalan
21%	Debenhams
15%	Primark
15%	Bhs
12%	Top Shop/Top Man
11%	Mail order/Internet/TV
10%	John Lewis
8%	Gap
6%	Sainsbury's
4%	H&M

● On average, the amount women spend in a lifetime on shoes and clothes that will never be worn is **£12,810.**

2 | How much do we spend?

● Clothing and footwear is now the largest category of consumer expenditure in the U.K., after food and non-alcoholic beverages, which in 2004 totalled £65.5 Bil.

● In 2004, clothing sales were worth £36.6 Bil. — a 19% increase since 2000.

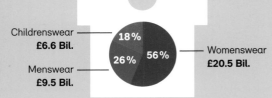

- Childrenswear £6.6 Bil. — 18%
- Menswear £9.5 Bil. — 26%
- Womenswear £20.5 Bil. — 56%

● In 2004, £5.95 Bil. was spent on footwear — a 34.3% increase since 2000.

● In 2004, the average weekly spend on clothing was £22.50 — 1999 was the record spending year at £23.50. The highest average spend by region was Northern Ireland at £30.20 a week. The lowest was in Wales at £17.80 a week.

Total weekly spending across the U.K. on clothes, by type, 2004, in Mio. £

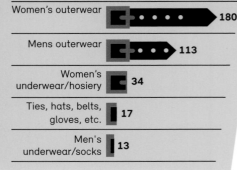

Category	£ Mio.
Women's outerwear	180
Mens outerwear	113
Women's underwear/hosiery	34
Ties, hats, belts, gloves, etc.	17
Men's underwear/socks	13

● In 2004, the retail price index showed that clothes were cheapest in **July** and most costly in September.

- Women in the U.K. spend, on average, **£31,680** on shoes over the course of their lifetime.

- 34% of 9–10 year olds spend some of their average weekly pocket money on clothing. This rises to **51%** of 11–16 year olds.

- In 2000, Marks & Spencer's share of the U.K.'s clothing market was 14.2%. In 2004, it was **11.8%**.

- When Kate Moss was seen carrying a quilted Marc Jacobs handbag in September 2005, sales of the bag rose **10**-fold at Selfridges in two days.

- In 2005, Primark sold nearly 250,000 of its £12 military-style jackets (available in 12 styles). Primark takes, on average, 6 weeks to get a concept design into its shops.

3 | What do we buy?

Top 5 clothing imports
by value, 2004

- Trousers £1.89 Bil.
- T-shirts £1.52 Bil.
- Pullovers £1.02 Bil.
- Blouses £0.5 Bil.
- Skirts £0.39 Bil.

Best-selling fashion at Marks & Spencer
December 2005 (not in order)

Gray wool military coat	£99
Black military jacket	£60
Black knee-length culottes	£35
Black and white-striped beatnik jumper	£29.50
Autograph black shift dress with bow	£89

Designer clothes

- In 2003, the global designer luxury goods market was worth $69.3 Bil.
  - Fashion and leather $29 Bil. — 42%
  - Perfumes & cosmetics $25.4 Bil. — 37%
  - Watches & jewelery $12.4 Bil. — 18%
  - Others $2.3 Bil. (3%)

- In 2002/03, Gucci derived 14% of its turnover from ready-to-wear clothing.

Global market share of designer labels (ready-to-wear and footwear only, 2003, %)

Polo Ralph Lauren	12.0
Tommy Hilfiger	10.6
Max Mara	6.9
Prada	6.0
Chanel	5.4

4 | Where are our clothes made?

Total value of clothing imports

- 1995 EU £1.57 Bil.
- Non-EU £3.72 Bil.
- 2004 EU £2.67 Bil.
- Non-EU £8.19 Bil.

Top 5 U.K. clothing suppliers

Importer	Import value	% change since 2003
Hong Kong	£1.53 Bil.	−7%
Turkey	£1.22 Bil.	+5%
China	£1.01 Bil.	+17%
Italy	£0.60 Bil.	+6%
Bangladesh	£0.50 Bil.	+12%

- The World Trade Organization estimates that China made 17% of the world's textiles in 2003. It expects the figure to reach 50% by the end of 2006.

- Within six months of the 31-year-old clothing quota between the EU and China ending on January 1, 2005, clothing imports from China to the EU had risen by 534%.

5 | Where does the money go?

Price breakdown of a £100 sport shoe

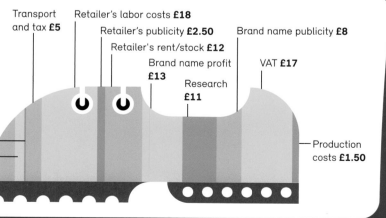

- Workers' labor costs £0.50
- Subcontractors' profits £3
- Materials £8.50
- Transport and tax £5
- Retailer's labor costs £18
- Retailer's publicity £2.50
- Retailer's rent/stock £12
- Brand name profit £13
- Research £11
- Brand name publicity £8
- VAT £17
- Production costs £1.50

- Debenhams spends more on advertising than any other clothing retailer. In 2004, it spent **£19.9 Mio.** — a 151% rise since 2000.

Fashion

Animal Attraction

This graphic looks at all angles of the British pet phenomenon — the demographic, the financial, the medical, and the moral.

Nothing completes an English country home quite like a fine pedigree mastiff hound. Just ask Sherlock Holmes. But the Hound of the Baskervilles is just the most famous pet dog in England — the country's pet dog population is actually well over six million. The great Sir Winston Churchill was very much his own man and in his affection for animals was no exception to the rule. Traditionally epitomized as the incarnation of the British Bulldog himself, the legendary leader of Britons declared "I like pigs. Dogs look up to us. Cats look down on us. Pigs treat us as equals."

1 | What kind of pets do we keep?

In 2003, 52.7% of U.K. households owned a pet — of these households, 24.4% owned a cat, 20.9% owned a dog and 8.6% owned goldfish.

Pet population in the U.K., 2002

Dogs	6.1 Mio.
Cats	7.5 Mio.
Budgerigars	0.75 Mio.
Rabbits	1.1 Mio.
Goldfish	14.7 Mio.
Tropical fish	9.3 Mio.
Marine fish	0.7 Mio.
Guinea pigs	0.73 Mio.
Hamsters	0.86 Mio.
Canaries	0.26 Mio.
Other birds	1.06 Mio.
Reptiles	0.14 Mio.
Total number of pets	**43.2 Mio.**

Cats and dogs

Least and most popular regions to own a cat and dog in U.K., 2003

Adults who own a dog: Scotland 30%, Southeast 18%
Adults who own a cat: Scotland 21%, Southeast 35%

Population of domestic cats and dogs, 1992–2002

% change 1992–2002 Cats: +7
% change 1992–2002 Dogs: −16

2 | How much do we spend on pets?

Pedigree pets
Percentage of dogs and cats that are non-pedigree

Dog: 24.7%
Cat: 92%

Average lifetime costs for dogs and cats, RSPCA estimate, 2004

Dog (12 yrs.) Total £9,844
- Food £5,664
- Pet insurance £1,200
- Vet £1,200
- Other costs (such as kennels) £1,780

Cat (14 yrs.) Total £9,459
- Food £5,664
- Vet £1,300
- Pet insurance £1,050
- Cat litter £2,184
- Other costs (such as cattery) £1,276

NB A Great Dane can cost up to £33,000 over its average 10-year life, according to Churchill Insurance.

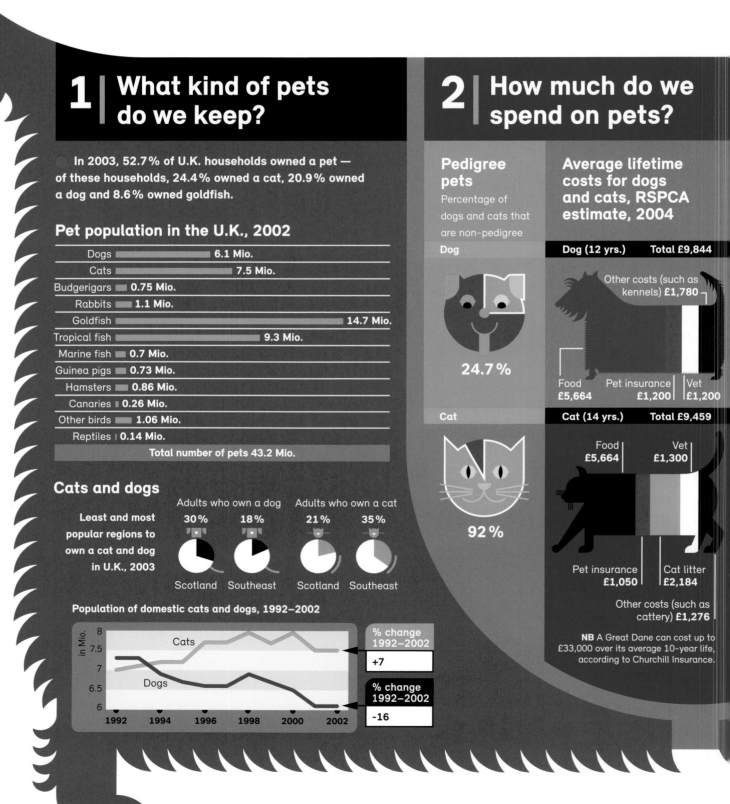

- The world's oldest cat is "Creme Puff" of Texas who turned 37 in 2004, according to the Guinness World Records.

- In 2003, 27% of women in the U.K. owned a cat, compared to **20%** of men.

- On average, each pet-owning household spent £2.90 a week on pets in 2002/03.

Most popular breeds of pedigree cats and dogs in U.K., 2002

Typical cost of puppy
- £300–£500 Labrador Retriever
- £400–£600 Cocker Spaniel
- £300–£600 English Springer Spaniel

Typical cost of kitten
- £250–£500 British Shorthair
- £250–£350 Siamese
- £250–£400 Persian

3 | What do we buy our pets?

Total household spending on pets in 2003
Total £3,640
- Pet purchase and accessories £832 Mio.
- Pet food £1,872 Mio.
- Veterinary and other services £936 Mio.

Pet accessories, by sector, 2003
- Grooming/toilet equipment £32.4 Mio.
- Other £4 Mio.
- Toys £61.0 Mio.
- Collars and leads £39.0 Mio.
- Cat litter £50.8 Mio.
- Baskets, bedding, and carriers £41.0 Mio.

Pet healthcare sales, 2003
- Flea treatments £85 Mio.
- Worming / Skin treatments £35 Mio.
- Vitamins, minerals, and supplements £24 Mio.
- £19 Mio.
- Others £8 Mio.

- The total market for dog and cat food was valued at an estimated £1.55 Bil. in 2003.

- In 2005, 34% of dogs and 21% of cats were insured. The pet insurance market was worth £265 Mio. in 2005 — a 157% increase since 1996.

4 | How many pets are poorly treated?

- In 2004, the RSPCA removed 157,482 animals from danger or abuse. It rehomed 69,787 animals, mostly through its network of 175 branches.

Top five animals groups of concern, 2004

- Dogs — 18,334
- Farm animals — 16,174
- Small domestic — 7,630
- Cats — 7,349
- Equines — 5,884

Top five dog breeds involved in RSCPA prosecution cases, 2004

Non-pedigree	268
German Shepherd	106
Staffordshire Bullterrier	80
Rottweiler	43
Jack Russell	39

Number of reptiles that passed through the animal reception center at Heathrow Airport

2001 **67,000** 2002 **100,000**

The RSPCA estimates that up to 15% of reptiles kept as pets are undernourished.

- In 2003, the average initial vet consultation fee in London was £18.72. In Scotland, it was **£11.92.**

- In 2003, Masterfoods (includes Cesar, Sheba, and Whiskas) and Nestlé Purina (includes Bakers Complete, Winalot, Arthur's, and Felix) between them controlled 79% of the cat food market and **68%** of the dog food market.

- In 1999, "single-serve" foods accounted for 16% of the cat food market. In 2005, they accounted for **54%**.

- The largest ever litter of puppies is **24**, born to "Tia," a Neopolitan Mastiff, in Cambridgeshire in 2004.

Pets
THE GOOD LIFE

The Covers Cobra

How some songs wind on through pop history.

There are songs we simply cannot shake off. This snaky visualization traces the history of the most covered artists, as their songs echo through the ages. Sadly, this visualization does not record how successful those reimaginings were. Too few musicians remember the golden rule when it comes to attempting a cover version for commercial release — do it differently or do it better. The Beatles dominate — their clean, pure melodies have become embedded in our DNA. But who would have thought that John Lennon, the spikiest of their members, would have inspired enough cover versions to make the chart in his own right? Of course, his closest collaborator, Paul McCartney still boasts one of the most covered songs of all time — *Yesterday*, with some 2,200 covers.

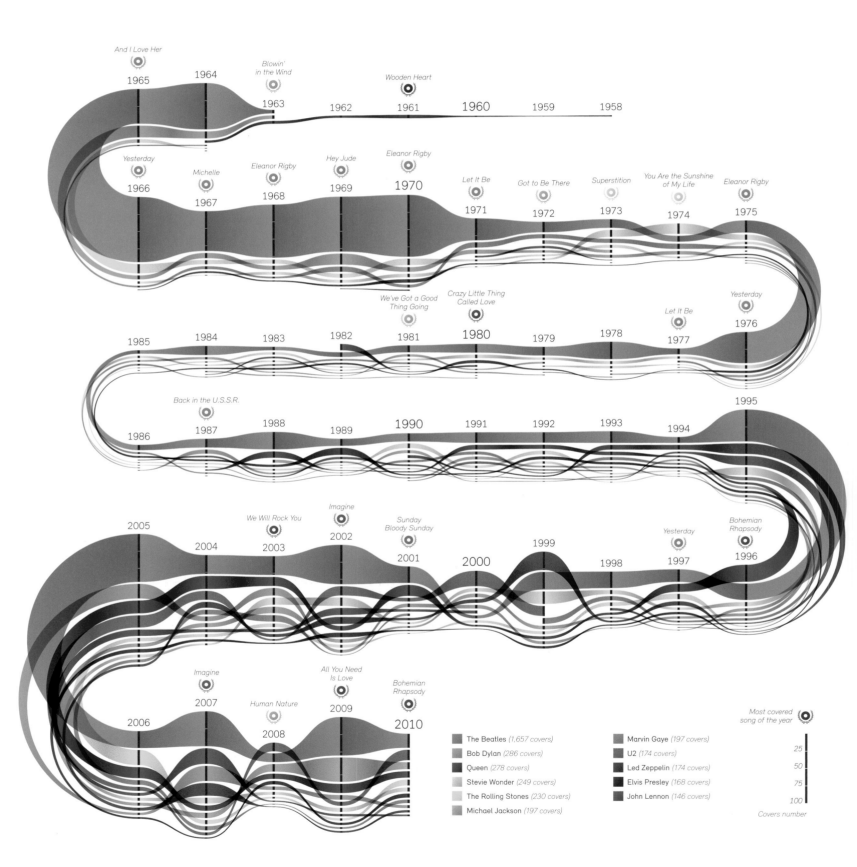

Bigger than Jesus

The Beatles records sales measured against those of their closest rivals.

By the time the Beatles split, Paul, John, George, and Ringo had broken every sales record going, changed pop music forever, provided the soundtrack for a cultural revolution, and written exactly 229 works of genius between them — and they were still barely 30 years old. They are still the biggest-selling artists of all time.

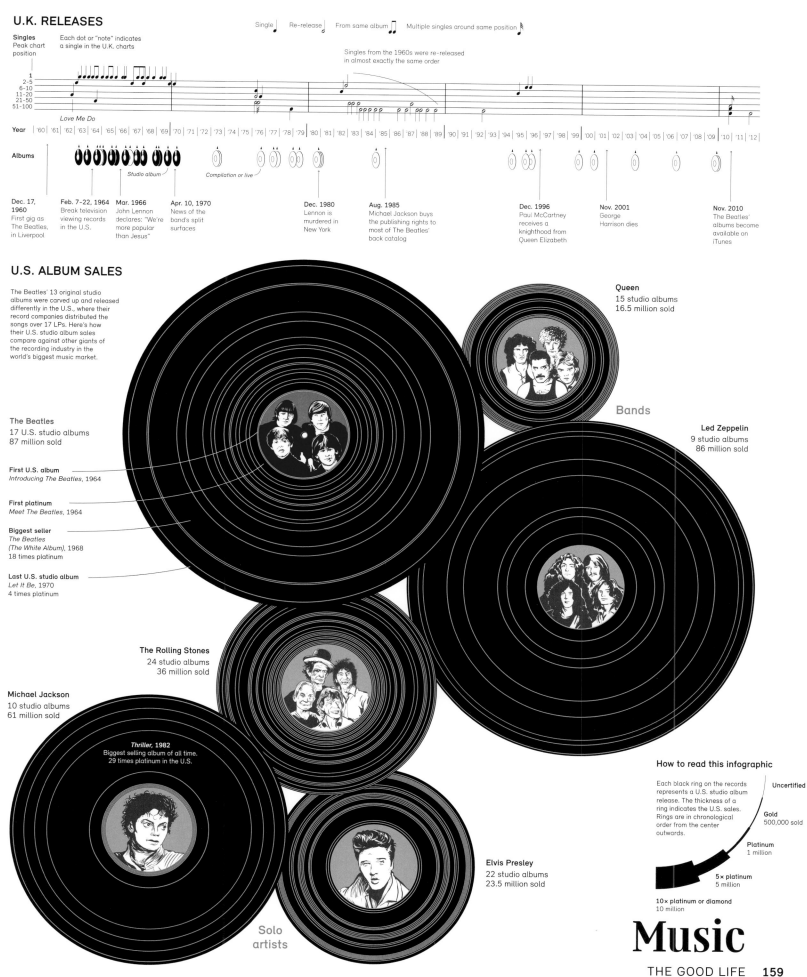

Music

THE GOOD LIFE 159

Square Eyes

These are 550 of the most important and influential (and mainly American) TV shows ever made.

You probably spend at least a decade of your life sitting in front of the TV. But don't let anyone ever tell you that it's wasted — especially if you've been watching any of these shows. For they represent over 70 years of our biggest cultural output. It is interesting to see how some genres, like the Western, have dwindled to a mere trickle, while others, like the Crime Drama, have swelled to a torrent. The whole TV phenomenon, of course, has widened to a tsunami of radio waves — or digital signals — out of that dark box in the corner. Just don't fight over the remote!

TV Shows

THE GOOD LIFE

Art Lives

The lives of the greatest painters of the twentieth century juxtaposed.

They were all brilliant individuals, but their lives were also intertwined — sometimes they knew each other personally, even formed friendships, but more importantly, their works were sometimes intimately related via different movements. Following the cultural fracturing after the First World War, the early twentieth century was to become a bewildering array of artistic "isms" — Cubism, Fauvism, Futurism, Cubo-Futurism, Expressionism, Impressionism. Some were imposed by critics searching for definitions with which to get a handle on this creative golden age, while others were self-imposed schools founded by like-minded geniuses. This graphic places the lives of these individuals in parallel, and traces their influences on each other. Events in their personal lives — lovers, children, travels — are also marked.

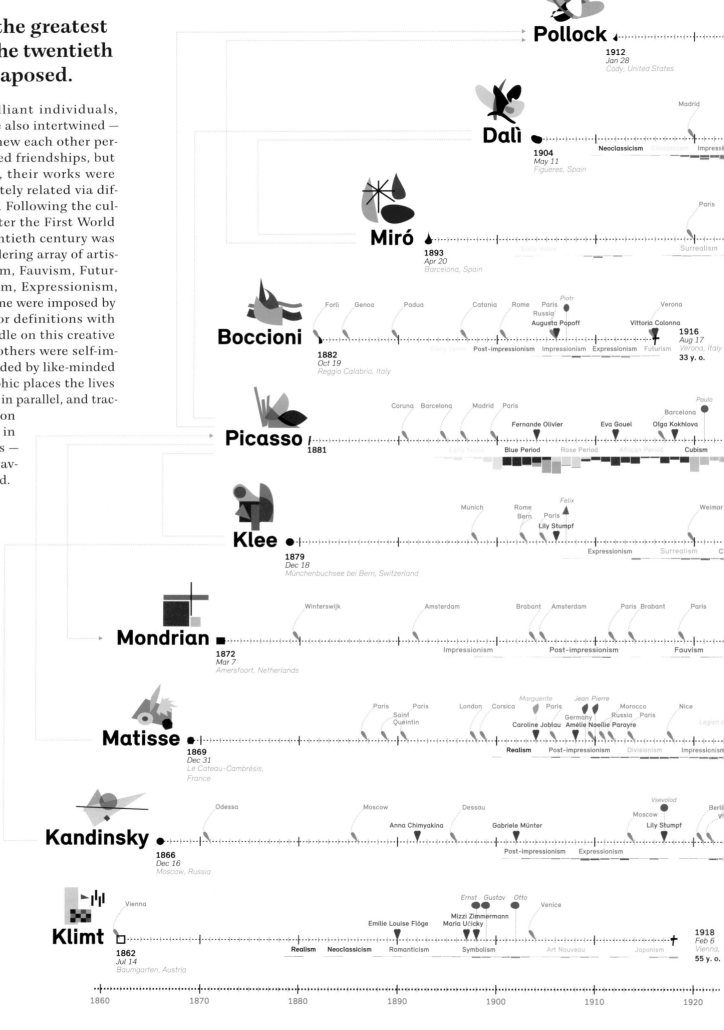

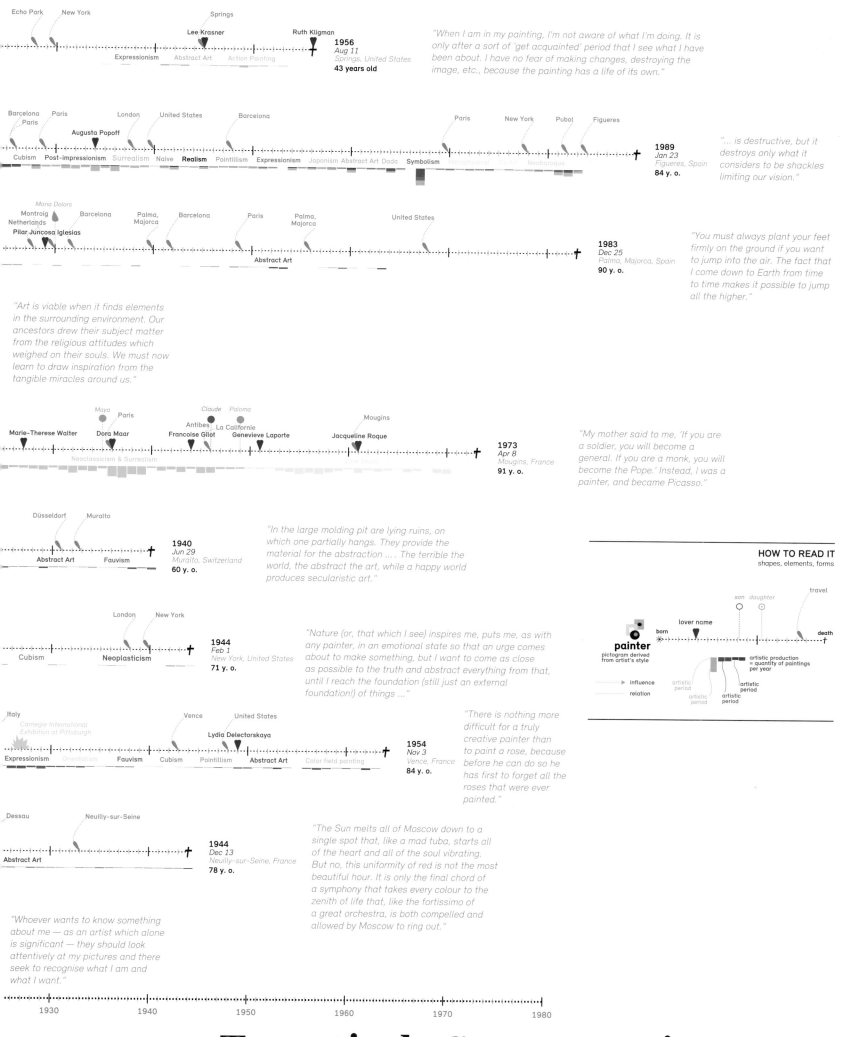

Twentieth-Century Painters

THE GOOD LIFE

The Art of the Deal

A lot of the greatest artists did not live to see the day their art was dicovered, praised, and sold at astronomical prices. These artists did.

Life is not fair and and art is no different. We sometimes nurture the romantic fantasy of artists as lonely geniuses, working their misunderstood talents to the bone in chilly attics with cracked windows, shivering over their easels with nothing but a puny coal stove to warm them. With paintbrushes strapped to their shivering, arthritic fingers, they keep themselves alive with the thought that suffering is essential to their creation and their art. Van Gogh may only have sold a single painting in his lifetime, but today art is commerce and artists like Damien Hirst or Banksy are marketing and communication geniuses. Here are the prices of the most expensive works by living artists who, unlike Vincent van Gogh, are still here to enjoy the material rewards of their art.

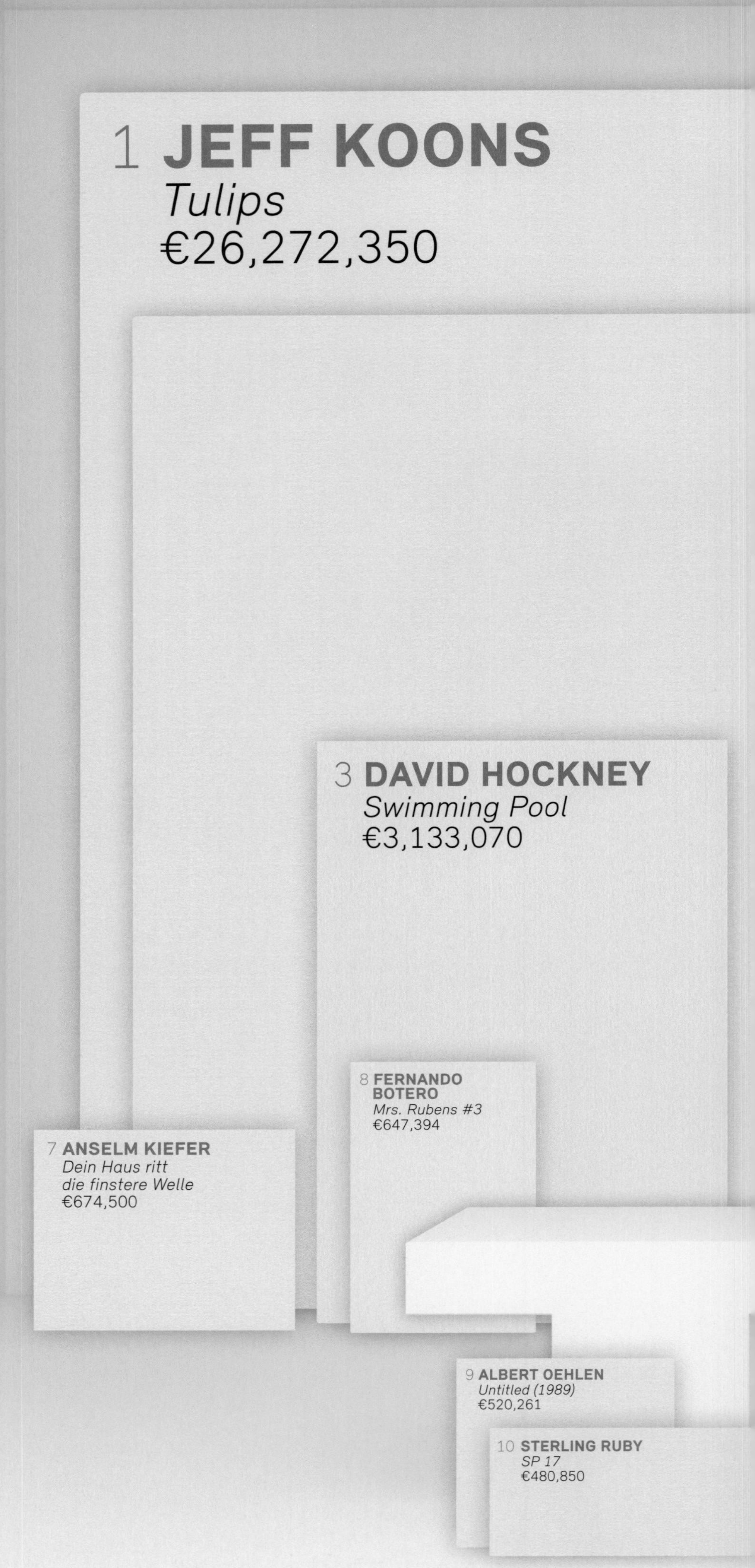

1 **JEFF KOONS**
Tulips
€26,272,350

3 **DAVID HOCKNEY**
Swimming Pool
€3,133,070

8 **FERNANDO BOTERO**
Mrs. Rubens #3
€647,394

7 **ANSELM KIEFER**
Dein Haus ritt die finstere Welle
€674,500

9 **ALBERT OEHLEN**
Untitled (1989)
€520,261

10 **STERLING RUBY**
SP 17
€480,850

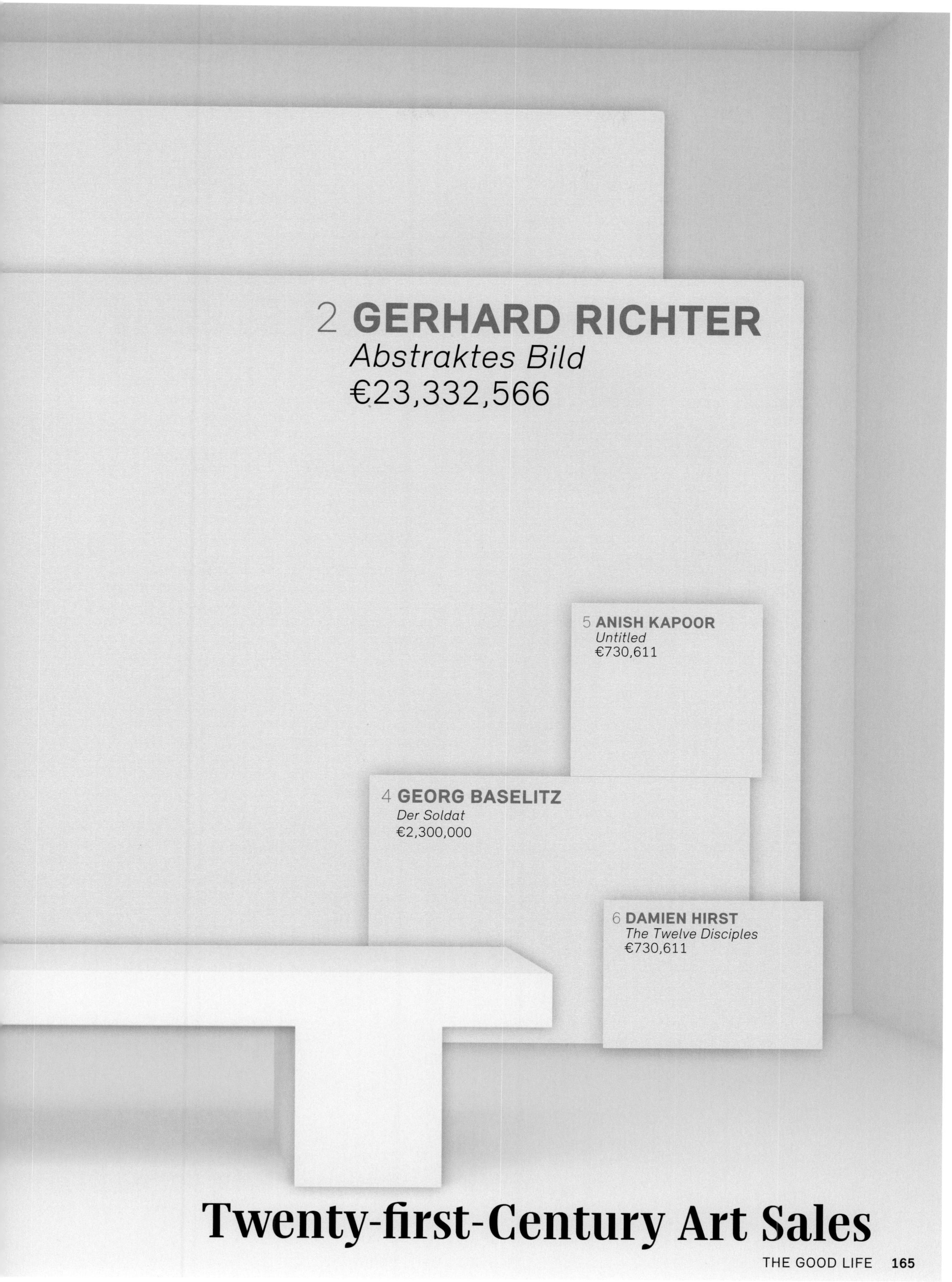

2 GERHARD RICHTER
Abstraktes Bild
€23,332,566

5 ANISH KAPOOR
Untitled
€730,611

4 GEORG BASELITZ
Der Soldat
€2,300,000

6 DAMIEN HIRST
The Twelve Disciples
€730,611

Twenty-first-Century Art Sales

Forging the Rings

How the Olympic Games began.

James Bond, the Sex Pistols, and a parachuting Queen — one wonders what the Ancient Greeks might have made of London 2012. Back in their day, it was all naked men racing round a dusty bowl for a leafy crown. But then again, our paltry 30 Summer Games spread over little more than a century would have looked feeble to them — the ancient Games were staged in Greece for nearly 1,200 years.

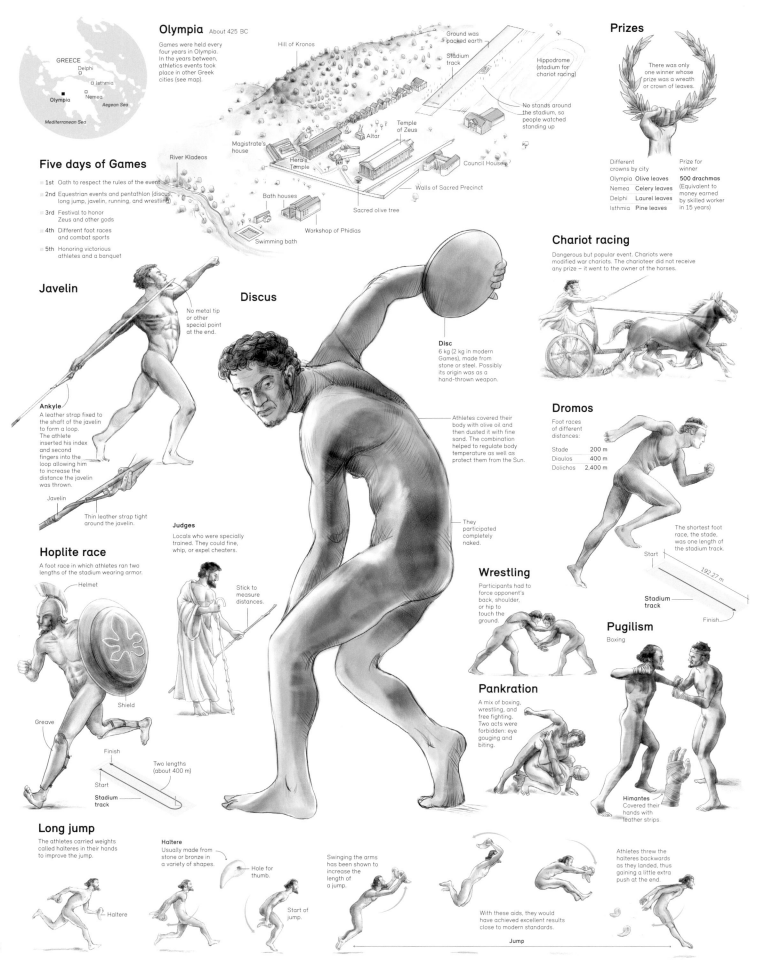

Bettering the Best

Records are made to be broken, but some are more fragile than others.

Is there a limit to what the human body can achieve? Can we stretch it that little bit further, make our flesh and blood and bones move faster and higher? Well, the Olympic Games are a good place to find out. Over the century in which the Games have developed, there has been a pattern to the graphs that chart world records: a steep rise in the first few decades, followed by incremental advances, with the line gradually flattening. And yet, at each Games, it seems like records keep tumbling, especially on the track and in the pool. Meanwhile in some disciplines, especially on the field, it looks like we have indeed hit the ceiling — the men's high and long jump records, for example, are both over 20 years old. Sports scientists analyze dozens of individual athletic performances to devise complex schematics that make projections about when new world records will be established. But their efforts are continually thwarted by certain athletes that seem to defy human limits — in 2008 one egghead pronounced that the men's 100-meter-sprint record of 9.68 seconds would never be bettered, only to be embarrassed a year later by a certain Jamaican named Usain Bolt, who took 14 hundredths of a second off it.

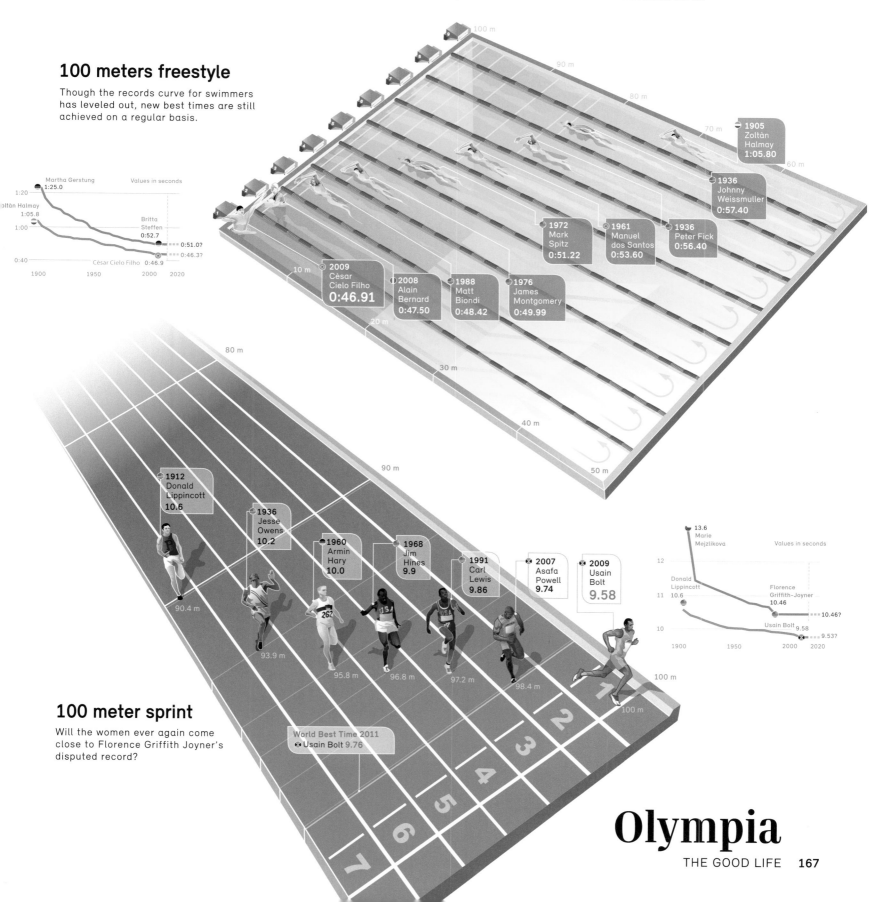

100 meters freestyle
Though the records curve for swimmers has leveled out, new best times are still achieved on a regular basis.

100 meter sprint
Will the women ever again come close to Florence Griffith Joyner's disputed record?

Olympia
THE GOOD LIFE

Any Given Saturday

Football is not only the most popular sport in the world — for millions it's a livelihood.

Crowds like the one pictured here gather all over Europe on any particular weekend (most games are played on Saturdays). This graphic shows all the people who converge regularly on the Allianz Arena, the home stadium of perennial German Bundesliga champions Bayern Munich. While 69,000 fans crane their necks dementedly to see 22 men chase an updated synthetic pig's bladder around a field, some 2,000 people are working on the sidelines, in the director's box, or in the bowels of the stadium to make sure the crowds get their football fix.

Firefighters (7), technical director (1), medical personnel (approx. 48) — of these: paramedics (8), doctors (6)

11,000 cars

350 buses

13,500 standing room

Mobile vendors (approx. 150)

Vendors in megastores & shops from esplanade (45)

Service personnel for fan meetings & arena à la carte (50)

Tier 4 (30)

Service & hospitality (approx. 350)

Kiosk vendors (approx. 400)

Tier 3 (18)

Cooks, assistant cooks (approx. 40)

Tier 2 (8)

Tier 1 (4)

At the turnstiles (8)

LEGO children's club (11)

Hostesses (72)

Opponents' coaching staff

Opponents' substitutes

Opposing team

6 flag wavers

4 referees

3 greenkeepers with 6 assistants

Escort kids

Stadium announcer

Mascot "Berni"

Ball boys (10-15)

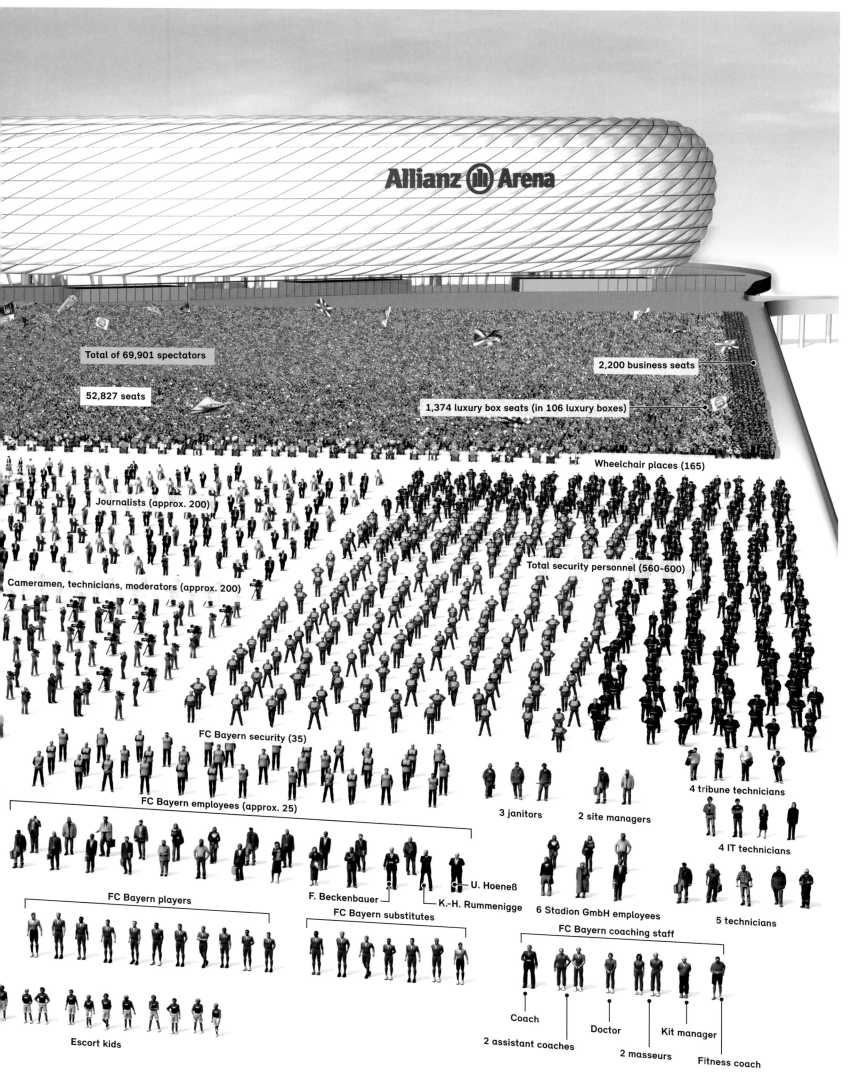

FEAR AND LOATHING

Beware, dear reader. The truth is, the good life does not come naturally.

Fear is perhaps our most basic emotion. It is the feeling that tries hardest to keep us alive, but this doesn't mean that it is a rational sensation — or that we are afraid of the right things. Our brains sometimes may tell us to fear the strangest of things, from squirrels (quilophobia) to flatulence (flatuphobia). Fear sells, but it isn't all bad: to have a rational fear of dying is just one way of demonstrating a love of life.

Nowadays humans are growing older and older, partly because some of the ancient scourges of mankind can be contained by recent progress in science and medicine, or because of understanding the relationship between hygiene and the spread of disease. We have all but eradicated smallpox, leprosy, tuberculosis, and polio — at least from the Western world.

The selection we have chosen here represents a wide variety of causes and victims. These causes and victims do not live in our collective consciousness in the ways that, say, the World Wars do. The Spanish Flu killed more people in 20 years than died on both sides in both World Wars, yet its name is hardly ever heard. But the impact of this and other contagions lives on in our subconscious, recalled every time a new outbreak occurs without a known cure. We have not removed our susceptibility to viruses like HIV and their mutations.

Behind the many graphics chosen from recent publications — a graph of fatal airliner crashes, a map showing the global path of smallpox contagions — are huge numbers that represent groups of real people who once lived, the ending of whom created a single data point to add to the others. But while we live, we feel, and the consumption of drugs, both legal and illegal, hack our brain's pleasure centers to create quick escape routes from unpleasant realities.

However much we reduce illness, violence and the proliferation of handguns require a different kind of cure. But thanks to great movie making, we can at least turn real fear into entertaining scares for a couple of hours of relief.

Not So Death Proof

Death by dog, sword, or five-point-palm-exploding-heart-technique.

Ever wondered what Quentin Tarantino's birthday cake might look like? Well, here's a start. This graphic, created especially for the influential U.S. director's 50th birthday, breaks down all the fictional deaths he is responsible for in eight octane-fueled movies, alongside which weapons were used. Unsurprisingly, small-arms fire is responsible for the most killings, so guns have been granted their own special balance sheet, so you can see exactly how many film fatalities were caused by head-shot, chest-shot, or gut-shot. Strange or unusual deaths have been granted their own special icons. Interesting that death tolls tend to have increased with the trajectory of the great man's career.

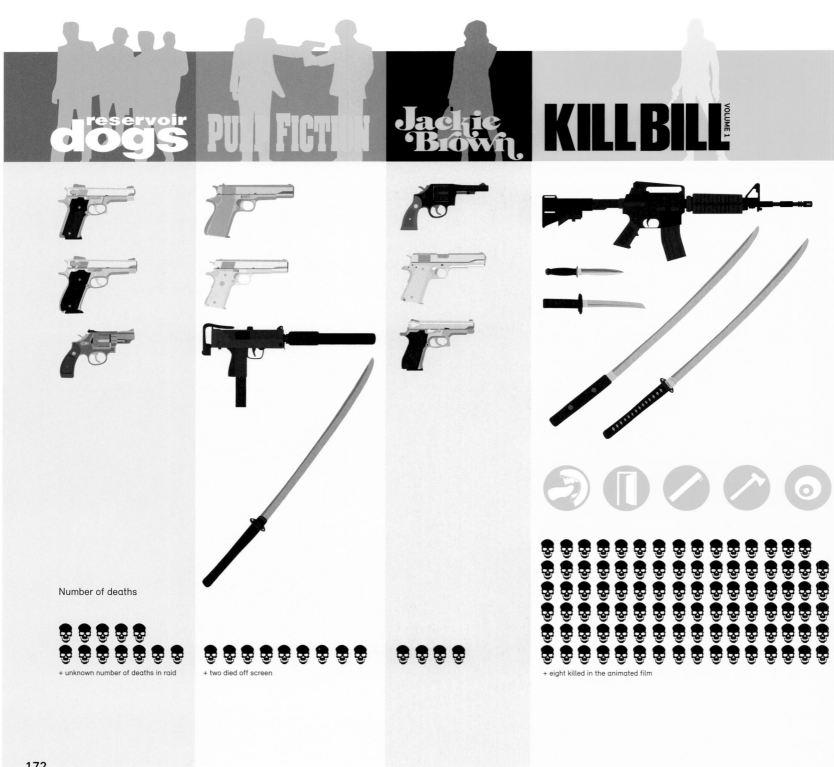

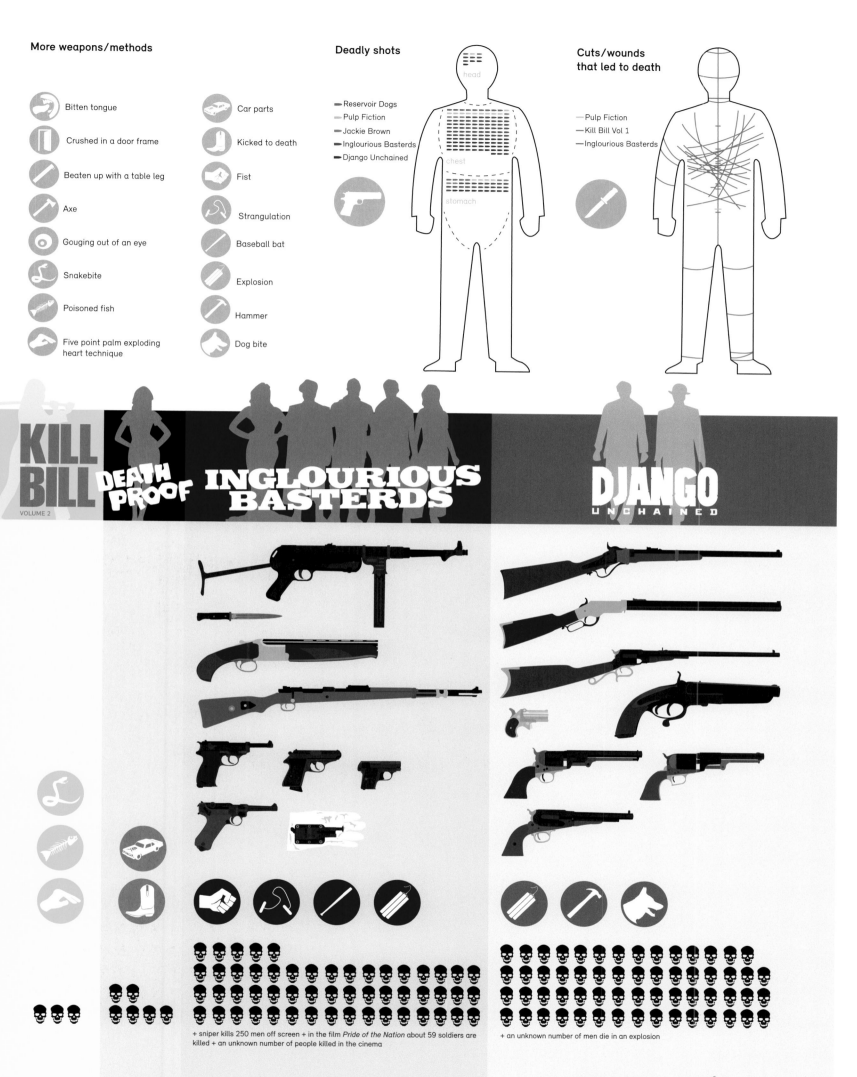

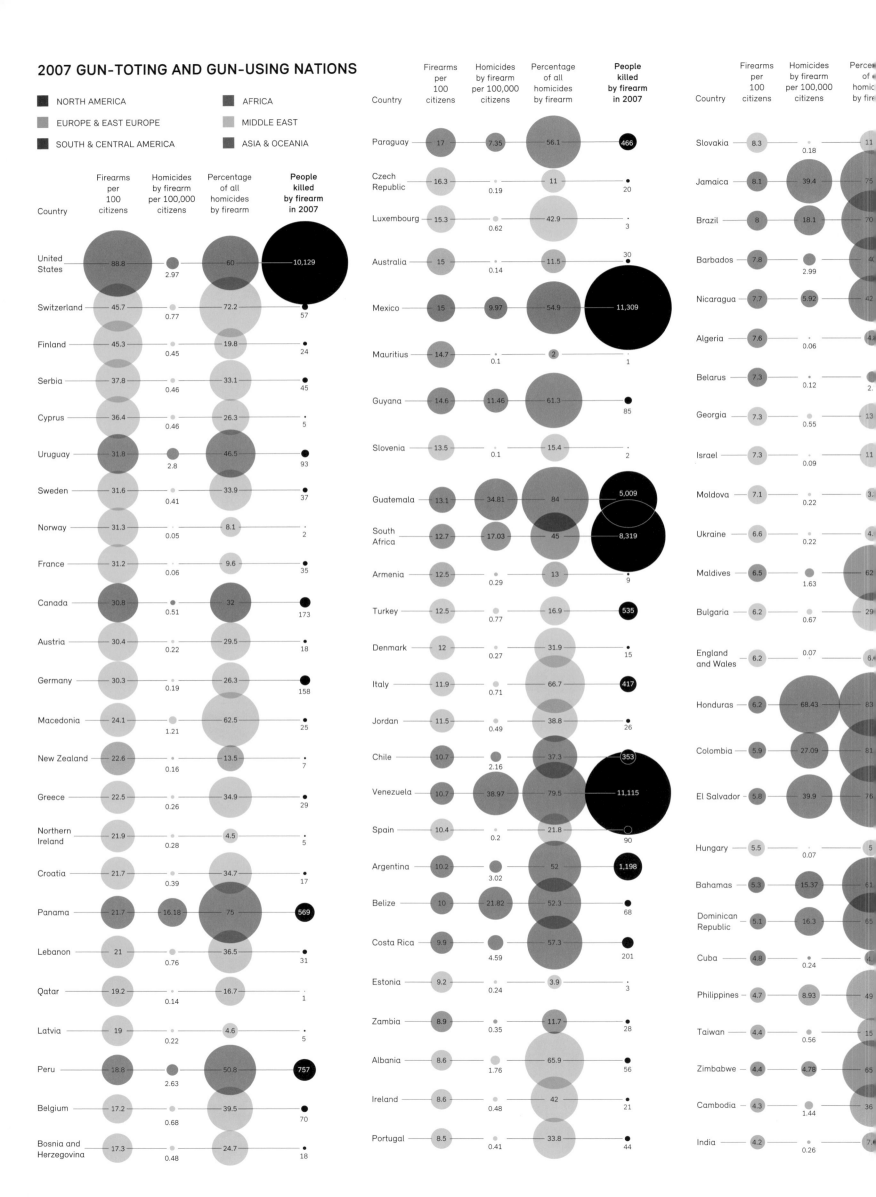

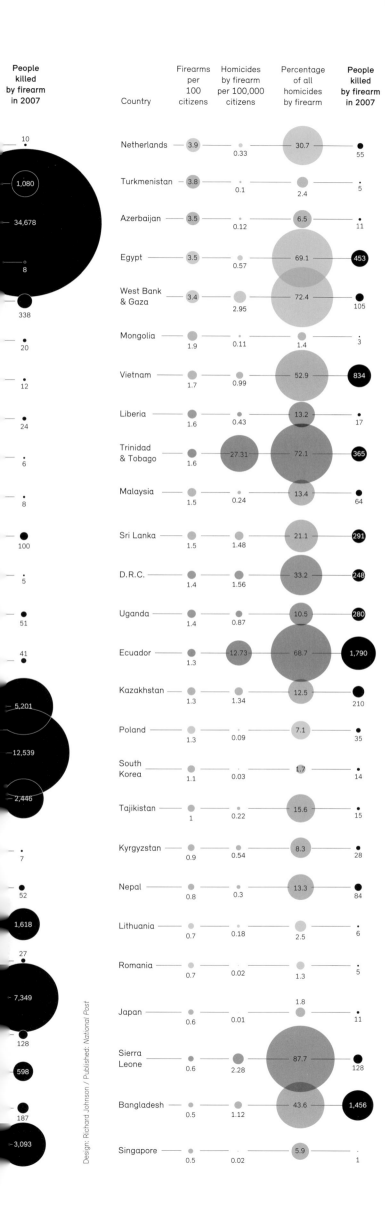

Where the Gun Rules

Firearms are the original weapon of mass destruction.

The endless debate over the United States' firearm laws consistently makes the news everywhere, but American gun culture is not the deadliest in the world. While the U.S. is the country with the highest gun ownership rate — an average of 88.8 per 100 people in 2007 — when it comes to firearms, several other countries have a worse murder rate. The most gun-related deaths happen in South and Central America, where Honduras, El Salvador, and Venezuela are particularly deadly. This list compares gun data from across the globe, starting with nations where guns are most common.

Guns

FEAR AND LOATHING

Irrational Psychological Mess

Phobias usually make little logical sense, but that doesn't mean they have no cause.

Phobias come in two basic kinds: specific and social. The specific are more commonly known: the fear of a specific animal (the spider and the snake are common), a specific sight (blood, injections), or a specific situation (enclosed spaces, heights). Social phobias, on the other hand, are more difficult to recognize, and yet hold a subtle power over many people. They are defined as a fear of how others will react to you — the fear of rejection, of being judged. If you consider how much we configure our own thoughts around other people's opinions of things, this is a powerful force indeed. But where do phobias come from? Obviously many phobias are acquired in traumatic childhood experiences — an early wasp attack can have untold consequences — which is known as "classical conditioning." Meanwhile, some research suggests that intelligent children are more likely to acquire phobias by observing them in others. Then there is a theory that some of our phobias are evolved — other apes tend to learn to fear snakes and spiders more quickly than other animals — something which may have been useful in our origins in the jungles of Africa.

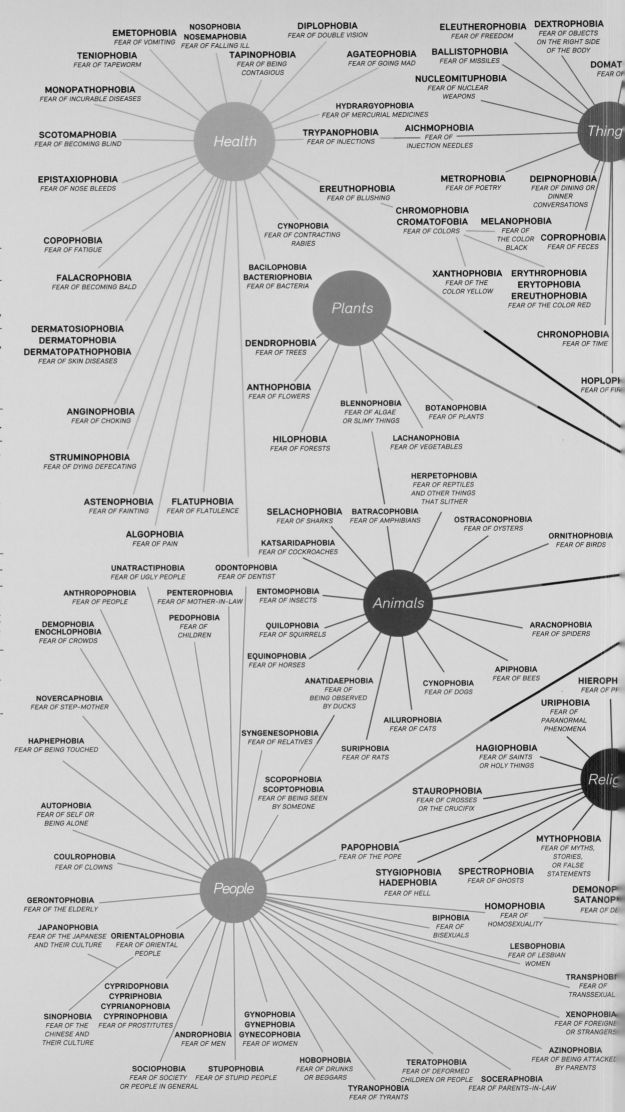

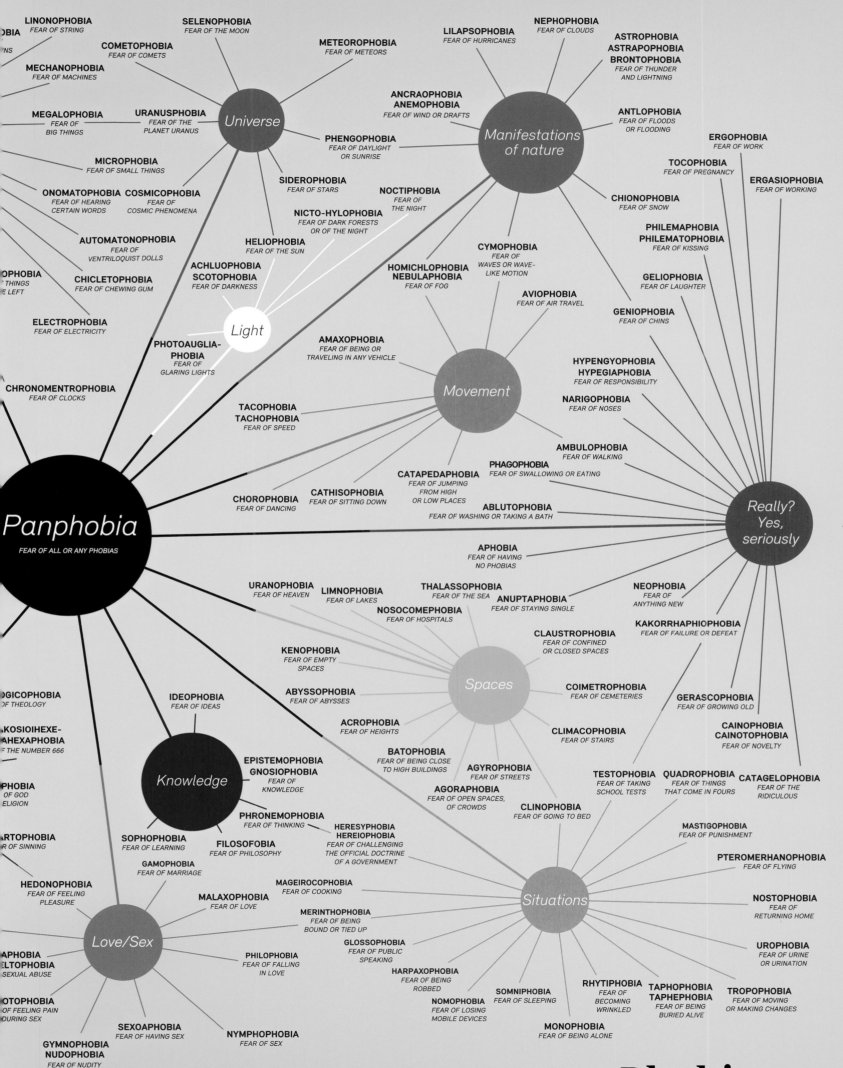

Phobias

FEAR AND LOATHING

Up in the Air

Planes have gotten much safer since the 1970s.

Statistically speaking, air travel is the safest means of transport there is — including escalators. In the past decade, the number of fatal plane crashes has dropped significantly, despite a massive increase in air travel. These days, even a crash

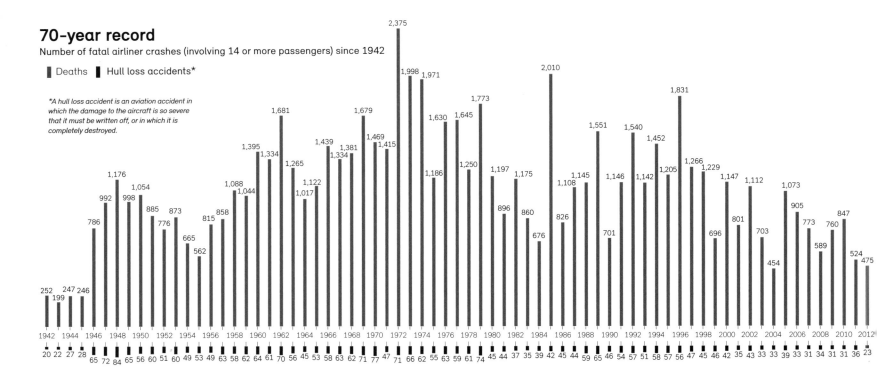

70-year record
Number of fatal airliner crashes (involving 14 or more passengers) since 1942

■ Deaths ■ Hull loss accidents*

*A hull loss accident is an aviation accident in which the damage to the aircraft is so severe that it must be written off, or in which it is completely destroyed.

What caused recent crashes ...
Different types of occurrences that led to fatalities within the worldwide commercial jet fleet from 2002 to 2011

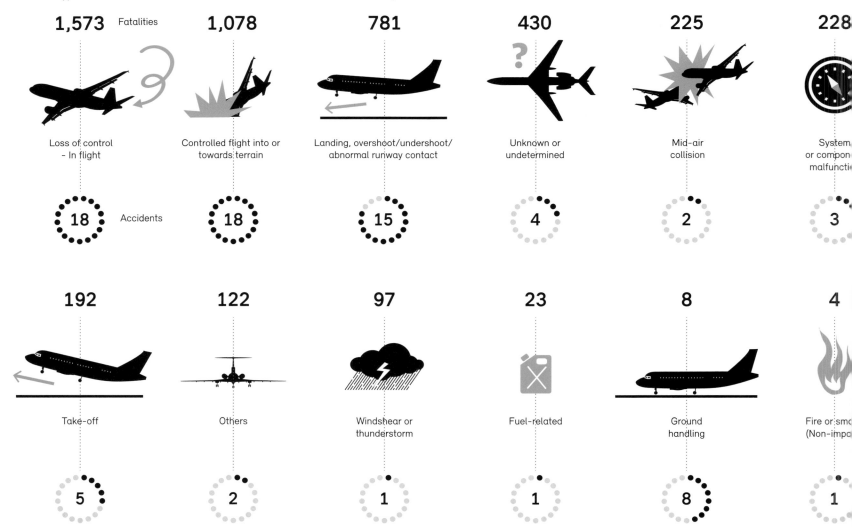

Fatalities	Category	Accidents
1,573	Loss of control - In flight	18
1,078	Controlled flight into or towards terrain	18
781	Landing, overshoot/undershoot/abnormal runway contact	15
430	Unknown or undetermined	4
225	Mid-air collision	2
228	System or component malfunction	3
192	Take-off	5
122	Others	2
97	Windshear or thunderstorm	1
23	Fuel-related	1
8	Ground handling	8
4	Fire or smoke (Non-impact)	1

does not mean everyone on board is necessarily doomed. For example, changes in the material used for cabin interiors provide crucial extra seconds to brace yourself in case of emergency. Moments that save lives.

and how many died?

e of circles indicates the number of fatalities

Since 1990   ○ Top 10 fatalities in history

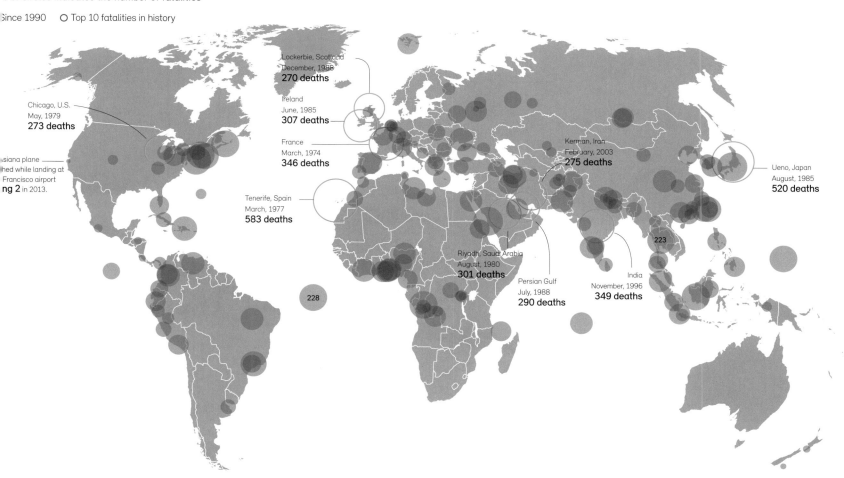

eaths by phase of flight

al accidents and onboard fatalities of the worldwide commercial jet fleet, 2002–2011 (%)

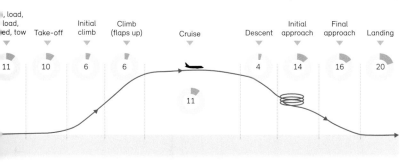

The hot seat

The survival rate is much lower for the seats six rows away from the emergency exit

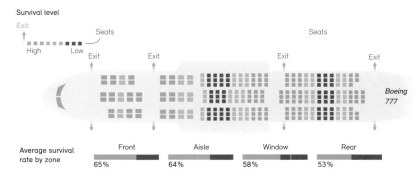

ack record

e 10 safest and least safe airlines out of 60 over the 30 years to 2012 based on the JACDEC Airline Safety Ranking 2012 and using the JACDEC Safety Index*

	Finnair	Air New Zealand	Cathay Pacific	Emirates	Etihad Airways	Eva Air	TAP Portugal	Hainan Airlines	Virgin Australia	British Airways	Scandi. Airlines	S. African Airways	Thai Airways	Turkish Airlines	Saudia	Korean Air	GOL Transp. Aér.	Air India	TAM Airlines	China Airlines
loss 1983	0	0	0	0	0	0	0	0	0	1	5	1	5	6	4	9	1	3	6	8
of ities	0	0	0	0	0	0	0	0	0	0	110	159	309	188	310	687	154	329	336	755

sed on Jet Airliner Crash Data Evaluation Center's annual safety calculations which include hull loss accidents and incidents in the past 30 years in relation to revenue passenger kilometers.

Plane Crashes

FEAR AND LOATHING

Trio Infernale

Containing the three deadliest diseases.

Our history books are full of stories of war and destruction, but some of the deadliest events were often treated as only footnotes: malaria, leprosy, and smallpox have killed millions over the millenia. For decades, the development of efficient and affordable vaccinations was an important challenge for medical scientists and research institutions. Thanks to them, leprosy and smallpox have largely been banished from the world. With strong backing from some of the world's richest and most influential people, the development of a vaccine against malaria is hopefully just around the corner.

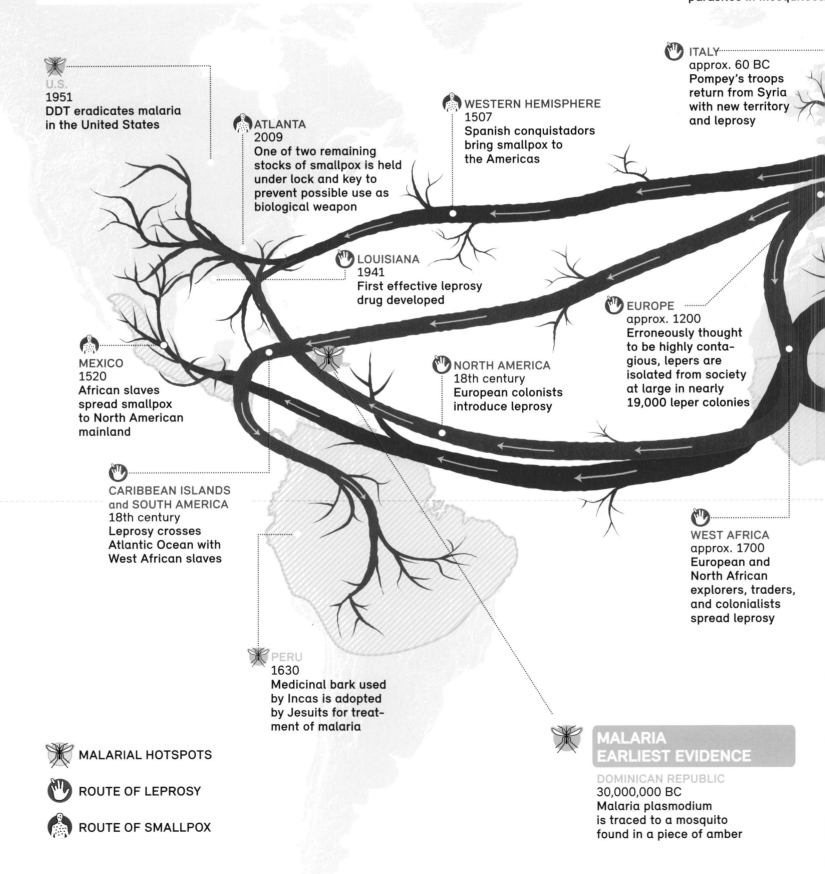

SWEDEN
1902
Ronald Ross is awarded the Nobel Prize in Physiology or Medicine for discovering malarial parasites in mosquitoes

ITALY
approx. 60 BC
Pompey's troops return from Syria with new territory and leprosy

U.S.
1951
DDT eradicates malaria in the United States

ATLANTA
2009
One of two remaining stocks of smallpox is held under lock and key to prevent possible use as biological weapon

WESTERN HEMISPHERE
1507
Spanish conquistadors bring smallpox to the Americas

LOUISIANA
1941
First effective leprosy drug developed

EUROPE
approx. 1200
Erroneously thought to be highly contagious, lepers are isolated from society at large in nearly 19,000 leper colonies

MEXICO
1520
African slaves spread smallpox to North American mainland

NORTH AMERICA
18th century
European colonists introduce leprosy

CARIBBEAN ISLANDS and SOUTH AMERICA
18th century
Leprosy crosses Atlantic Ocean with West African slaves

WEST AFRICA
approx. 1700
European and North African explorers, traders, and colonialists spread leprosy

PERU
1630
Medicinal bark used by Incas is adopted by Jesuits for treatment of malaria

- MALARIAL HOTSPOTS
- ROUTE OF LEPROSY
- ROUTE OF SMALLPOX

MALARIA EARLIEST EVIDENCE

DOMINICAN REPUBLIC
30,000,000 BC
Malaria plasmodium is traced to a mosquito found in a piece of amber

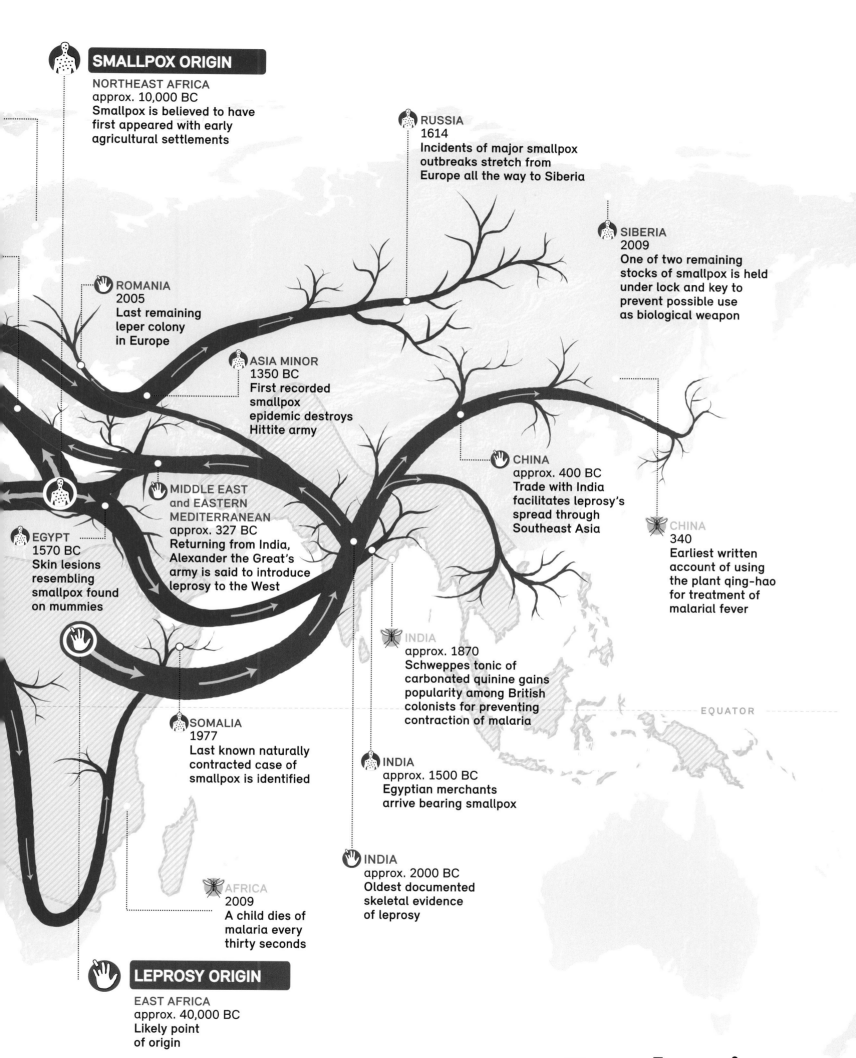

Pandemics

FEAR AND LOATHING

Pandora's Pandemic

There are so many things you could die of, but so few kill so many. Here are the biggest serial killers in the world.

Nothing breeds better and faster than a virus. If you are nothing but a string of random cells, you can evolve much quicker and better than anything else — and you can kill whatever you want, regardless of race, social status, or nationality. This is a graphic portraying history's biggest, and tiniest, killers. You'll notice their symptoms look surprisingly similar — have you tested for buboes recently?

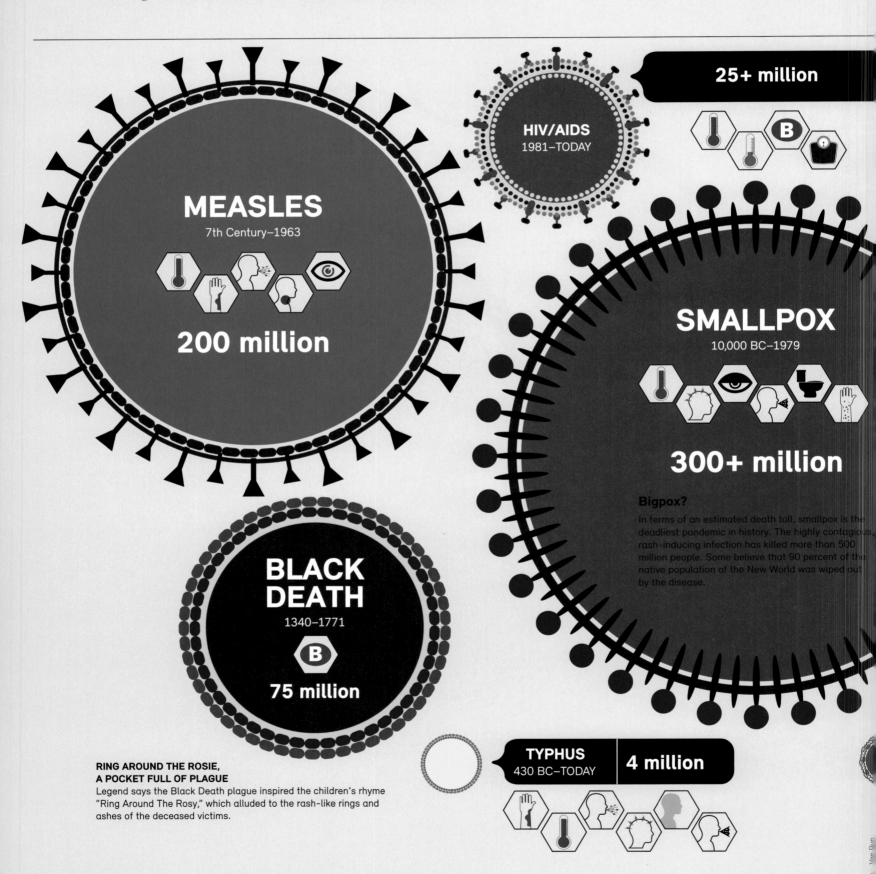

MEASLES
7th Century–1963
200 million

HIV/AIDS
1981–TODAY
25+ million

SMALLPOX
10,000 BC–1979
300+ million

Bigpox?
In terms of an estimated death toll, smallpox is the deadliest pandemic in history. The highly contagious, rash-inducing infection has killed more than 500 million people. Some believe that 90 percent of the native population of the New World was wiped out by the disease.

BLACK DEATH
1340–1771
75 million

RING AROUND THE ROSIE, A POCKET FULL OF PLAGUE
Legend says the Black Death plague inspired the children's rhyme "Ring Around The Rosy," which alluded to the rash-like rings and ashes of the deceased victims.

TYPHUS
430 BC–TODAY
4 million

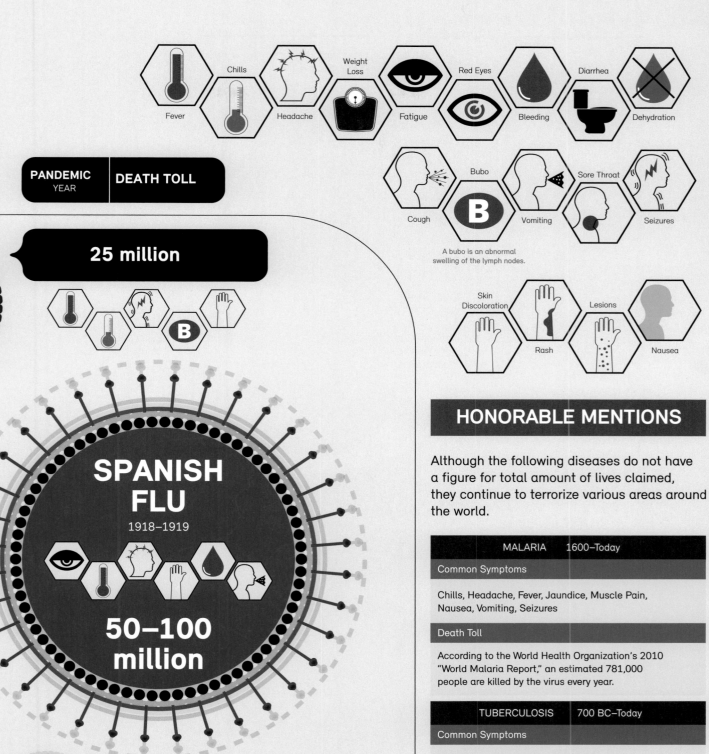

Pandemics

FEAR AND LOATHING

Nicotine: The Fake Neurotransmitter

In the chemical world inside your brain, nicotine works because it is one of the best imitators of all.

It doesn't matter how big they make the health warnings on the packet, no smoker is going to see them any better. Not a single smoker is going to wake up one day and say, spot the black and white square and say, "Thank God I saw this — I thought these things were full of vitamins." That's because nicotine is so addictive that it switches off your rationality. It's so addictive that 50 percent of people who have survived lung cancer surgery continue smoking. 50 percent! That's because of one thing only — nicotine, a molecule richly present in a certain family of plants, happens to exactly mimic the neurotransmitters in your brain that tell you you're happy.

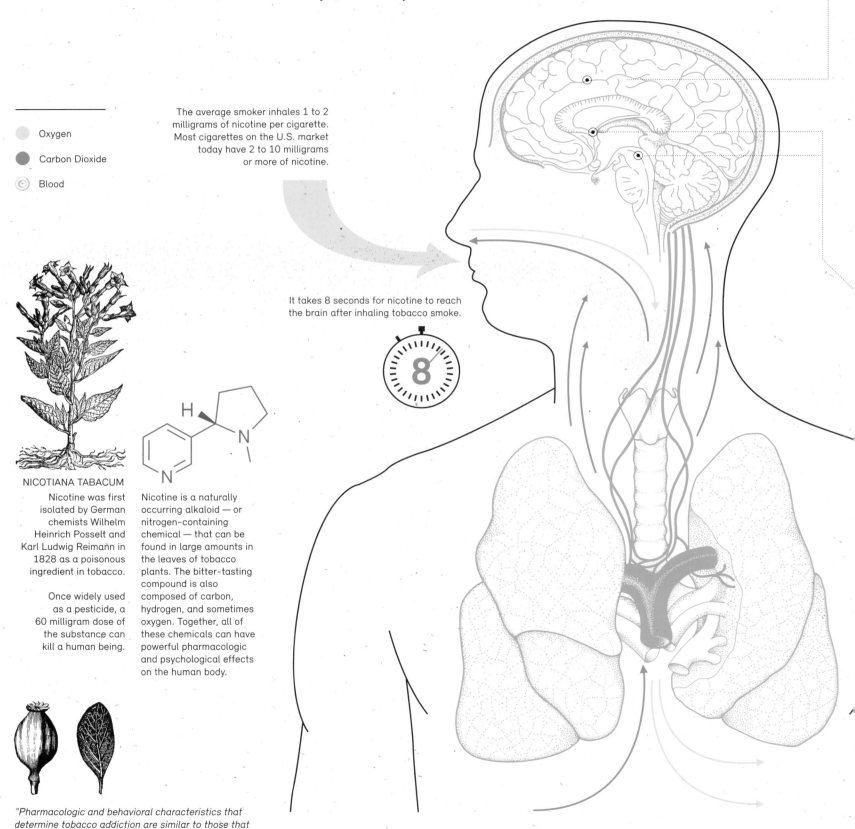

Oxygen

Carbon Dioxide

Blood

The average smoker inhales 1 to 2 milligrams of nicotine per cigarette. Most cigarettes on the U.S. market today have 2 to 10 milligrams or more of nicotine.

It takes 8 seconds for nicotine to reach the brain after inhaling tobacco smoke.

NICOTIANA TABACUM

Nicotine was first isolated by German chemists Wilhelm Heinrich Posselt and Karl Ludwig Reimann in 1828 as a poisonous ingredient in tobacco.

Once widely used as a pesticide, a 60 milligram dose of the substance can kill a human being.

Nicotine is a naturally occurring alkaloid — or nitrogen-containing chemical — that can be found in large amounts in the leaves of tobacco plants. The bitter-tasting compound is also composed of carbon, hydrogen, and sometimes oxygen. Together, all of these chemicals can have powerful pharmacologic and psychological effects on the human body.

"Pharmacologic and behavioral characteristics that determine tobacco addiction are similar to those that determine addiction to drugs such as heroin and cocaine."
American Heart Association

Inhalation of tobacco smoke.

THE PLEASURE PRINCIPLE

Research has shown that nicotine is the main silent agent in cigarettes. More than just a physical craving, it's also a psychological dependency.

1/
Our brain is made up of billions of nerve cells. They release chemical messengers called neurotransmitters that activate the receptor's nerve cell in order to communicate with other cells.

2/
The nicotine molecule is shaped like the neurotransmitter called acetylcholine, which is involved in everything from muscle movement and heart rate to memory.

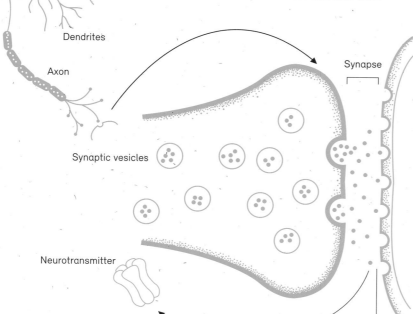

3/
Once the nicotine reaches the brain, it's able to attach to the acetylcholine receptors and mimic its actions. The body then responds to the nicotine as if they were natural neurotransmitters.

4/
Simultaneously, nicotine also raises dopamine and endorphin levels, which control feelings of pleasure and euphoria. At night, when you're asleep, these pathways shut down and cravings often occur immediately upon waking up.

5/
As smokers increase the amount of cigarettes they smoke, more nicotine is absorbed by their lungs and into their bloodstream. They have to maintain this level of nicotine in order to keep the nervous system stable and prevent withdrawal symptoms.

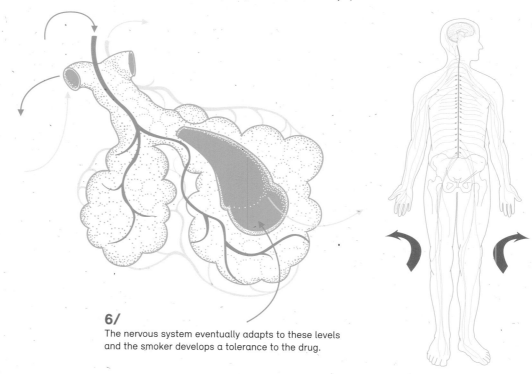

6/
The nervous system eventually adapts to these levels and the smoker develops a tolerance to the drug.

Smoking
FEAR AND LOATHING

America's Favorite Drugs

Some drugs are bigger than others.

America's "War on Drugs" is fought everywhere, and it's a different battle in different places. But whether it's the struggle to shut down Californian meth labs, or a desperate fight against cocaine fiends in Florida, it looks like guerrilla tactics are the order of the day. Sometimes it's hard to tell who is really in control — do the police represent the reigning order and the drug dealers the rebels, or is it the other way around? Either way, this is a map of America based on what local law enforcement officials see as the greatest dangers to their communities.

WEST

WEST/ISLANDS
ALASKA, AMERICAN SAMOA, CENTRAL CALIFORNIA, GUAM, HAWAII, IDAHO, NEVADA, NORTHERN CALIFORNIA, NORTHERN MARIANA ISLANDS, OREGON, WASHINGTON

NORTH/MIDWEST
COLORADO, IOWA, KANSAS, MISSOURI, MONTANA, NEBRASKA, NORTH DAKOTA, SOUTH DAKOTA, SOUTHERN ILLIONOIS, UTAH, WYOMING

MIDWEST
INDIANA, KENTUCKY, MICHIGAN, MINNESOTA, NORTHERN ILLINOIS, OHIO, WISCONSIN

SOUTHWEST
ARIZONA, NEW MEXICO, OKLAHOMA, SOUTHERN CALIFORNIA, TEXAS

- PHARMACEUTICALS
- MARIJUANA
- METHAMPHETAMINE
- HEROIN
- COCAINE

Design: Kiss Me I'm Polish – Agnieszka Gasparska, Joshua Covarrubias / published: GOOD.is

THEAST	FLORIDA/ISLANDS	MID-ATLANTIC	NORTHEAST	NEW ENGLAND	E A S T
...MA, ARKANSAS, ...IA, LOUISIANA, ...SIPPI, NORTH CAROLINA, ... CAROLINA, TENNESSEE	FLORIDA, PUERTO RICO, THE U.S. VIRGIN ISLANDS	DELAWARE, MARYLAND, PENNSYLVANIA, VIRGINIA, WASHINGTON, D.C., WEST VIRGINIA	NEW JERSEY, NEW YORK	CONNECTICUT, MAINE, MASSACHUSETTS, NEW HAMPSHIRE, RHODE ISLAND, VERMONT	

Drugs

FEAR AND LOATHING

Best Laid Plans

Why can't we keep our New Year's resolutions?

The euphoria kicks in when that ball drops, or that clock chimes midnight. You've got a whole year ahead of you, clean and pure like virgin snow. Imagine all the wonderful things you can fill it with — *finally* you have *time*. But whatever it is you want most — quit smoking or drinking, lose weight, get a better job, write a novel — according to the cold, hard figures, the unfortunate truth is that you have only a slightly better than one in ten chance of keeping your resolutions in the next twelve months. In fact, statistically, one in four people have abandoned their dreams of a better organized, more productive life by January 7. So here are some tips from the experts: Stage 1 — Imagine your "big picture" life — where do you want to be in ten years? Stage 2 — Break it down into short-term goals. And finally, Stage 3 — get started. IMPORTANT NOTICE: you have to make SMART goals. That is, Specific, Measurable, Attainable, Relevant, and Time-bound. The only thing stopping you is you!

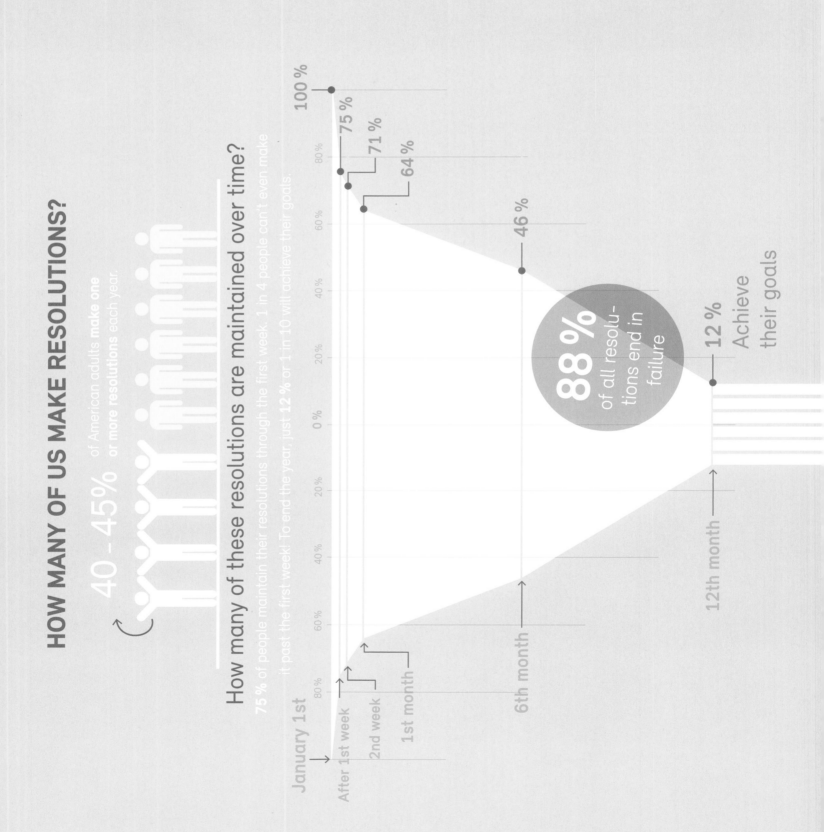

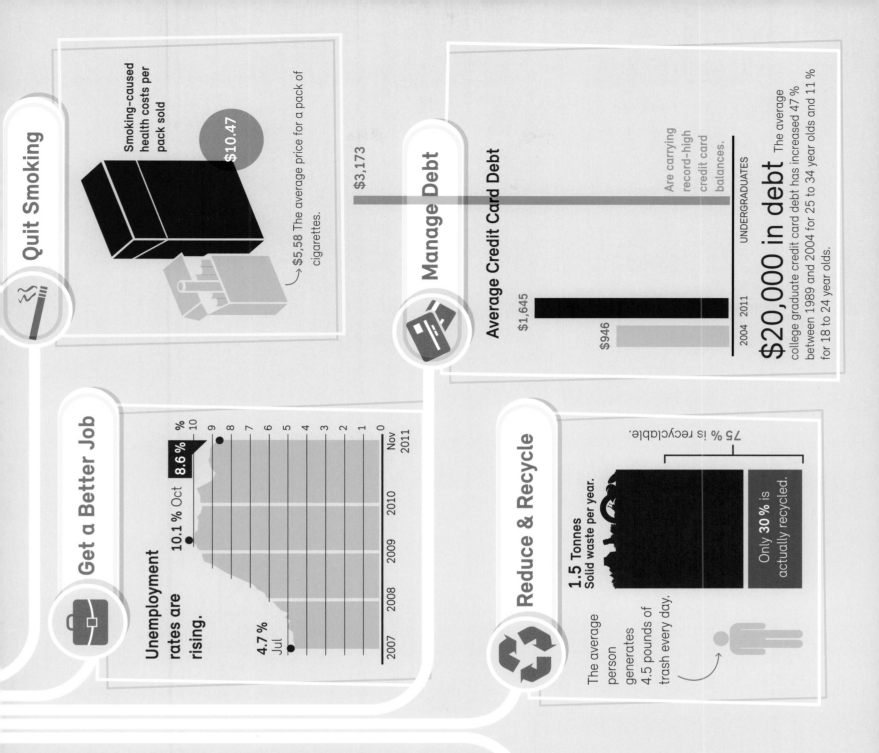

New Year's Resolutions

MONEY MAKES THE WORLD GO ROUND

Economic events in recent years have led to some hard, almost existential questions: What IS money? Who has it? Who needs it? And where the heck did it all go?

No matter what its politics, every country has its own currency, and its money is unequally distributed. The very existence of money demands that this is so — like the famous fact about some sharks, money is only alive as long as it keeps moving. The money is nothing without the marketplace, a strange mechanism with its own internal logic, filled with self-proclaimed experts yet so complex that nobody can be certain about its future.

The recent financial crash showed how connected the world's financial systems have become. This is a global marketplace now, and its rules keep on shifting. Are these booms and busts an inevitable part of having a functional economy? If we all want to win, how can we minimize losses? Is it possible for a person or a country to win and lose at the same time? How do we adjust for desirable outcomes that the market won't supply?

One thing is clear: almost everything we produce is now connected to the needs of another country, from meat to cars, illegal drugs to uranium. Yet this leads to more difficult questions, as the balance remains unequal between what the market dictates and who has the greatest need, a gap currently filled by international aid. But for how long will it make sense to have the majority of the world's population be so dependent on the richest 1 percent, when the global networks ensure that the poorer countries are also hugely important links in the chain? By being so interdependent, are the existing relationships sustainable? Are we due for a shift in priorities, and if so, how will that be carried out?

As politicians and non-governmental actors constantly negotiate such ever-changing dynamics, thousands of boats, planes, and trucks continue to travel across borders, moving products and minerals, money and aid. Brands stretch and mutate as they journey across borders, while cables beneath the oceans shoot intangibles from screen to screen at thousands of miles per second. Sometimes these connections work smoothly, to balance demand and supply on a vast and unimaginable scale. Other times there is an unforeseen glitch, market assumptions are proven unstable, and the world enters a crisis of readjustment.

Throughout all of this, our thirst for energy grows and grows, increasing everyone's need for oil, for sunshine, for wind, and for water. These resources are real and tangible, used to generate power to sustain the technological growth to which we seem addicted. Whoever controls the supply of power becomes the powerful.

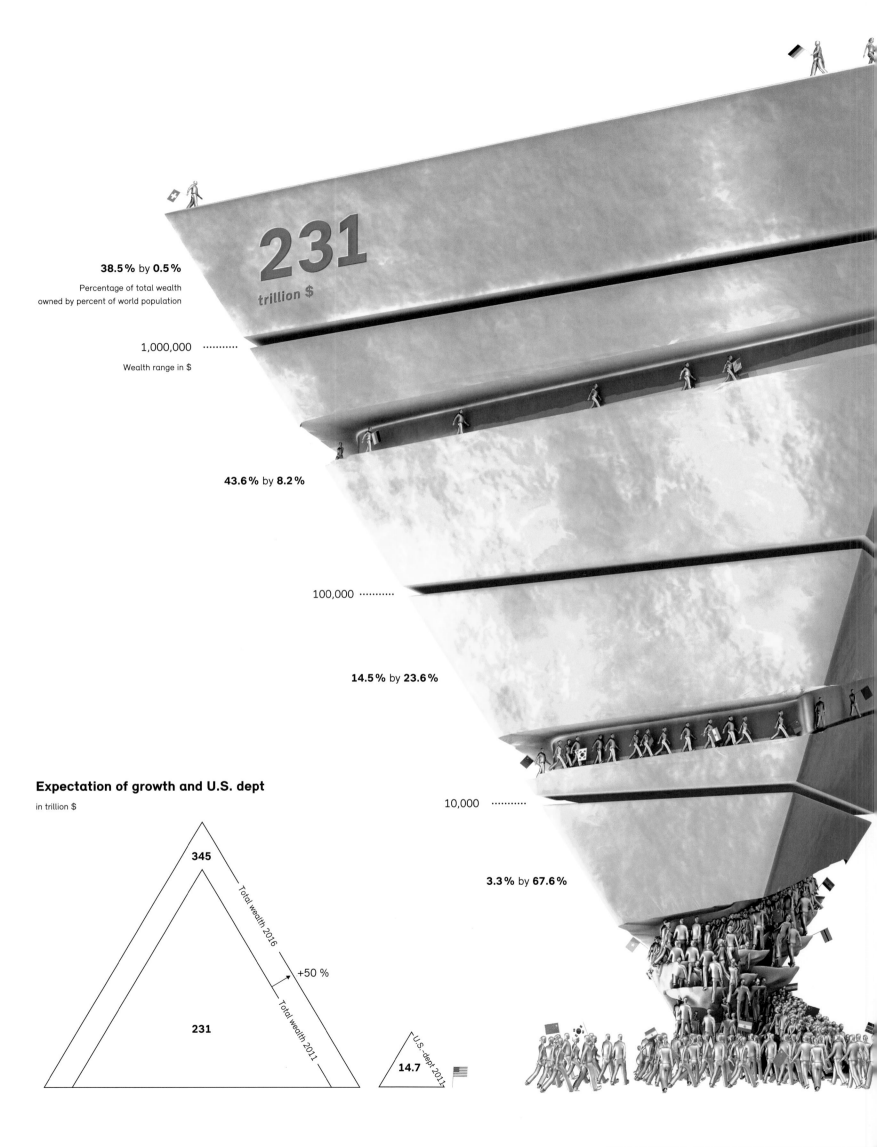

Wealth shares in 2011
per regions

Inhabitants in billion		Wealth shares in percent
1.526 Asia-Pacific		22
1.339 China		9
1.211 India		2
1.129 Africa		1
0.754 Europe		34
0.549 North America		28
0.469 Latin America		4

Supporting the Rich

Why being middle class is the new being poor.

Despite the Lehman Brothers catastrophe, bankers' bonuses, and the economic crisis ravaging the southern half of the European Union, the idea of capitalism is still as powerful as ever. It's easy to see why. It carries plenty of moral weight: most people believe people should work to support themselves and their families. And there's sound practical evidence too. Look at the world: the richest countries are capitalist, and immigrants, legal and illegal, are desperate to get in. There's no doubt that, measured by the sheer amount of money there is, capitalism has been extraordinarily successful — if our ancestors saw us today they'd think they'd fathered a line of monarchs. But this relentless faith in accumulation for its own sake has created a shocking imbalance in the world. Something can't be right, many say, when more than one in seven people in the world are below the poverty line — that is, living on less that $1.25 a day and surviving without basic human needs. Even in the United States, one of the richest countries in the world, a 2012 report found that 50 million Americans, more than 16 percent of the population, were living in conditions familiar to populations in sub-Saharan Africa. More interestingly, one survey carried out recently in the U.S. by a Harvard business professor found that the gap between people's ideals of wealth distribution, their perceptions, and the reality, were massive. In short: asked to envision a fair, healthy capitalist world, Republicans and Democrats alike thought that money should be distributed a lot more equally than it is now.

Distribution of wealth in 2011
per inhabitant and country in $

■ > 100,000   ■ 100,000–25,000   ■ 25,000–5,000   ■ < 5,000   ■ no data

Pyramid of Wealth

MONEY MAKES THE WORLD GO ROUND

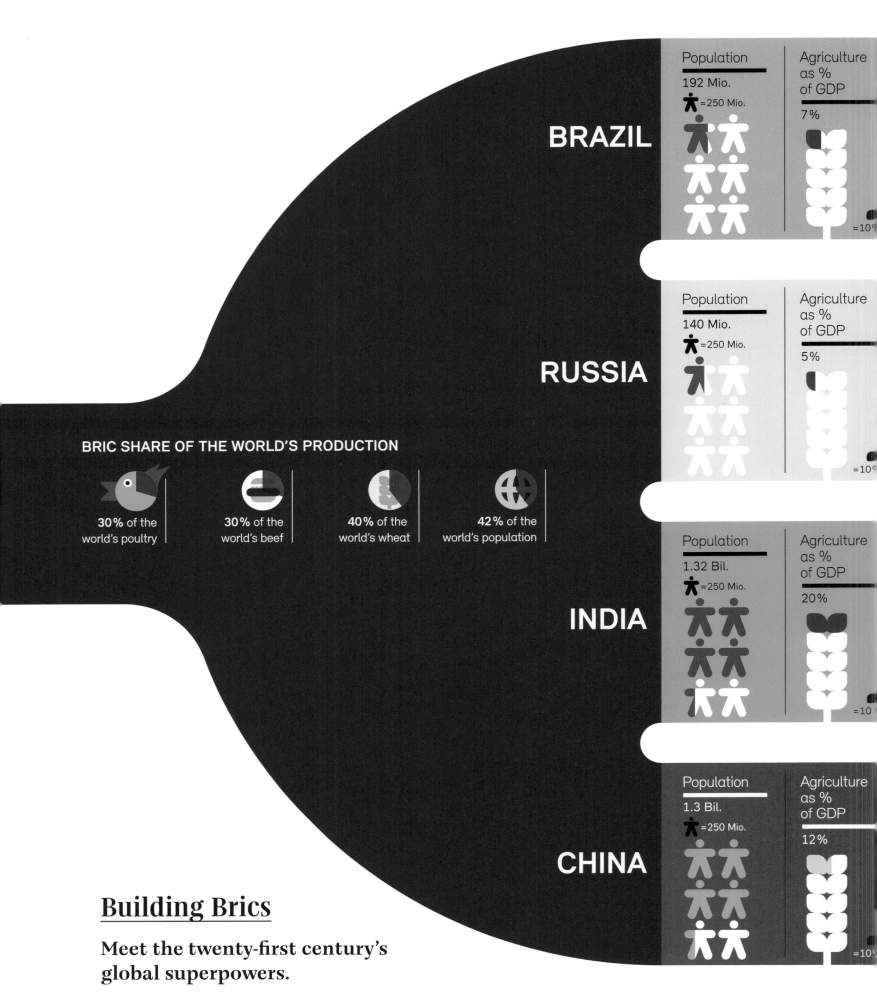

Building Brics

Meet the twenty-first century's global superpowers.

The four countries pictured here are home to 42 percent of the world's population, with vast reservoirs of natural resources and land, not to mention the power to change the world. These are the so-called BRIC countries, and their diplomats are increasingly throwing their weight around on the international stage, claiming the right to intervene in international crises and make peace between warring peoples, just as the U.S. has in the past century. But, as this graphic shows, each of these countries also faces considerable problems at home — the biggest one being the struggle to keep their own people from starving.

of land
dedicated to
agriculture

Average farmland values
£660 per acre
 = 100 acres

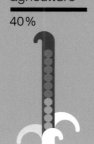

Expansion in agriculture
40%

● **KILLER FACT**

Between 2000 and 2006, Brazil lost 150,000 km² of rainforest, an area larger than Greece.

○ **BIGGEST HURDLE**

Increasing agricultural output while protecting the Amazon rainforest. Cleared land is worth **5 to 10** times more than forested land, but deforestation could be responsible for up to **10%** of global CO_2 emissions.

○ **BRIGHT IDEA**

One of Brazil's largest beef exporters, Bertin, has signed a pact with Greenpeace to refuse purchases of cattle reared in recently deforested parts of the Amazon jungle. As well as using satellite imagery to map ranches and detect where illegal logging is taking place, the company actively trains suppliers to improve their management of the land.

---

of land
dedicated to
agriculture

Average farmland values
£500 per acre
 = 100 acres

Expansion in agriculture
26%

● **KILLER FACT**

Between 1999 and 2009, Russia turned from a grain importer to the **third largest exporter after the USA and the EU.**

○ **BIGGEST HURDLE**

Though Russia is the **largest country in the world**, much of its land area is affected by permafrost and is unusable as farmland. Increasing instances of drought and escalating debts are also hampering farmers.

○ **BRIGHT IDEA**

Russian president Dmitry Medvedev has signed a new food security doctrine, which aims to boost domestic production of basic foodstuffs to 80% by 2020. Domestic grain supplies have already achieved their 95% target, while production of meat should increase to 85% — meat from the U.S. currently accounts for the highest value of imports at around U.S. $630 Million.

---

of land
dedicated to
agriculture

Average farmland values
£660 per acre
 = 100 acres

Expansion in agriculture
21%

● **KILLER FACT**

India's population includes **43%** of **children who are underweight due to malnutrition.**

○ **BIGGEST HURDLE**

India's population is growing faster than its ability to produce rice and wheat. Improving productivity is a key issue, due to poorly maintained irrigation systems and lack of access to markets and modern equipment.

○ **BRIGHT IDEA**

More farmers in remote areas now have access to modern equipment, thanks to a business venture that brings agricultural supermarkets into the countryside. The Hariyali Kisaan Bazaar sell quality fertilizers, seeds, and tools, as well as financial products such as crop insurance. The chain aims to add another 300 stores around rural India to its existing 300 by 2012.

---

of land
dedicated to
agriculture

Average farmland values
£660 per acre
 = 100 acres

Expansion in agriculture
26%

● **KILLER FACT**

China has **22%** of the world's population, but only **7%** of its arable land.

○ **BIGGEST HURDLE**

The country's limited space for farming is hampering expansion. Land in the west and north is generally colder and drier than traditional farmlands to the east, leading to a greater reliance on imports and on farm concessions in other parts of Asia and Africa.

○ **BRIGHT IDEA**

To limit the damage caused to crops by flooding, scientists at the International Rice Research Institute are attempting to develop a waterproof variety that can withstand being submerged for two weeks. The government is also attempting to change people's habits to eating more potatoes, which need less water to grow than rice or wheat, and yield more calories per acre.

B.R.I.C. States

MONEY MAKES THE WORLD GO ROUND

More than Just Money

Providing aid to developing countries is a complex challenge.

Insular aid workers driving around in white air-conditioned SUVs buying the most expensive food the country has to offer. Aid agencies building wells that are then abandoned because the funds run out to maintain them. If you ask the people living in some of Africa's poorest countries about the international Non-Governmental Organizations (NGOs) operating among them, they often give a startlingly negative answer. In many cases, only the smallest charities trying to solve the most local problems win the trust of inhabitants and succeed in making a difference. At the same time, only the major NGOs can help coordinate the global strategies needed to achieve the United Nations' development targets. And there has been some success in recent years. There are still a billion people in the world who live in extreme poverty — defined as earning less than one US$1 a day — but that number is steadily falling. At the same time, there is a growing realization that pouring money into an economy to help it grow does not necessarily help. If the UN's goals are to be met, then tackling global inequality is vital. The graphics on this page summarize the answers that major aid organizations gave in a survey on the future of development aid.

Which factors are most likely to increase humanitarian need in the coming years?

Agencies were asked to choose three from a list of nine factors, and rank them in terms of importance, 1 being the most important. Points were then awarded to each answer: 1 = 3 points, 2 = 2 points, 3 = 1 point

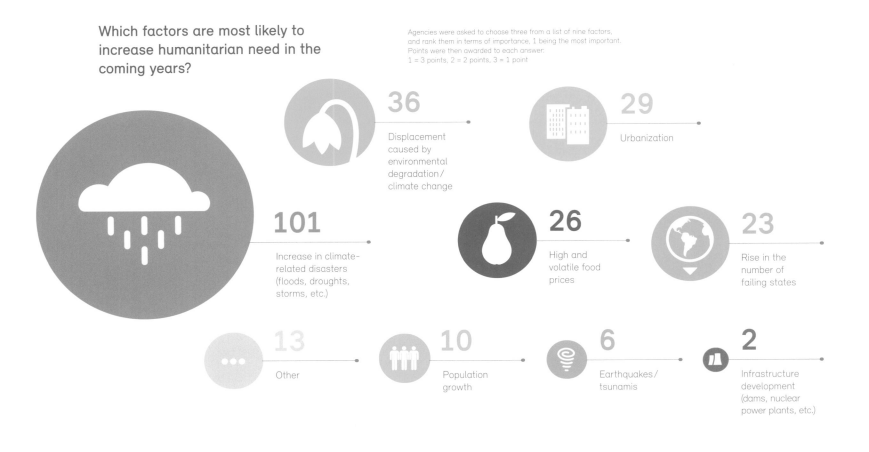

- **101** Increase in climate-related disasters (floods, droughts, storms, etc.)
- **36** Displacement caused by environmental degradation/climate change
- **29** Urbanization
- **26** High and volatile food prices
- **23** Rise in the number of failing states
- **13** Other
- **10** Population growth
- **6** Earthquakes/tsunamis
- **2** Infrastructure development (dams, nuclear power plants, etc.)

What are the biggest challenges to the delivery of humanitarian aid?

Aid agencies were asked to select three from a list of eight challenges.

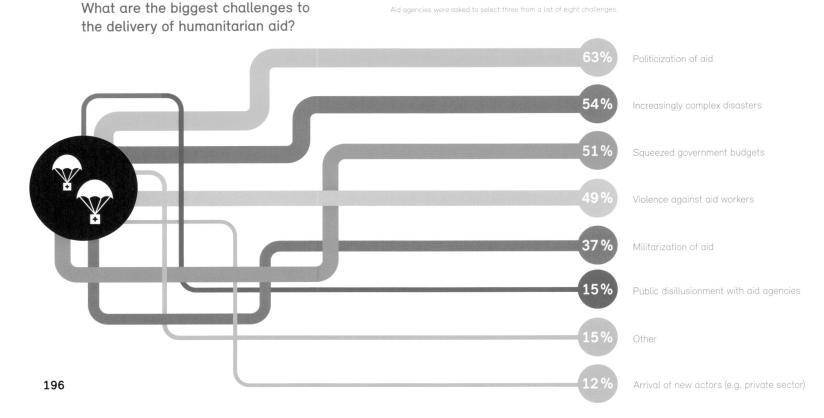

- 63% Politicization of aid
- 54% Increasingly complex disasters
- 51% Squeezed government budgets
- 49% Violence against aid workers
- 37% Militarization of aid
- 15% Public disillusionment with aid agencies
- 15% Other
- 12% Arrival of new actors (e.g. private sector)

What is the single most important obstacle to adapting?

Aid agencies were asked to select one obstacle from a list of six.

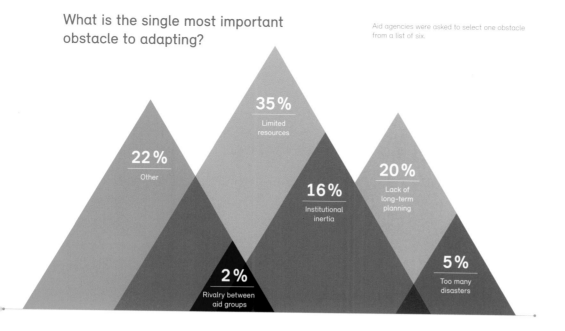

- 22% Other
- 35% Limited resources
- 2% Rivalry between aid groups
- 16% Institutional inertia
- 20% Lack of long-term planning
- 5% Too many disasters

How will the sources of humanitarian funding most likely look in five years' time?

Aid agencies were asked to select one response from four options. 54 percent (22 agencies) indicated they believe that government funding for humanitarian aid will fall.

46% Governments will still provide the bulk.
19 agencies

29% There will be a growing gap between funding and needs, as government and private donations fall.
12 agencies

25% Governments will have cut aid budgets heavily, with donations from individuals and companies filling the gap.
10 agencies

0% A major proportion will be funded by international taxes (e.g. on financial transactions).
0 agencies

How will the distribution of humanitarian funding most likely look in five years' time?

Aid agencies were asked to select one response from three options.

  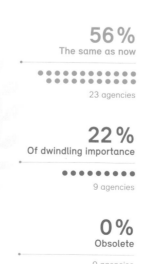

5% Most aid will go through a central fund for emergencies managed by the UN.
2 agencies

17% Donors will bypass the UN, giving money directly to national governments and charities.
7 agencies

78% The system will continue to use a range of channels, as now.
32 agencies

Do you agree with the following statement: "The international humanitarian aid system delivers value for money"?

Aid agencies were asked to select one response from five options.

- 7% Strongly agree — 3 agencies
- 59% Agree — 24 agencies
- 25% Disagree — 10 agencies
- 2% Strongly disagree — 1 agency
- 7% Don't know — 3 agencies

How important do you think the UN aid system will be to humanitarian response over the next decade?

Aid agencies were asked to select one response from five options.

- % Indispensible — agencies
- 7% Of growing importance — agencies
- 56% The same as now — 23 agencies
- 22% Of dwindling importance — 9 agencies
- 0% Obsolete — 0 agencies

Development Aid
MONEY MAKES THE WORLD GO ROUND

Forever Blowing Bubbles

How the economy punishes us for our foolish fads.

What if you woke up one morning and all the cashpoints had stopped dispensing, the banks had all defaulted on their debts, and there was no more money. This nightmare scenario was narrowly averted with one late-Sunday-night government decision in Great Britain in 2008, when the collapse of the financial services giant Lehman Brothers began to spread to "system-relevant" banks in Europe. Since then, complex legislation is being debated and passed in parliaments across the world in an attempt to impose restrictions on banking practices and market trading to ensure that financial institutions do not run up insane, unguaranteed debts by trading theoretical money that, effectively, only ever exists in the future. But all the legislation in the world cannot regulate human greed — as history shows, the smell of money triggers strange chemical reactions in our brains and we pay the price for gambling on hope against reality.

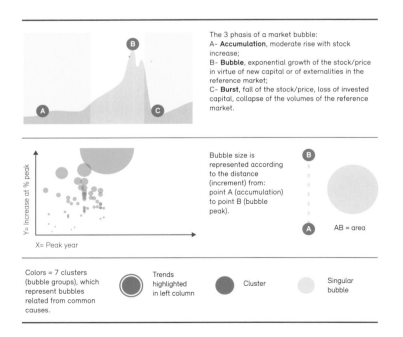

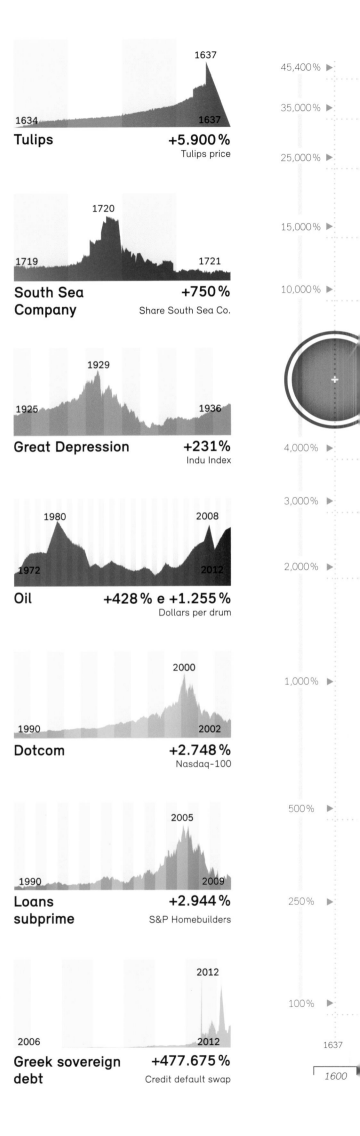

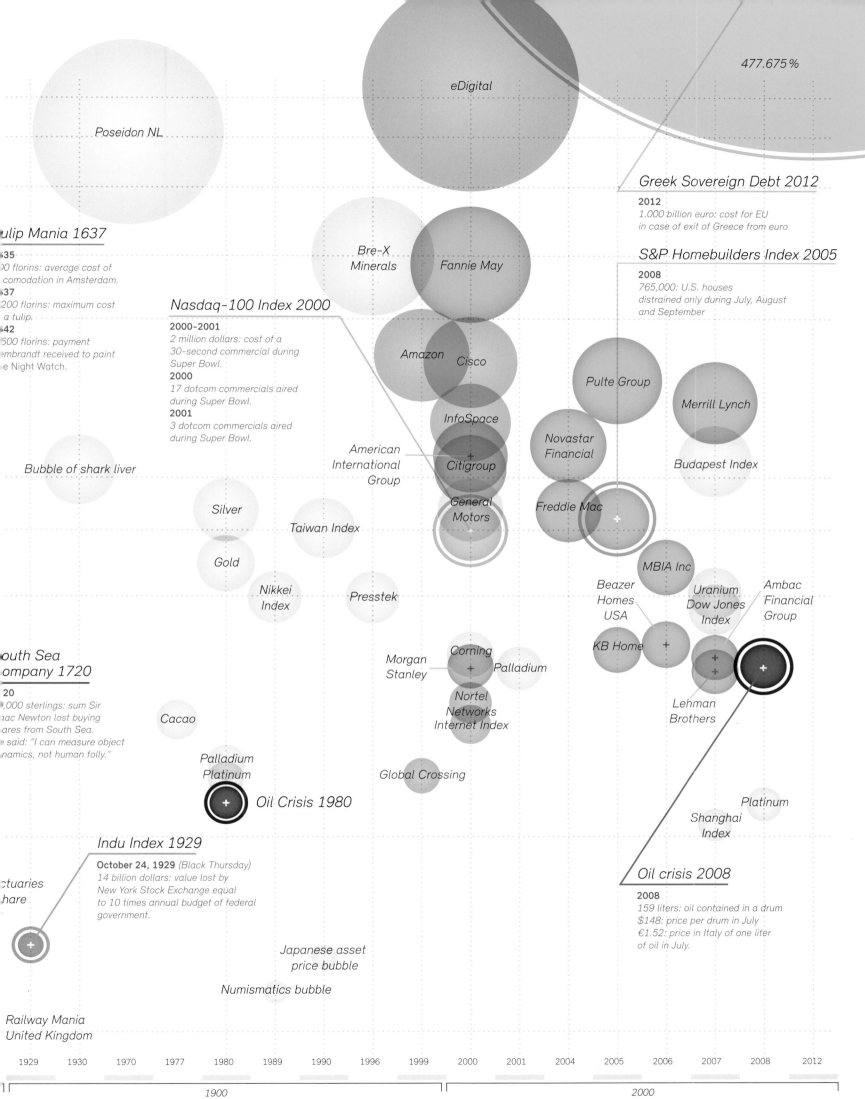

Spreading the Load

Ships, trains, planes, and trucks circle the world every day just to bring you your apples — and as for your apples, well, they came from everywhere.

Pick up the nearest object to you — chances are that at least one component of it traveled halfway around the world, from some strange exotic place, to be in your house. If the thing you picked up is high-tech, chances are that many of its parts have been sourced from several spots across the globe. Thanks to globalization and that vast army of dockworkers, shippers, and logistics companies, everything from mines, to factories, to stores, to shopping malls, to street vendors, is connected in a vast network of consumption. And some of it is not exactly legal.

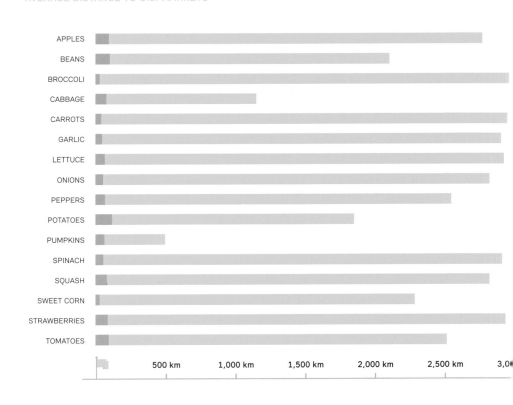

HOW FAR FOOD TRAVELS
AVERAGE DISTANCE TO U.S. MARKETS

CARGO CHANNELS
U.S. TRADE BY WEIGHT

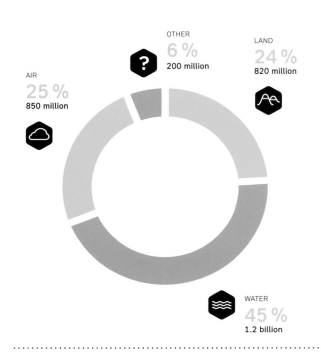

- AIR 25% — 850 million
- OTHER 6% — 200 million
- LAND 24% — 820 million
- WATER 45% — 1.2 billion

THE WORLD'S TOP PORTS
SHIPPING CONTAINERS HANDLED

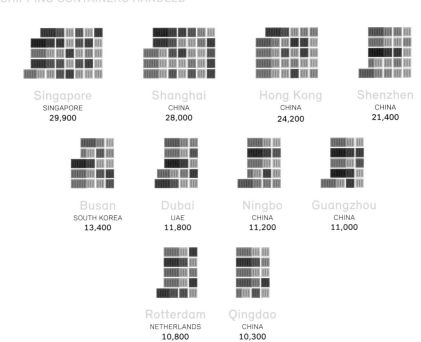

Port	Country	Containers
Singapore	SINGAPORE	29,900
Shanghai	CHINA	28,000
Hong Kong	CHINA	24,200
Shenzhen	CHINA	21,400
Busan	SOUTH KOREA	13,400
Dubai	UAE	11,800
Ningbo	CHINA	11,200
Guangzhou	CHINA	11,000
Rotterdam	NETHERLANDS	10,800
Qingdao	CHINA	10,300

FREIGHT INFRASTRUCTURE

	ROADWAYS	RAILWAYS	WATERWAYS	PIPELINES	AIRPORTS
USA	6,500,000 km	230,000 km	41,000 km	790,000 km	5,200 airports
India	3,300,000 km	63,000 km	15,000 km	23,000 km	251 airports
China	1,900,000 km	78,000 km	110,000 km	58,000 km	413 airports

MAKING A LAPTOP

A typical laptop manufacturer sources its raw materials from over 35 locations around the world, fabricates and assembles its parts in over 15, and undergoes final assembly in China before shipping to your retailer or doorstep.

RAW MATERIALS
FABRICATION — 1 WAFERS — 2 SILICON PROCESSING — 3 PLASTICS — 4 LCD SCREEN — 5 BATTERY
CHIP ASSEMBLY
LAPTOP ASSEMBLY

HOW GOODS ARE SOLD

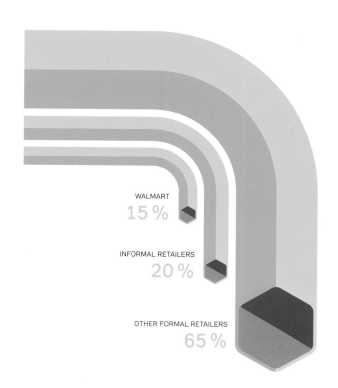

WALMART 15 %
INFORMAL RETAILERS 20 %
OTHER FORMAL RETAILERS 65 %

HOW TO SMUGGLE GOODS

Hide Shipments
Trade from China to underground markets often goes underreported. The number of shipping containers doesn't have to match the invoice.

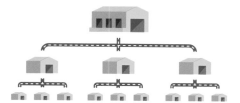

Use Middlemen
If you're an established producer like Procter & Gamble, move your goods through a string of wholesalers with access to informal channels.

Target the Ports
Go through ports with less supervision and reduced customs fees. Informal smugglers, truck drivers, and loaders will take it from here.

Sweeten the Deal
If all else fails, bribe an official or worker to look the other way at various checkpoints, such as a port, customs area, or weigh station.

Globalization
MONEY MAKES THE WORLD GO ROUND

Where the Drug Mules Trek

Two plants, a few chemical concoctions — and a worldwide industry.

Despite drastic punishments for drug dealing, up to and including dea[th] in many countries, the worldwide illegal drug trade continues to flo[ur]ish — partly because, no doubt, those who are making the real mon[ey] rarely get caught. While governments invest vast sums of taxpayer[...]

PRODUCTION

Farmers and chemists form the basis of the drug business as producers. They operate in the underground. The dealers work hand in hand with pilots, accountants, legal advisors, and financial experts.

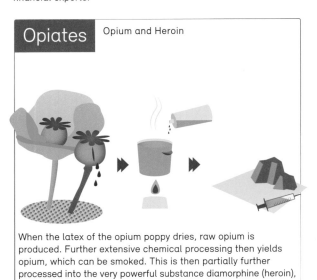

Opiates — Opium and Heroin

When the latex of the opium poppy dries, raw opium is produced. Further extensive chemical processing then yields opium, which can be smoked. This is then partially further processed into the very powerful substance diamorphine (heroin), which can be smoked, sniffed, or injected.

Cocaine — Cocaine and Crack

The leaves of the lightly narcotic coca plant are processed into a white powder in illegal laboratories. This cocaine hydrochloride can be sniffed in either pure or diluted form. When mixed with sodium bicarbonate and water and then heated, crack is produced, the vapors of which can be inhaled.

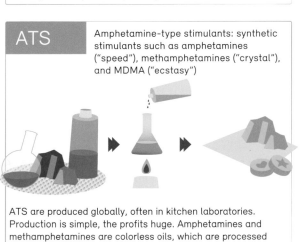

ATS

Amphetamine-type stimulants: synthetic stimulants such as amphetamines ("speed"), methamphetamines ("crystal"), and MDMA ("ecstasy")

ATS are produced globally, often in kitchen laboratories. Production is simple, the profits huge. Amphetamines and methamphetamines are colorless oils, which are processed into pastes or salts, often diluted and mixed. They appear in the market in the form of powders, pills, or liquids.

DISTRIBUTION

Hard drugs are especially popular in the USA, Europe, and Asia. Crossing all borders, consumers are supplied with these illegal products by truck, ship, or small plane. Opium originates primarily in Afghanistan or Myanmar, while cocaine comes mostly from Colombia, Peru, or Bolivia.

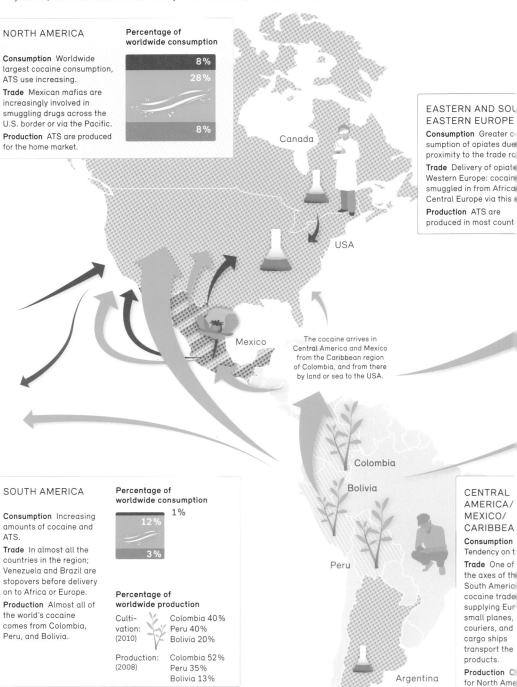

NORTH AMERICA — Percentage of worldwide consumption: 8%, 28%, 8%
- **Consumption** Worldwide largest cocaine consumption, ATS use increasing.
- **Trade** Mexican mafias are increasingly involved in smuggling drugs across the U.S. border or via the Pacific.
- **Production** ATS are produced for the home market.

The cocaine arrives in Central America and Mexico from the Caribbean region of Colombia, and from there by land or sea to the USA.

EASTERN AND SOU[TH] EASTERN EUROPE
- **Consumption** Greater c[on]sumption of opiates due [to] proximity to the trade ro[ute]
- **Trade** Delivery of opiate[s to] Western Europe; cocain[e] smuggled in from Africa [to] Central Europe via this [...]
- **Production** ATS are produced in most count[ries]

SOUTH AMERICA — Percentage of worldwide consumption: 1%, 12%, 3%
- **Consumption** Increasing amounts of cocaine and ATS.
- **Trade** In almost all the countries in the region; Venezuela and Brazil are stopovers before delivery on to Africa or Europe.
- **Production** Almost all of the world's cocaine comes from Colombia, Peru, and Bolivia.

Percentage of worldwide production
- Cultivation (2010): Colombia 40%, Peru 40%, Bolivia 20%
- Production (2008): Colombia 52%, Peru 35%, Bolivia 13%

CENTRAL AMERICA/ MEXICO/ CARIBBEA[N]
- **Consumption** Tendency on t[he rise]
- **Trade** One of the axes of th[e] South Americ[an] cocaine trade supplying Eur[ope]; small planes, couriers, and cargo ships transport the products.
- **Production** C[enter] for North Ame[rica]

PRODUCTION in tonnes per year
Gross amounts of the drugs produced

4,860* (of which 3,132 for the production of heroin)
1,111
494**

TURNOVER in billion U.S. dollars per year

Almost 1 percent of the global gross domestic product can [be] assigned to the drug trade. Dealers and sellers in the consuming countries collect most of the money, and the farm[ers] receive only 1 percent of the opiate and cocaine turnover.

68   85   63

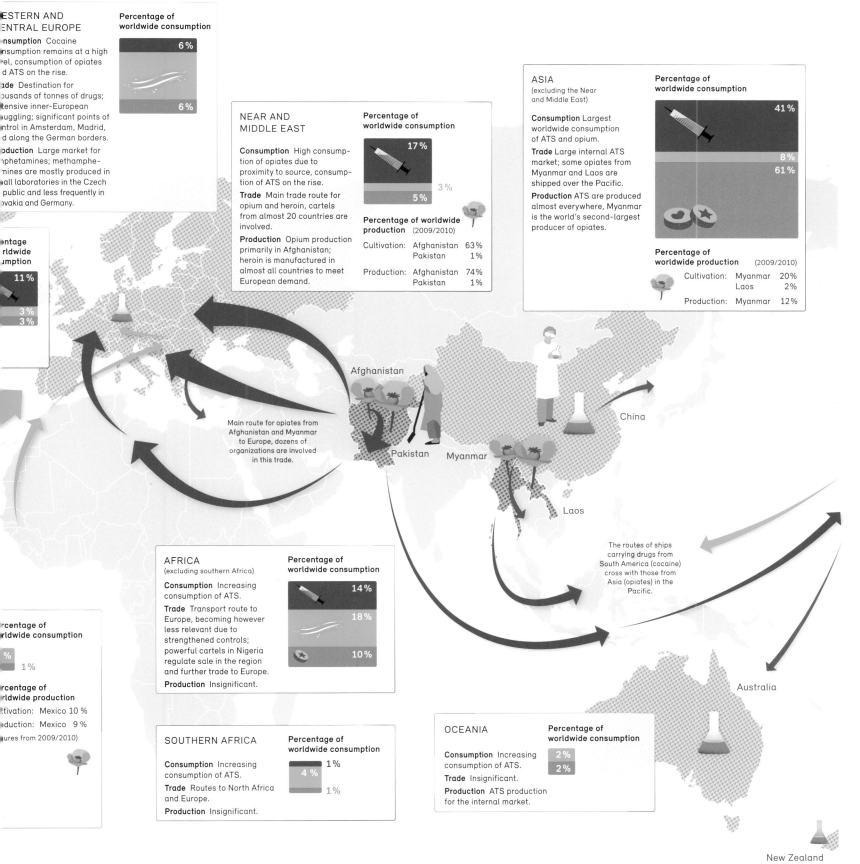

Drug Trafficking

MONEY MAKES THE WORLD GO ROUND

Eating It All Up

How our bottomless appetite is devouring the world's biodiversity.

You're taking a bite of your favorite burger, and a frog dies in the rainforest. Not only are the two events intimately connected — but that frog really matters. Humanity's impact on the Earth is deeper and more complex than we can perceive, and researchers are only beginning to understand it. Under the influence of natural climate change, species have the time to adapt — they can either migrate or evolve — and in turn, the ecosystems they are a part of have the breathing space to alter with them. But the human population is expanding inexorably, and its relentless consumption is severely impacting natural adaptation and lifecycles. This is partly due to the mass industrialized cultivation of certain in-demand produce — particularly soybean, cattle, and palm oil — which is putting massive pressure on ecosystems. This graphic focuses on the most biodiverse and most precious country in the world: Brazil. It is home to two vast ecosystems of millions of species of plants and animals — the Amazon rainforest and the Cerrado savannah. They are both under threat as Brazil's booming economy expands.

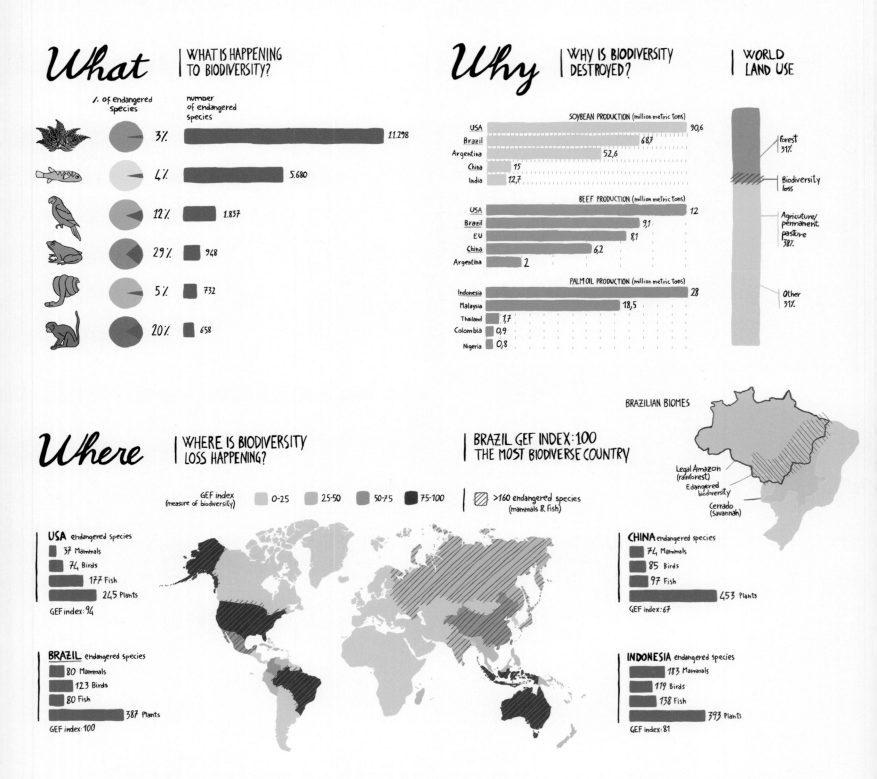

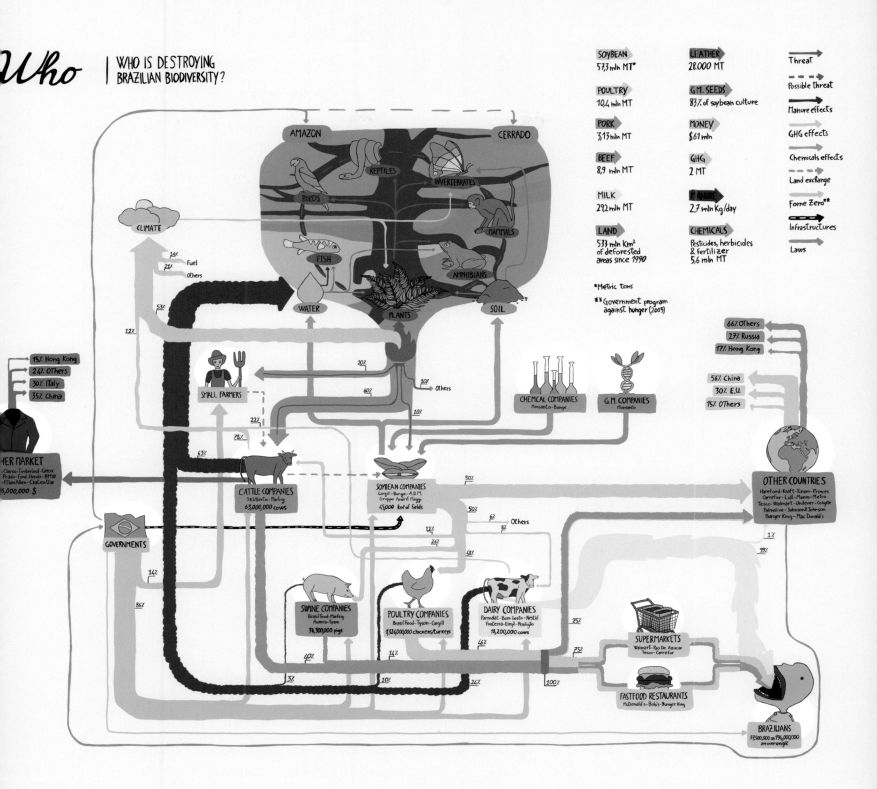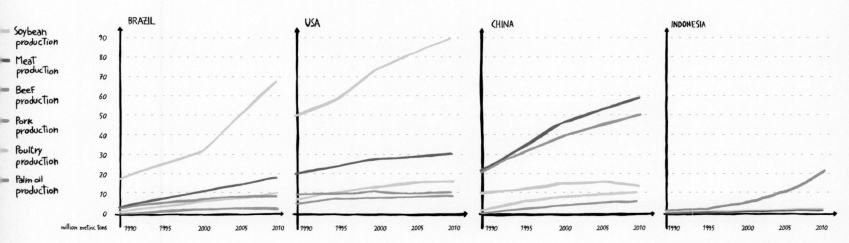

Feeling Peckish?

Our eating habits change the climate, which changes our food.

This graphic uncovers the symbiotic link between our food and the Earth's climate. The production of some foods, especially meat and butter, requires more carbon emissions than, say, fruit and vegetables — and that doesn't even include the extra energy you need to cook meat. On top of that, massive amounts of energy are lost on the journey to your dinner table. And then of course there's all the food that simply gets thrown away every day. Increased industrialization means that food, especially meat, is getting cheaper and cheaper across the globe, and meat consumption is expected to rise sharply in the southern hemisphere over the coming decades.

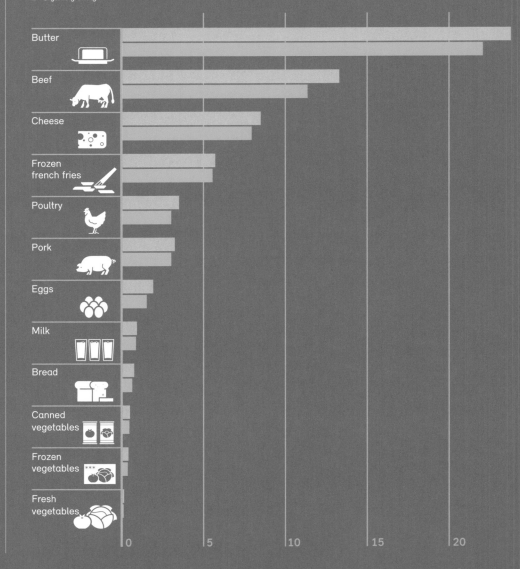

THE CLIMATE BURDEN OF BUTTER, MEAT & CO.

The production of various foodstuffs releases a certain amount of greenhouse gas (in kilograms of CO_2 equivalent per kilogram of product).

- Conventional growing
- Organic growing

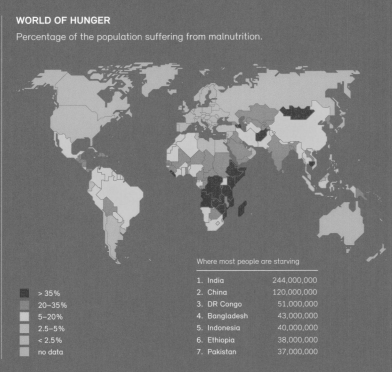

HUNGER DUE TO CLIMATE CHANGE
Forecast changes in agricultural productivity up to 2080.

- −50%
- −15%
- 0%
- +15%
- +35%
- no data

WORLD OF HUNGER
Percentage of the population suffering from malnutrition.

- > 35%
- 20–35%
- 5–20%
- 2.5–5%
- < 2.5%
- no data

Where most people are starving

1.	India	244,000,000
2.	China	120,000,000
3.	DR Congo	51,000,000
4.	Bangladesh	43,000,000
5.	Indonesia	40,000,000
6.	Ethiopia	38,000,000
7.	Pakistan	37,000,000

JST FOR MEAT

at consumption per head and per year up to 2050
world region (in kilograms).

E WORLD OF MCDONALD'S

proximately 31,000 McDonald's restaurants around the world provide 47 million people
more than 100 countries with meat and fatty foods.

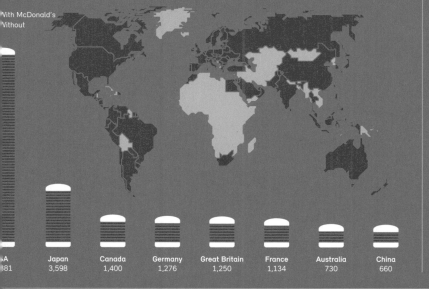

NOTONY IN THE STABLE

cause few turbo-breeds can survive, most domestic farm animal breeds
threatened with extinction.

Farm animal breeds in Germany:
- Endangered
- Not endangered

orses 12/14 | Cattle 15/19 | Sheep 19/21 | Pigs 3/5 | Goats 3/4

LOSS OF FARMERS

Amount of the working population in agriculture of Germany.

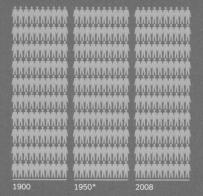

1900 | 1950* | 2008

* Former territory of the Federal Republic

CHEAPER AND CHEAPER FOOD

Statistics for foodstuffs – figures for private households (in percent).

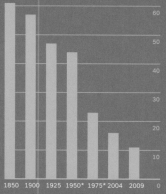

1850 | 1900 | 1925 | 1950* | 1975* | 2004 | 2009

* West Germany only

This is how long employees must work to buy the following foods (in minutes).

- 1970
- 2008

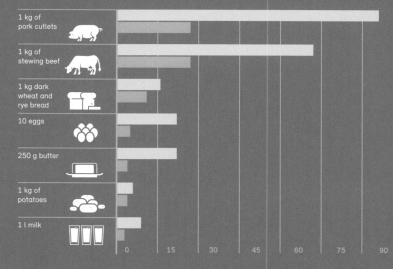

WHAT REMAINS OF THE HARVEST

56 percent of the calories produced on the field are lost – due to harvest losses, inefficient meat production, and foodstuffs thrown away (Basis: 2,000 kilocalories — daily requirement for one person).

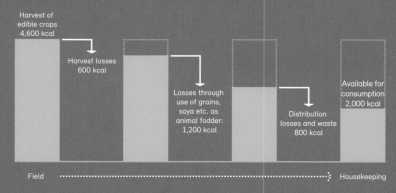

HARDLY USED VARIETY

Wheat, rice, and corn are the main sources of food for humans.

- 3 — 3 plant species cover 50 percent of the human calorie requirement
- 30 — 30 plant species produce 95 percent of the calorie requirement
- 7,000 — 7,000 plant species are used
- 250,000 — 250,000 plant species are known worldwide

Food

MONEY MAKES THE WORLD GO ROUND

Brand Octopuses

These are the companies that feed, wash, and shave us.

If you just fed your cat, brushed your teeth, used cotton buds, had breakfast, or had your inevitable after-lunch cigarette outside the office building, then chances are that almost 100 percent of what you consumed belongs to one of the corporations in these pages. The world is filled with millions of brands. But behind those household names are only a few corporations that control the global consumer goods market for food, beverages, cleaning agents, personal care products, and tobacco. They control the production of industrial sugar, fat, starch, and other ingredients essential for the world's favorite snacks, soft drinks, chewing gum, and anything else you buy in a supermarket. For some local flavor, corporations buy domestic brands and eventually sell the same products in different countries under the same or different names. Germany is one of the most competitive markets for consumer products in the world, and food is relatively cheap. In 1955, Germans spent 50 percent of their income on food, and in 2010 they spent only 12 percent. The rise of the conscious consumer and organic products will lead to higher prices but better food. Pictured here are the brands on sale and their revenues in Germany alone.

*Nestea and Lipton Ice Tea are joint ventures of Coca-Cola and Nestlé/PepsiCo/Unilever.

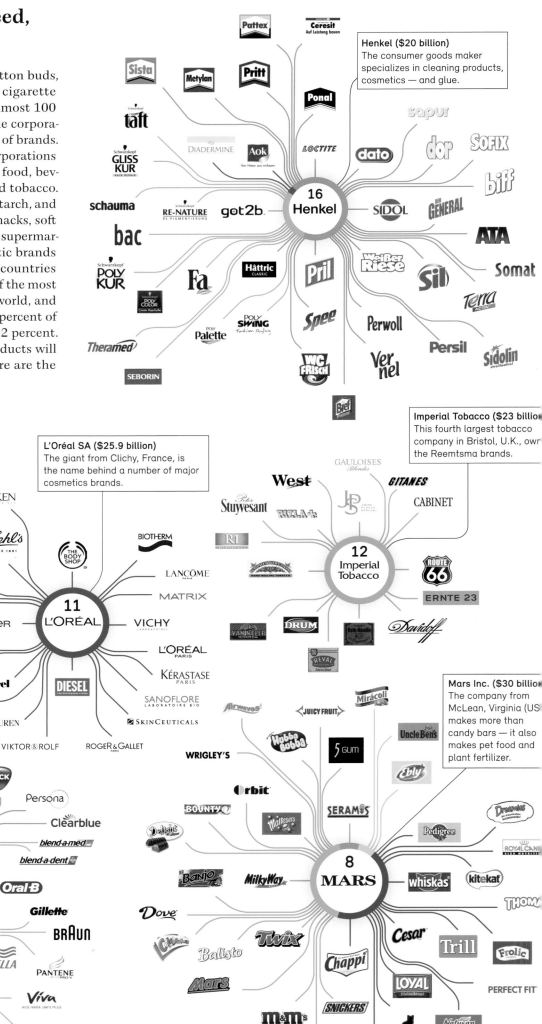

Henkel ($20 billion)
The consumer goods maker specializes in cleaning products, cosmetics — and glue.

L'Oréal SA ($25.9 billion)
The giant from Clichy, France, is the name behind a number of major cosmetics brands.

Imperial Tobacco ($23 billion)
This fourth largest tobacco company in Bristol, U.K., owns the Reemtsma brands.

Procter & Gamble ($82.6 billion)
The consumer goods multinational is based in Cincinnati, Ohio (USA), specializes in toiletries.

Mars Inc. ($30 billion)
The company from McLean, Virginia (US) makes more than candy bars — it also makes pet food and plant fertilizer.

208

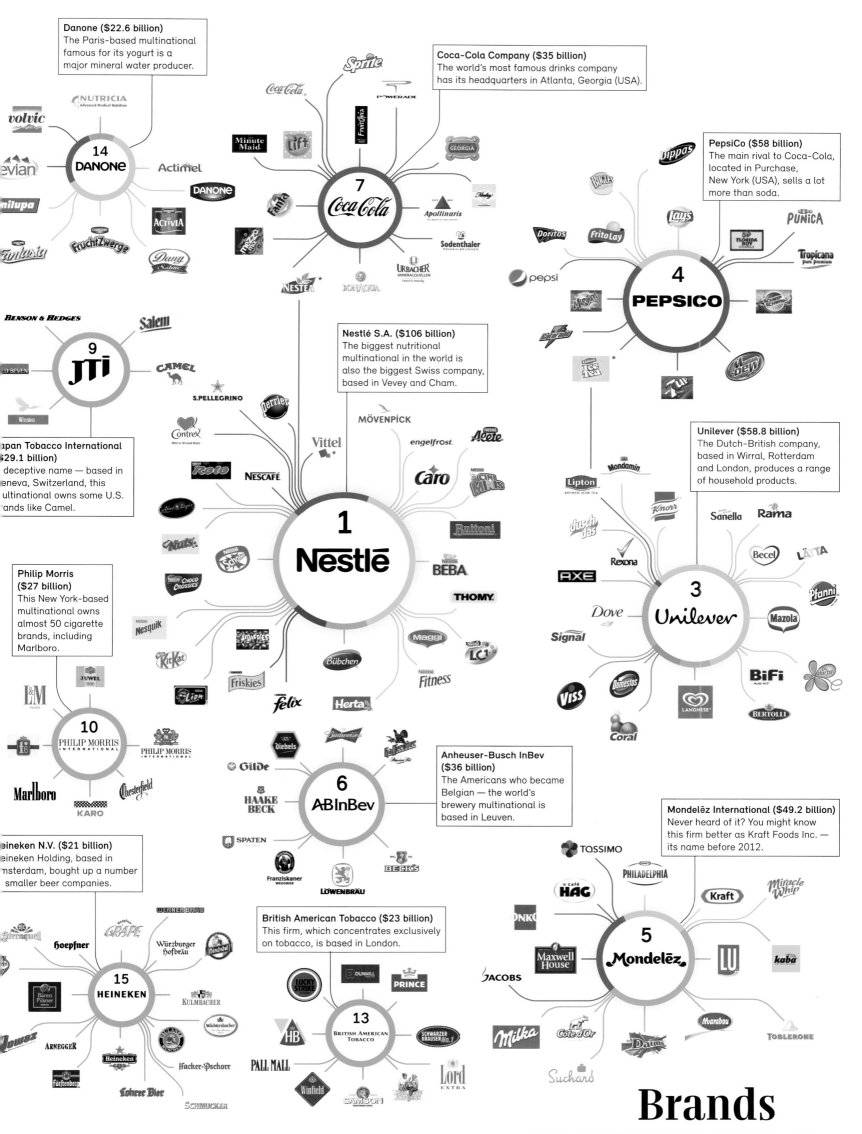

A Carmaker's Family Tree

Who are Chrysler's ancestors, and exactly how many parents does General Motors have?

This graphic offers the chance to excavate the ancient strata of the world's major car companies. Of course, it is in a way the opposite of a regular genealogy. Due to a history of mergers and acquisitions, rather fanning out, this family tree flows together into vast rivers of cars united. In other words, our roads are gradually being taken over by a few giant carmakers hiding behind a myriad of brands. The thickness of the line represents each company's share in the North American automobile market — which might explain why world giants like Volkswagen and Mitsubishi have been reduced to rather sorry trickles.

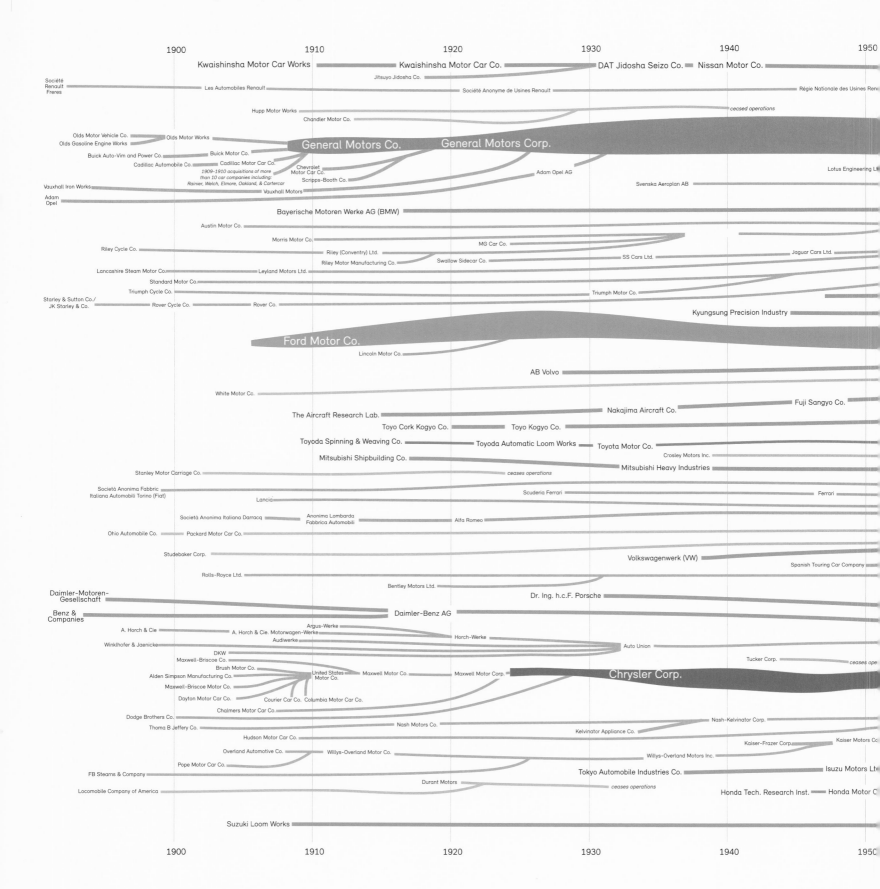

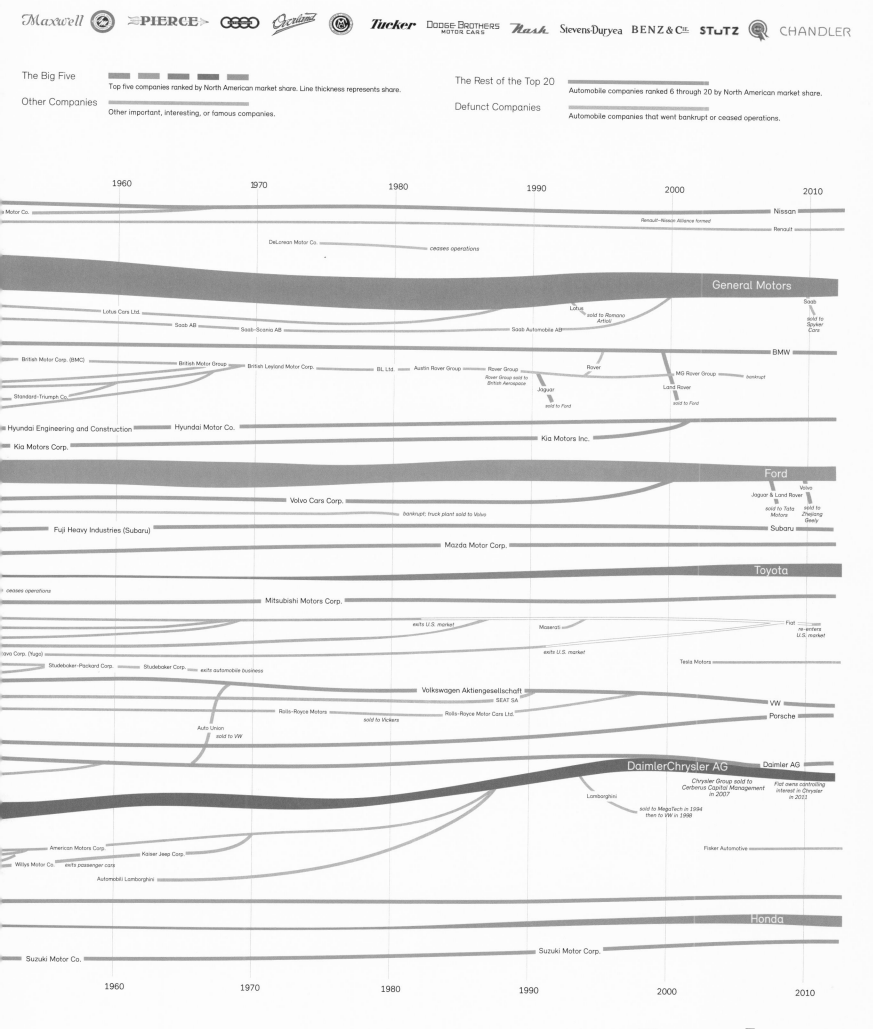

Brands

MONEY MAKES THE WORLD GO ROUND

Nuclear-Fueled World

Uranium-235: a long-term relationship.

It's hard to tell whether nuclear power is a relic of the past or the way of the future. While European nations, especially Germany, are hoping to wind down their dependence on one of the most poisonous elements in the periodic table — particularly in the wake of the 2011 Fukushima disaster in Japan — developing countries everywhere are drawing up blueprints for new reactors as their thirst for electricity rises. But there are two main problems — for one thing, most of the world's reactors are over 25 years old and so becoming more dangerous to run. And then there's the waste, which can remain radioactive for tens of thousands of years. No economically viable and environmentally safe method of disposal has yet been found. But hey, we all need our toasters, right?

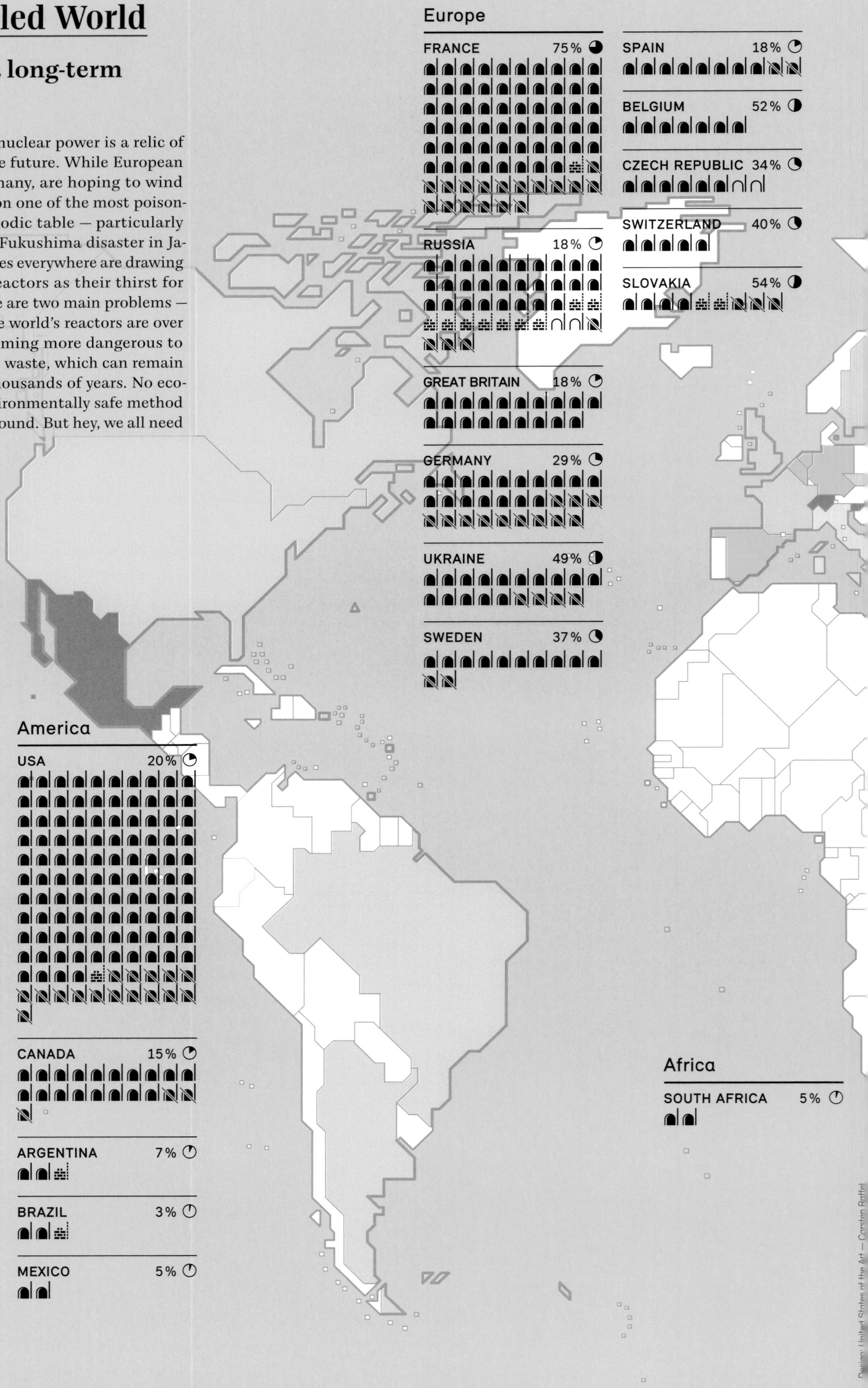

Nuclear reactors for power generation
- In operation
- Under construction
- Definite planning in progress
- Decommissioned or dismantled
- Share of power generation in percentage terms

The future of nuclear energy production
- Expansion planned*
- Exit postponed
- Re-entry planned
- At a stand-still

* This at least refers to Europe and North America, primarily because of political statements of intent, which say nothing about the financial viability of NPP new builds.

Europe

- FRANCE 75%
- RUSSIA 18%
- GREAT BRITAIN 18%
- GERMANY 29%
- UKRAINE 49%
- SWEDEN 37%
- SPAIN 18%
- BELGIUM 52%
- CZECH REPUBLIC 34%
- SWITZERLAND 40%
- SLOVAKIA 54%

America

- USA 20%
- CANADA 15%
- ARGENTINA 7%
- BRAZIL 3%
- MEXICO 5%

Africa

- SOUTH AFRICA 5%

212

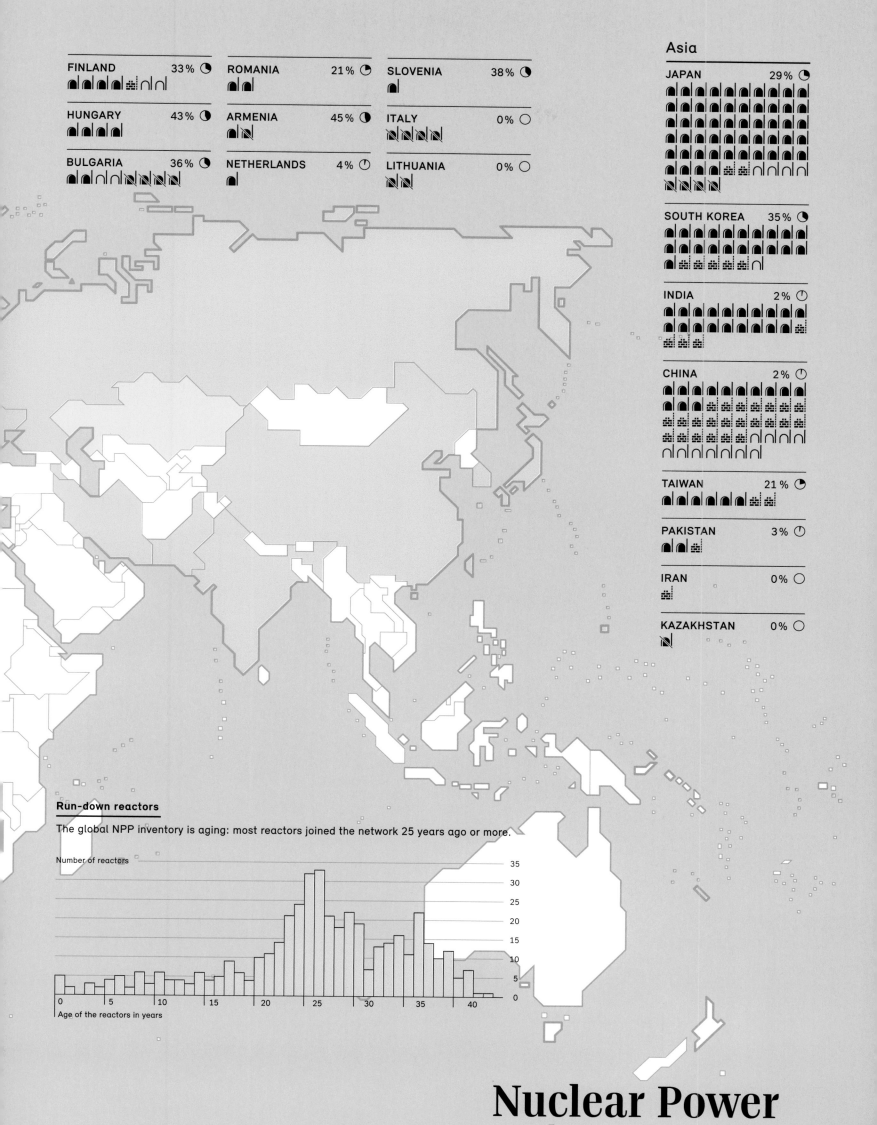

Here Comes the Sun

Our original power station.

An afternoon on a cloudless summer's day is all it takes to understand the power of the Sun. An unguarded hour of exposure is enough to fry your skin, and yet that energy has already traveled impossible distances across a vast cold void to do that to you. No wonder that the first human cultures worshipped the Sun. Everything we do is dependent on the star just eight and a half short light-minutes away at the center of our solar system. Not only that, we are formed from the same stuff — a molecular cloud that collapsed and spun and flattened to create the Sun and all the planets, asteroids, and comets in it. Incredibly, the Sun's mass accounts for 99.86 percent of the solar system. And the Sun still conceals many mysteries — the biggest of which is the fusion of hydrogen nuclei into helium, the nuclear reaction that powers the Sun. If we could repeat that trick on Earth, all of our energy worries would be resolved. For now though, we will have to make do with trying to harness the Sun's energy by the cruder means of photovoltaic technology. The graphics on this page explore how one Sun-kissed country — Italy — is doing its best to supplement its power supply with whatever spare rays the clouds deign to let pass through.

THE LONG ROAD OF SOLAR POWER IN ITALY: A LEVEL OF PRODUCTION STILL INSIGNIFICANT COMPARED TO TOTAL DOMESTIC DEMAND.

In Italy in 2008 the energy produced from renewable sources (59,244 GWh) was equal to **16.5%** of total domestic consumption (357,460 GWh). In 2000 the figure was **16%**. Despite the developments of recent years, among renewable sources, solar energy does not reach **1%**.

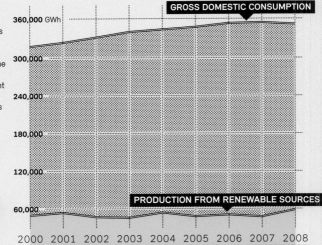

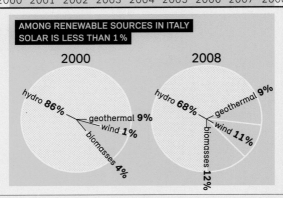

WATCH OUT FOR SUN STROKE
Forecast of world energy production according to the different sources.
1 Exajoule = 278 billion kilowatt hours / 100 Exajoules is the energy consumption in the United States.

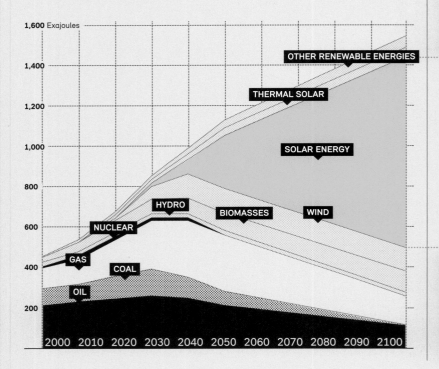

Every second the Sun produces an amount of energy.

Solar surface
6.0877×10^{18} m²

Solar mass
1.9891×10^{30} kg

Surface temperature
5,504.85 °C

Luminosity
3.827×10^{26} W

~ 150,000,000 km

Sun

8 minutes is the time it takes for sunlight to reach the Earth.

4 billion 100 watt lightbulbs is the equivalent of solar energy radiated per second on the Earth.

Man has always tried to portray the star at the center of our solar system. Adrian Frutiger has collected a number of representations:

1. The basic symbol of the Sun and the universe. *2. Primitive Sun symbol.* *3. Sun wheel with internal radiation.* *4. The Sun shining on the Earth.* *5. Circular swastika, symbol of the Sun and of life.* *6, 7, 8. The curved divisions do not represent the rays but the movement of the Sun in the sky.* *9. Rising Sun.* *10. Sun at sunset.*

WHO ARE THE PRODUCERS WHO PLAY WITH LIGHT?

There are many companies engaged in the production of solar-thermodynamic energy. Their projects are born from technological development (a discovery, the development of an idea: **technology push**) or from the request coming from a given market (**energy pull**).

BUSINESS MODELS
- ● energy pull
- ○ technology push

THE PRICE? EVER LOWER.

Grid parity is the moment in which solar energy will cost as much as that produced via traditional sources. Will be reached between 2014 and 2016.

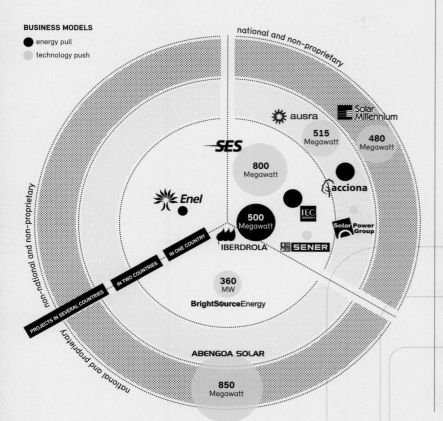

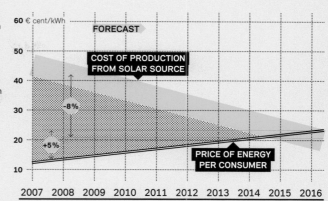

PHOTOVOLTAIC POWERS.

United States
- Installed in 2008: 400 Megawatt
- Cumulative up to 2008: 1,231 Megawatt

Spain
- Installed in 2008: 2,661 MW
- Cumulative up to 2008: 3,316 MW

Germany
- Installed in 2008: 1,350 MW
- Cumulative up to 2008: 5,212 MW

Japan
- Installed in 2008: 300 MW
- Cumulative up to 2008: 2,219 MW

HOW IMPORTANT IS THE OIL TREND?

- Brent dated index
- Energy price

Sufficient to satisfy the entire demand of the Earth for 500 years.

On March 20, 2009 the last solar panel was mounted on the **International Space Station** (ISS) that will allow a total output of 124 kilowatts of electricity to be achieved, equivalent to the consumption of approx. 42 homes in the United States.

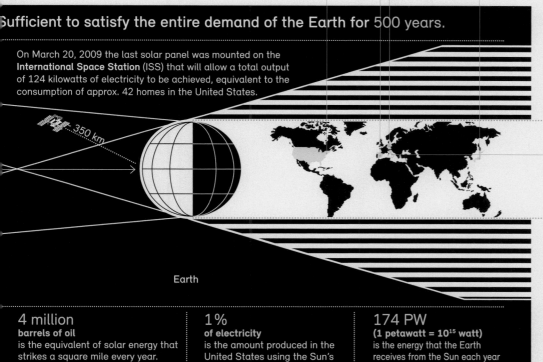

4 million barrels of oil is the equivalent of solar energy that strikes a square mile every year.

1% of electricity is the amount produced in the United States using the Sun's rays in 2007.

174 PW (1 petawatt = 10^{15} watt) is the energy that the Earth receives from the Sun each year (30% is reflected).

Italy's rays beam down on Lombardy.
Photovoltaic systems by region and energy production (March 09).

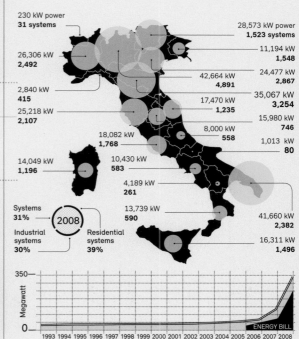

Systems 31% | Industrial systems 30% | Residential systems 39% (2008)

---

11. Traditional image of the Sun with rays. **12.** From an Indian coin. **13.** Relief on metal, Crete. **14.** Sun with flame rays. **15.** Indian sign painted on leather. **16.** Sun wheel, rural painting of the Alps. **17.** From a Roman tomb. **18.** Sun in a closed world, in Knossos. **19.** Ancient Chinese symbol of the forces of nature. **20.** Sand drawing, in India. **21.** Typical Celtic symbol concerning worship of the Sun. **22.** Sun with feet. **23.** From an Indian coin. **24.** Terracotta from Hopi, Arizona.

Solar Power
MONEY MAKES THE WORLD GO ROUND

Recharge the Renewables

Comparing the markets, and potential, of various energy sources.

As the old debates over renewable energy play out, technological developments speed on regardless. On one particularly sunny, windy day in April 2013, the European Energy Exchange announced that Germany had managed to supply half the country's workday needs from renewable energy sources. It amounted to 35,902 megawatts, equivalent to the output of 26 nuclear power stations. Germany, the European Union's biggest economy and most populous country, remains one of the biggest enthusiasts for renewables. But while that world record impressed many, the fact remains that most people are still skeptical that renewable resources will ever be able to quench our thirst for electricity. For better or worse, the cynics insist, we will always be dependent on the dirtier energy — coal, gas, or nuclear power. This graphic compares the figures for each viable energy resource in Great Britain.

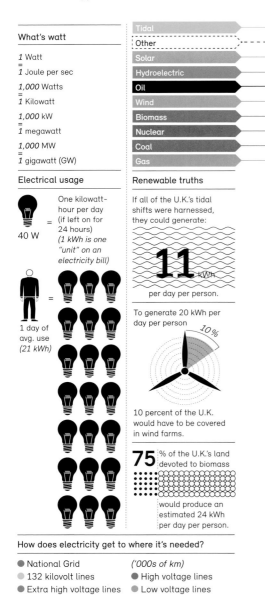

What's watt

1 Watt = 1 Joule per sec
1,000 Watts = 1 Kilowatt
1,000 kW = 1 megawatt
1,000 MW = 1 gigawatt (GW)

Electrical usage

40 W = One kilowatt-hour per day (if left on for 24 hours) (1 kWh is one "unit" on an electricity bill)

1 day of avg. use (21 kWh)

Renewable truths

If all of the U.K.'s tidal shifts were harnessed, they could generate: **11** kWh per day per person.

To generate 20 kWh per day per person 10% — 10 percent of the U.K. would have to be covered in wind farms.

75 % of the U.K.'s land devoted to biomass would produce an estimated 24 kWh per day per person.

How does electricity get to where it's needed?

- National Grid
- 132 kilovolt lines
- Extra high voltage lines
- ('000s of km)
- High voltage lines
- Low voltage lines

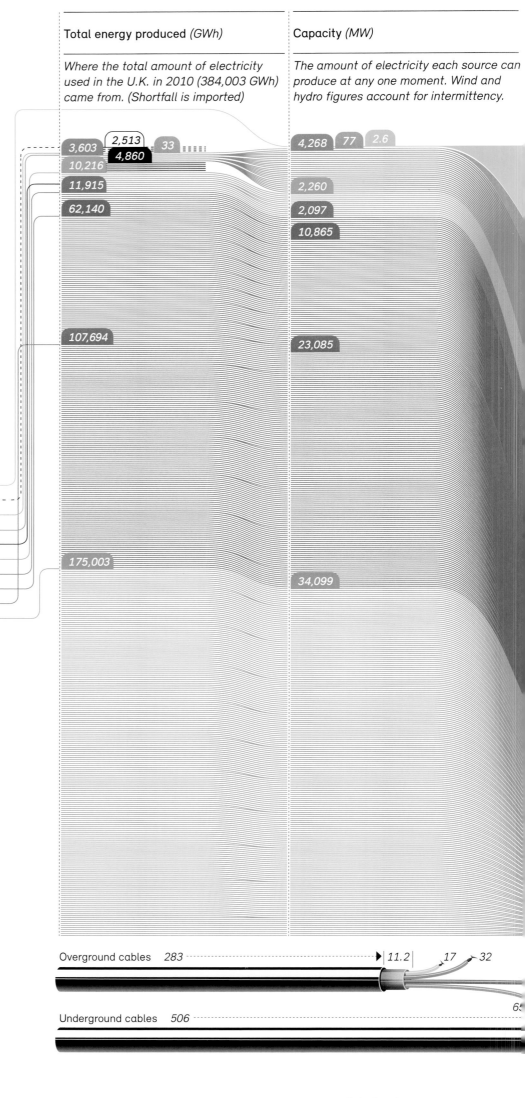

Total energy produced (GWh) — Where the total amount of electricity used in the U.K. in 2010 (384,003 GWh) came from. (Shortfall is imported)

Capacity (MW) — The amount of electricity each source can produce at any one moment. Wind and hydro figures account for intermittency.

Tidal · Other · Solar · Hydroelectric · Oil · Wind · Biomass · Nuclear · Coal · Gas

Total energy produced: 3,603 · 2,513 · 33 · 4,860 · 10,216 · 11,915 · 62,140 · 107,694 · 175,003

Capacity: 4,268 · 77 · 2.6 · 2,260 · 2,097 · 10,865 · 23,085 · 34,099

Overground cables 283 — 11.2 · 17 · 32
Underground cables 506

216

Design: Fraser Lyness
Published: Eureka Magazine — The Times Newspaper

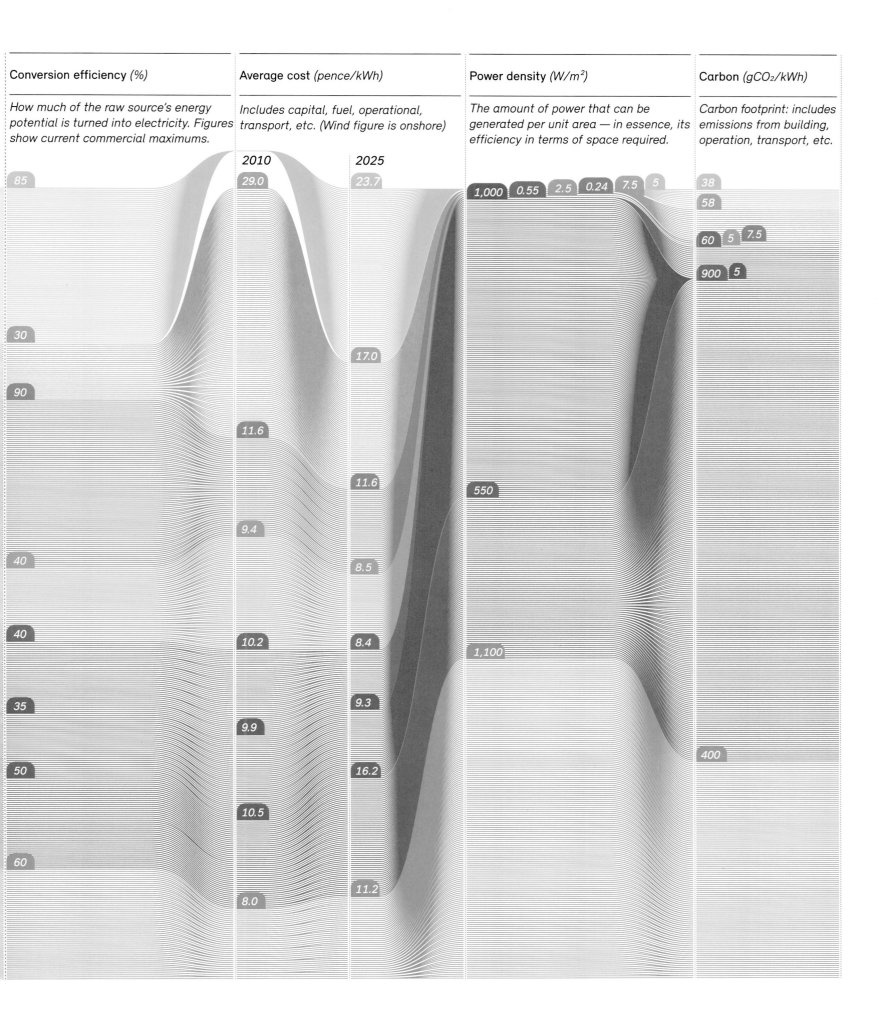

Conversion efficiency *(%)*

How much of the raw source's energy potential is turned into electricity. Figures show current commercial maximums.

85
30
90
40
40
35
50
60

Average cost *(pence/kWh)*

Includes capital, fuel, operational, transport, etc. (Wind figure is onshore)

2010   *2025*

29.0   23.7
17.0
11.6   11.6
9.4   8.5
       8.4
10.2   9.3
9.9   16.2
10.5   11.2
8.0

Power density *(W/m²)*

The amount of power that can be generated per unit area — in essence, its efficiency in terms of space required.

1,000   0.55   2.5   0.24   7.5   5

550

1,100

Carbon *(gCO₂/kWh)*

Carbon footprint: includes emissions from building, operation, transport, etc.

38
58
60   5   7.5
900 5

400

169
0.97 3 21
154
328

Energy Market

MONEY MAKES THE WORLD GO ROUND

Growing Thirsty

How China became the world's biggest gas-guzzler.

When the United States invaded Iraq in 2003, some critics muttered that the real reason for the war was to slake America's eternal thirst for oil and to establish permanent U.S. control over some of the biggest oil reserves in the Middle East. But if that's true, two factors intervened in U.S. plans. China's state-guaranteed oil and gas companies have begun to edge American corporations out of lucrative contracts. Second, the U.S. has discovered fracking opportunities on home soil, and the dependence on the Middle East is expected to slacken in the coming years. China's oil companies have a key advantage over their rivals when bidding for oil contracts — they are more interested in securing a steady supply than extracting large profits. That is because China is desperate to fuel its breakneck economy, and though it is building new nuclear and gas shale power plants, they are yet to go online. And China's economic growth is prodigious. In 2013, it was due to grow by "only" seven percent — its lowest in 13 years. Back in 2011, China's economy expanded by 9.3 percent. The graph shows how rapidly the country's thirst for oil has developed.

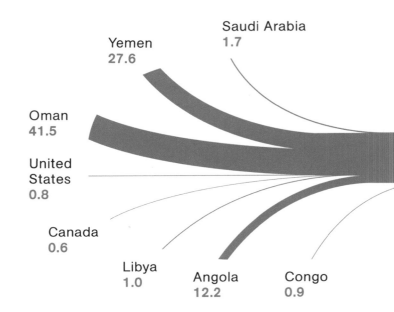

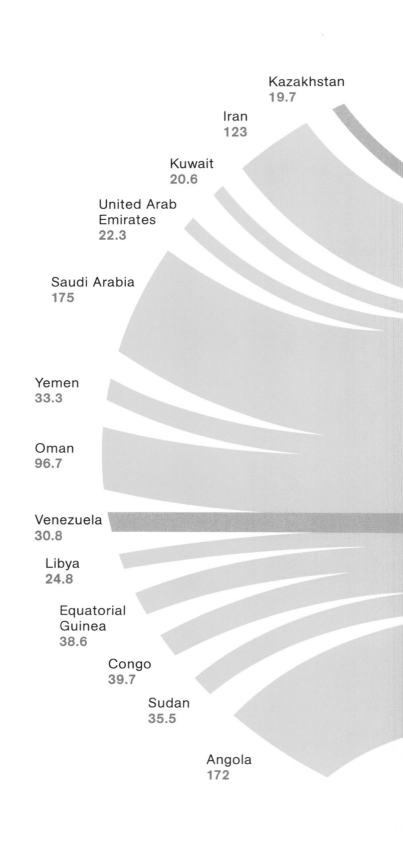

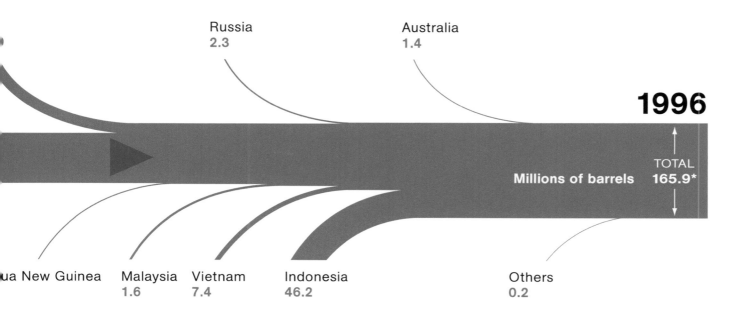

1996

Russia 2.3
Australia 1.4
Papua New Guinea
Malaysia 1.6
Vietnam 7.4
Indonesia 46.2
Others 0.2

TOTAL Millions of barrels 165.9*

COUNTRIES EXPORTING OIL TO CHINA

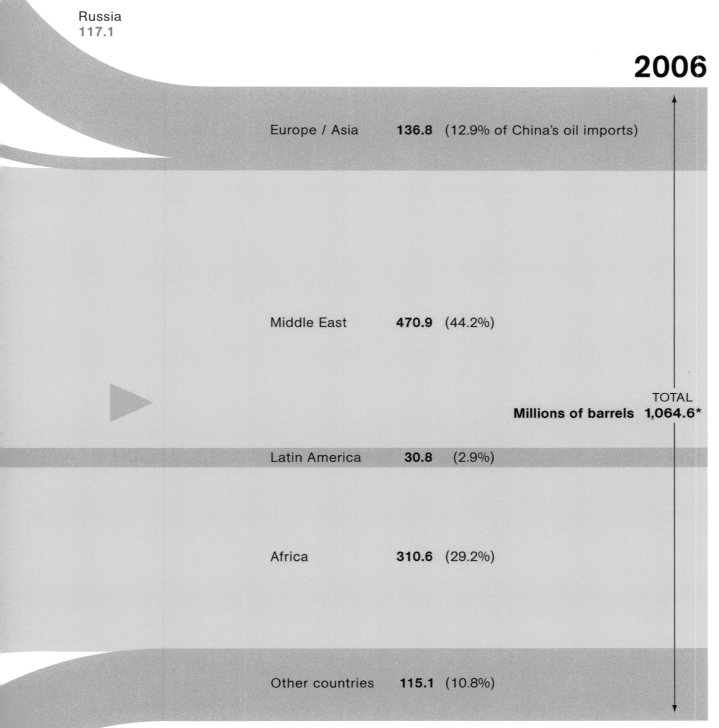

2006

Russia 117.1

Europe / Asia 136.8 (12.9% of China's oil imports)

Middle East 470.9 (44.2%)

TOTAL Millions of barrels 1,064.6*

Latin America 30.8 (2.9%)

Africa 310.6 (29.2%)

Other countries 115.1 (10.8%)

*Numbers are rounded and do not equal totals.

Oil

MONEY MAKES THE WORLD GO ROUND

Power Choices

How do we turn the right way on the energy decision chain?

The problem of providing the world with enough energy is likely to be a headache for the rest of this century, if not the next one. Unfortunately, the ancient Greeks ruled out the perpetual motion machine, so the problem is this: while there are plenty of ways to keep a wheel spinning, they all come at a cost. Either they are too inefficient, too expensive, too risky, too inflexible, too finite, or too environmentally destructive. Many governments have now decided that our best bet is to ignore the risk and ecological factors and go nuclear. Nuclear power is certainly cheap, and as long as you keep your safety standards up and you don't live near the edge of a tectonic plate, safe enough. But then again, if something does go wrong, the consequences can be more devastating and terrifying than any of the alternatives. This graphic attempts to find a way through the labyrinth of choices in our eternal quest for power.

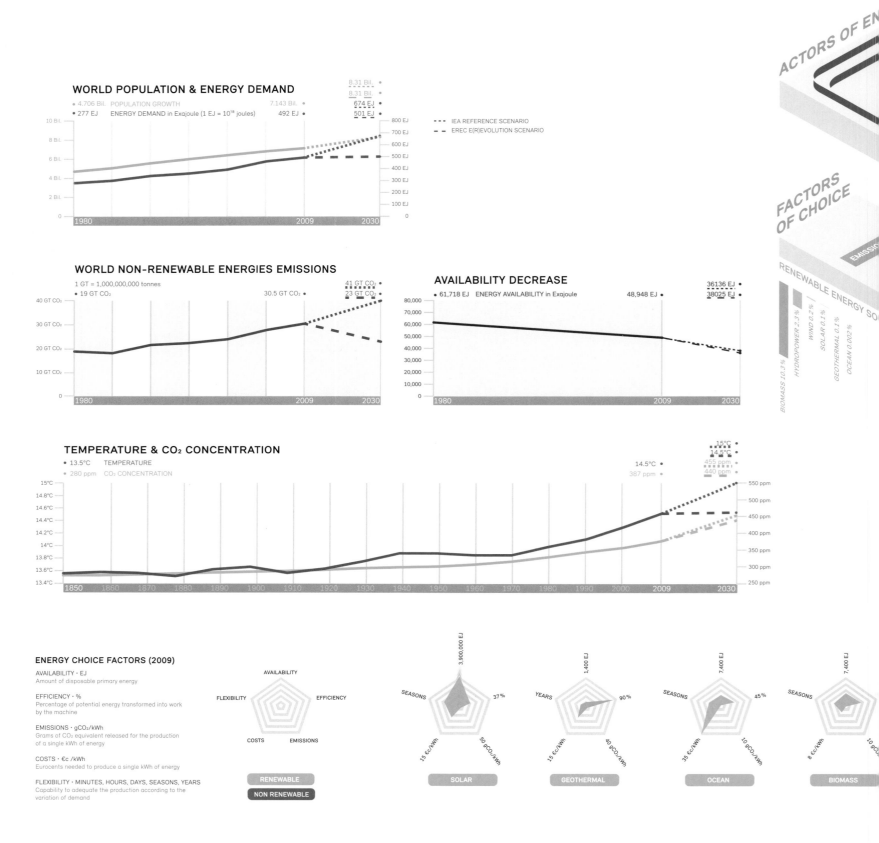

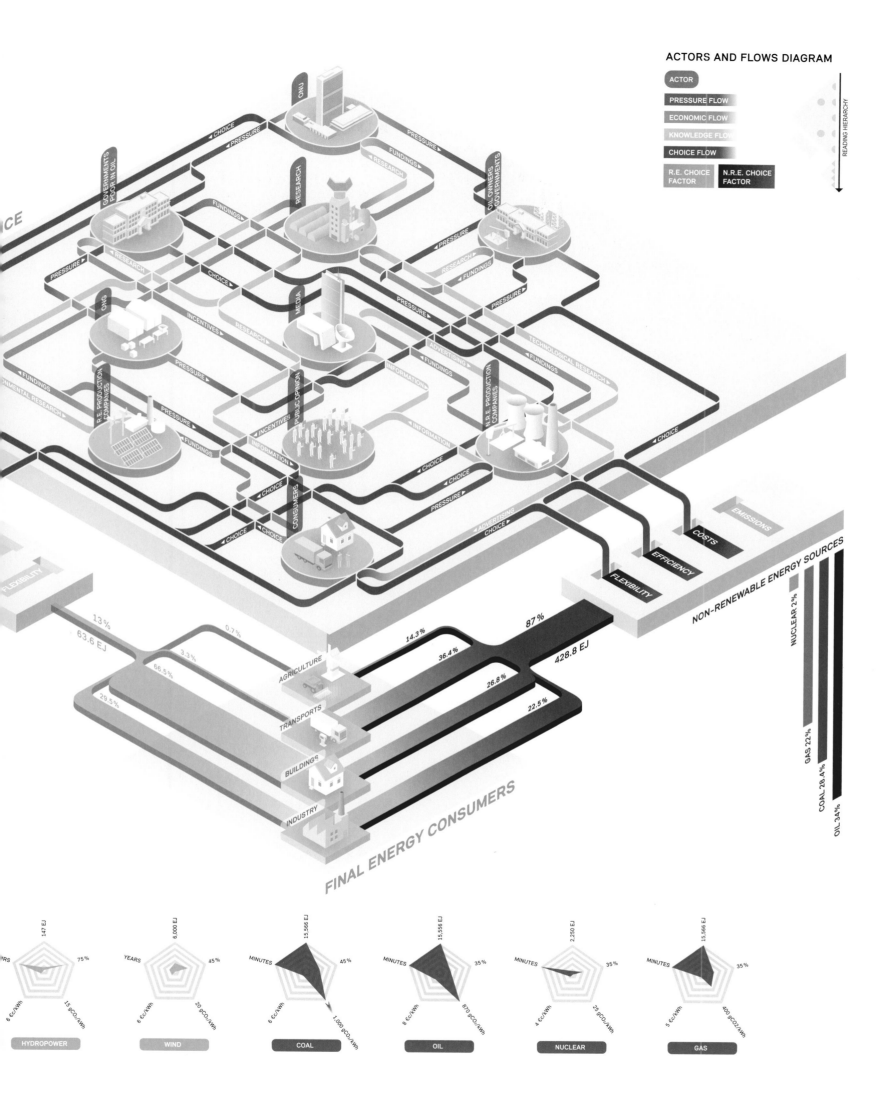

Energy Flow
MONEY MAKES THE WORLD GO ROUND

THE WORLD IS NOT ENOUGH

Are you living a responsible life? Do you care about the natural world around you? Is your behavior helping to preserve it for generations to come?

These are big questions, but they come down to countless small decisions that we make every day. For example, people from other countries often remark on the size of the portions served in American restaurants. But is it really such a surprise? The United States is a vast country founded by Europeans who came from lands of scarcity to what seemed like an infinitely large land mass with unending abundance. Why would they settle for less?

But there are consequences to this mindset, and the USA had to face up to this fact when the buffalo population of the country dropped from several hundred million in the 1700s to just 800 remaining buffalo within one hundred years. It had never occurred to anyone until then that having buffalo to kill would ever be anything other than a constant fact of life.

Just in time, the American buffalo were saved thanks to a reality check, and some careful breeding programs. The same might not be true about the world's fossil fuels, our minerals, and our air. As the graphics in this chapter demonstrate, we have less than a decade's worth of antimony left to dig out. Coal and copper will likely be used up in our lifetime, and titanium and aluminium in that of our children's. Workable replacements have yet to be properly deployed. Some may never be found.

In the meantime, as other images here show, our carbon emissions show no signs of slowing, the planet is slowly heating up, the Great Pacific Garbage Patch is four times the size of Germany, fresh water remains scarce, there seem to be more tornados and earthquakes due to our interventions in the climate and fracking, and hundreds of animals and insects are becoming seriously endangered while most politicians look the other way.

We have created this selection to remind ourselves of these and other important points. You were probably aware of at least some of these facts, and are probably unsurprised about the rest of them. But it isn't too late. If we want to change the world, we need to change nations' and companies' priorities, and that means continuing to state the facts in as many new and different ways as possible, in the hope that at least one of these methods might be enough to scare people into action.

Sometimes advocacy groups use the rhetoric of "We are killing the planet." The truth is, the planet doesn't care. This beautiful planet will keep spinning around the Sun, contain a couple thousand if not millions of species, and keep revolving regardless of whether its surface remains habitable for humans.

We should not make serious changes to our culture and lifestyle only because it's good for the planet. We should do it because otherwise humanity will soon be as depleted as everything else that it has consumed in its journey. As the following chapter will show, we can be incredible, resourceful, and brilliant. The time has come to apply more of that genius to our most pressing problem: getting a grip on ourselves.

Energy Vampires

What gets sucked out of your system at night.

The clock on your microwave, the little red light on your TV, the digital recorder waiting to record its show — most of the electrical appliances in your home are essentially undead beings, quietly leeching your electricity and sapping you of your cash while you're asleep. This graphic shows how much power is bled from your metaphorical veins each year, in kilowatt hours, and what it costs you — assuming 11 U.S. cents per kilowatt hour. The red lines show passive standby mode, while the blue lines show active standby mode.

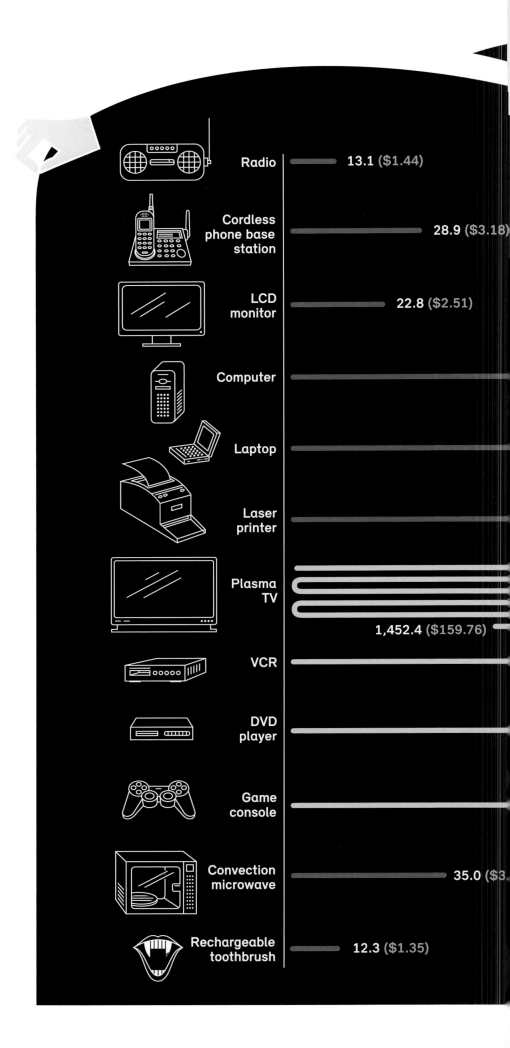

Design: Nigel Holmes
Published: *GOOD Magazine*

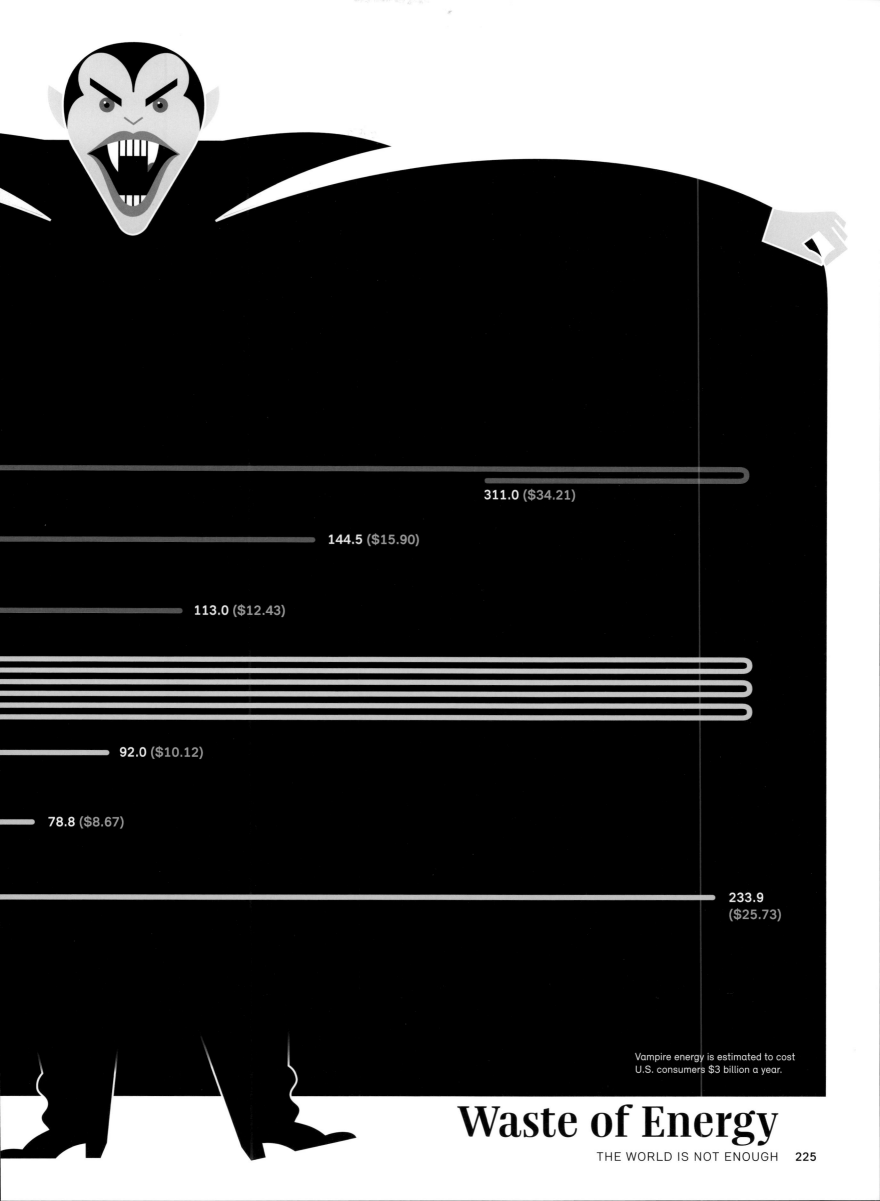

311.0 ($34.21)
144.5 ($15.90)
113.0 ($12.43)
92.0 ($10.12)
78.8 ($8.67)
233.9 ($25.73)

Vampire energy is estimated to cost U.S. consumers $3 billion a year.

Waste of Energy
THE WORLD IS NOT ENOUGH

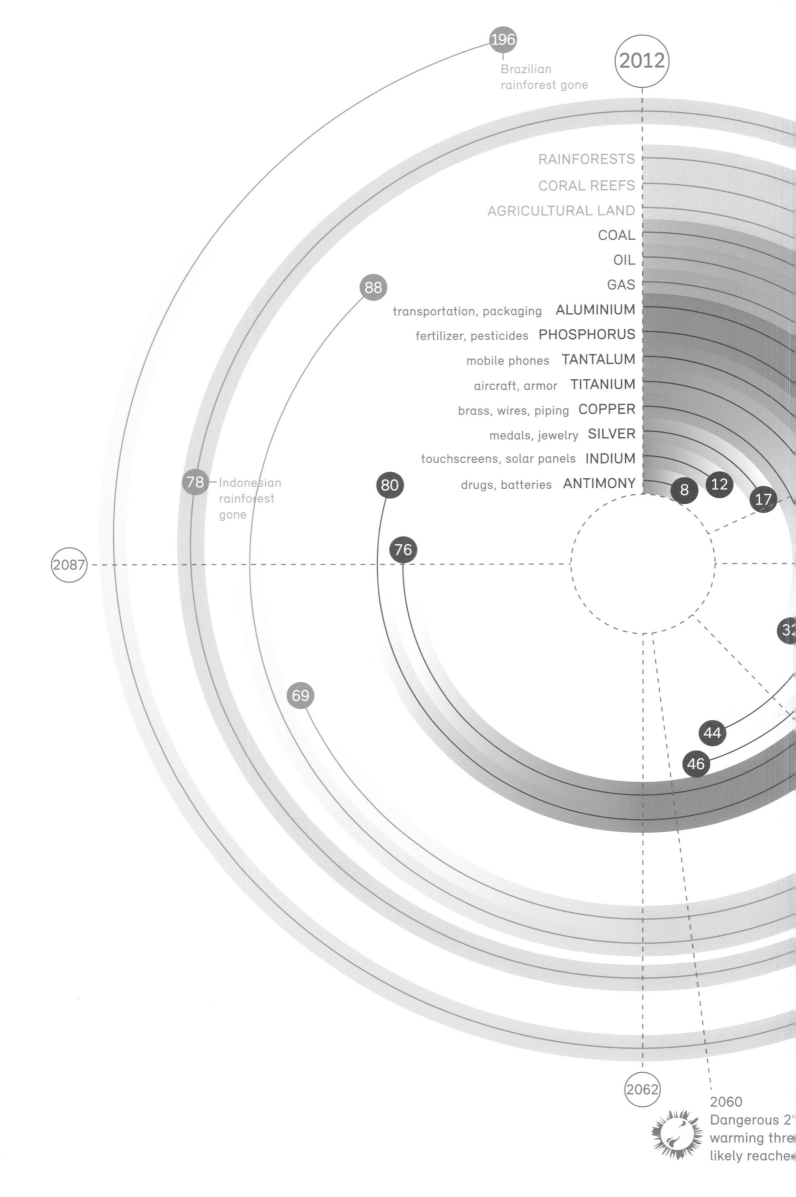

What's Left in Store

You know all those things you love doing?

Natural resources are what mankind and modern civilization are built upon. Some of those ressources are renewable and we can keep building our future with them. But because others are not, finding replacements and adjusting our lifestyle are probably the most fundamental challenges that western society currently faces. In some cases, this limitation might lead to great ideas and innovations. In other cases, we all might be reminiscing about the good old days when oil and polar ice were still around.

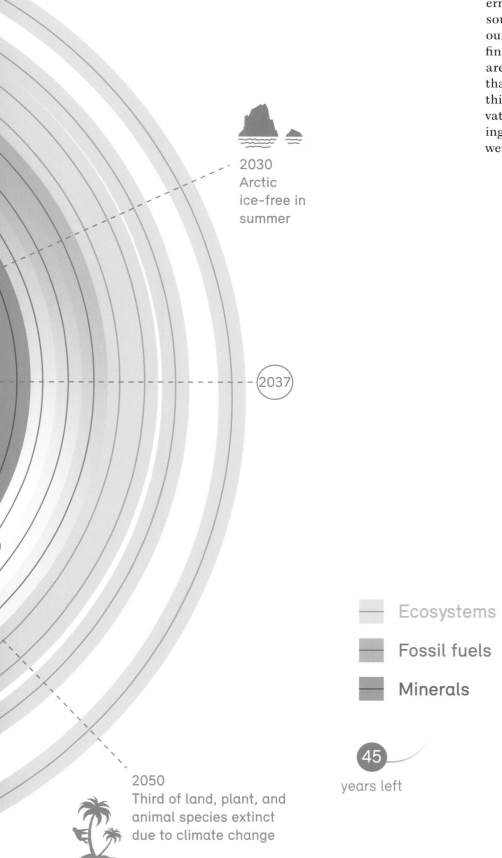

2030 Arctic ice-free in summer

2037

Ecosystems
Fossil fuels
Minerals

45 years left

2050 Third of land, plant, and animal species extinct due to climate change

Resources

THE WORLD IS NOT ENOUGH

We Need a New Earth

The bottom line: using the current rate of consumption, we need 1.2 Earths to satisfy our needs.

Somehow, we are just going to have to muddle through. One way or another, the human population of the world will likely struggle and survive, like we always have done when faced with famine, scarcity, and ruin. But the fact is that our current addictions — to fossil fuels, to meat, to plastic, among others —

ECOLOGICAL FOOTPRINT
The size of each area represents the ecological footprint of each country in million hectares.
The color scale represents the ecological footprint per capita in hectares (100 × 100 m).

<1   1.1–2   2.1–3   3.1–4   4.1–5   5.1–6   6.1–7   7.1–8   8.1–9   >9

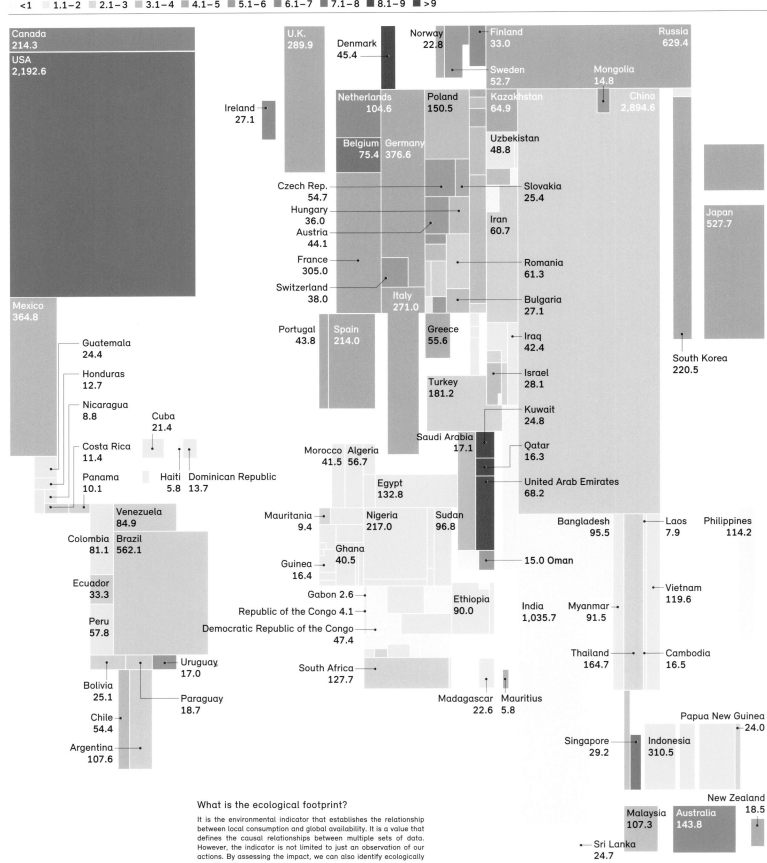

What is the ecological footprint?

It is the environmental indicator that establishes the relationship between local consumption and global availability. It is a value that defines the causal relationships between multiple sets of data. However, the indicator is not limited to just an observation of our actions. By assessing the impact, we can also identify ecologically sound practices.

are moving us towards a global crisis. Up until now, it is the developing world that has had to suffer and subsist while the rest of us thrive. But the largest of the poor nations — China and India — are already asserting themselves economically, developing a widening consumer class that wants the same things that everyone wants — the distraction of material goods. The same is true elsewhere too — by 2050, it has been estimated, there will be one billion Africans with a significant disposable income. So, has anyone got a spare Earth we could borrow?

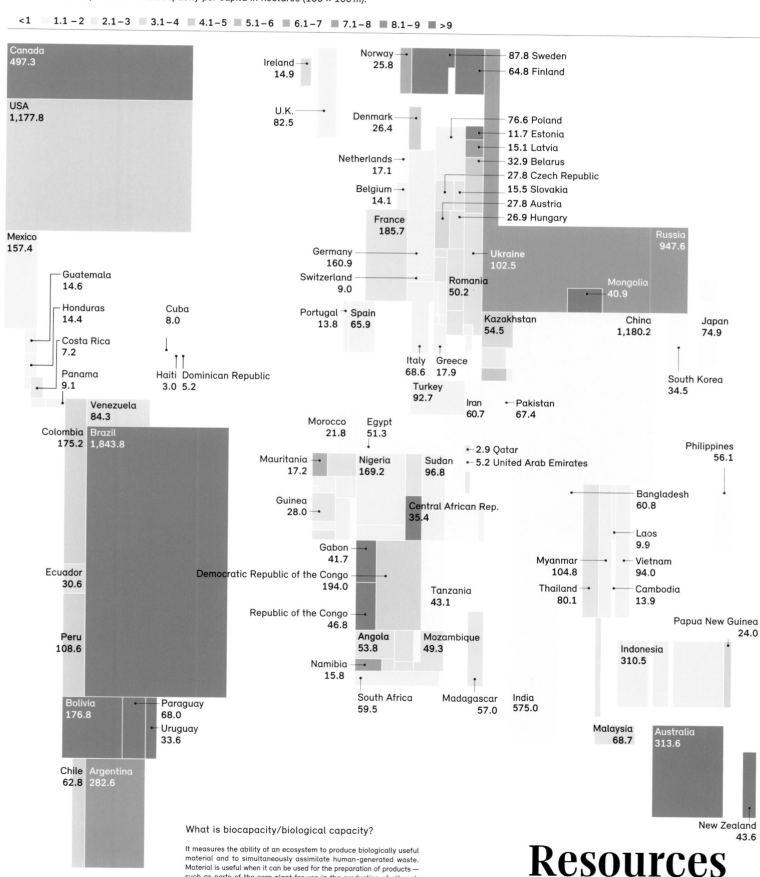

BIOCAPACITY
The size of each area represents the biocapacity of each country in million hectares.
The color scale represents the biocapacity per capita in hectares (100 × 100 m).

<1   1.1–2   2.1–3   3.1–4   4.1–5   5.1–6   6.1–7   7.1–8   8.1–9   >9

What is biocapacity/biological capacity?

It measures the ability of an ecosystem to produce biologically useful material and to simultaneously assimilate human-generated waste. Material is useful when it can be used for the preparation of products — such as parts of the corn plant for use in the production of ethanol. Thus, the rating changes from year to year depending on the further utilization of the raw material.

Resources
THE WORLD IS NOT ENOUGH

TOTAL CARBON DIOXIDE EMISSIONS FROM THE CONSUMPTION OF ENERGY
BY MAJOR NATION STATE 2009 IN MILLIONS OF METRIC TONNES

WHOLE WORLD (2009)
30,313.25

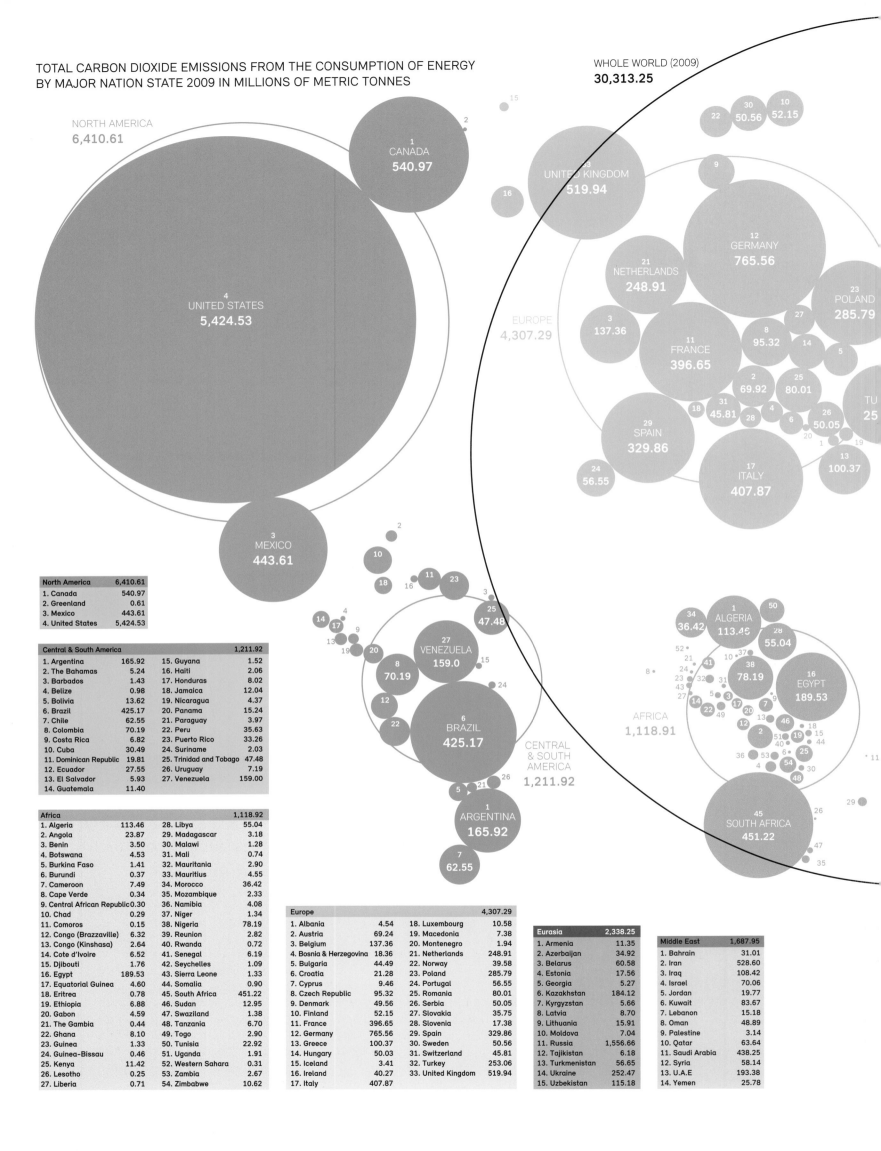

North America		6,410.61
1. Canada		540.97
2. Greenland		0.61
3. Mexico		443.61
4. United States		5,424.53

Central & South America				1,211.92
1. Argentina	165.92	15. Guyana	1.52	
2. The Bahamas	5.24	16. Haiti	2.06	
3. Barbados	1.43	17. Honduras	8.02	
4. Belize	0.98	18. Jamaica	12.04	
5. Bolivia	13.62	19. Nicaragua	4.37	
6. Brazil	425.17	20. Panama	15.24	
7. Chile	62.55	21. Paraguay	3.97	
8. Colombia	70.19	22. Peru	35.63	
9. Costa Rica	6.82	23. Puerto Rico	33.26	
10. Cuba	30.49	24. Suriname	2.03	
11. Dominican Republic	19.81	25. Trinidad and Tobago	47.48	
12. Ecuador	27.55	26. Uruguay	7.19	
13. El Salvador	5.93	27. Venezuela	159.00	
14. Guatemala	11.40			

Africa				1,118.92
1. Algeria	113.46	28. Libya	55.04	
2. Angola	23.87	29. Madagascar	3.18	
3. Benin	3.50	30. Malawi	1.28	
4. Botswana	4.53	31. Mali	0.74	
5. Burkina Faso	1.41	32. Mauritania	2.90	
6. Burundi	0.37	33. Mauritius	4.55	
7. Cameroon	7.49	34. Morocco	36.42	
8. Cape Verde	0.34	35. Mozambique	2.33	
9. Central African Republic	0.30	36. Namibia	4.08	
10. Chad	0.29	37. Niger	1.34	
11. Comoros	0.15	38. Nigeria	78.19	
12. Congo (Brazzaville)	6.32	39. Reunion	2.82	
13. Congo (Kinshasa)	2.64	40. Rwanda	0.72	
14. Cote d'Ivoire	6.52	41. Senegal	6.19	
15. Djibouti	1.76	42. Seychelles	1.09	
16. Egypt	189.53	43. Sierra Leone	1.33	
17. Equatorial Guinea	4.60	44. Somalia	0.90	
18. Eritrea	0.78	45. South Africa	451.22	
19. Ethiopia	6.88	46. Sudan	12.95	
20. Gabon	4.59	47. Swaziland	1.38	
21. The Gambia	0.44	48. Tanzania	6.70	
22. Ghana	8.10	49. Togo	2.90	
23. Guinea	1.33	50. Tunisia	22.92	
24. Guinea-Bissau	0.46	51. Uganda	1.91	
25. Kenya	11.42	52. Western Sahara	0.31	
26. Lesotho	0.25	53. Zambia	2.67	
27. Liberia	0.71	54. Zimbabwe	10.62	

Europe				4,307.29
1. Albania	4.54	18. Luxembourg	10.58	
2. Austria	69.24	19. Macedonia	7.38	
3. Belgium	137.36	20. Montenegro	1.94	
4. Bosnia & Herzegovina	18.36	21. Netherlands	248.91	
5. Bulgaria	44.49	22. Norway	39.58	
6. Croatia	21.28	23. Poland	285.79	
7. Cyprus	9.46	24. Portugal	56.55	
8. Czech Republic	95.32	25. Romania	80.01	
9. Denmark	49.56	26. Serbia	50.05	
10. Finland	52.15	27. Slovakia	35.75	
11. France	396.65	28. Slovenia	17.38	
12. Germany	765.56	29. Spain	329.86	
13. Greece	100.37	30. Sweden	50.56	
14. Hungary	50.03	31. Switzerland	45.81	
15. Iceland	3.41	32. Turkey	253.06	
16. Ireland	40.27	33. United Kingdom	519.94	
17. Italy	407.87			

Eurasia		2,338.25
1. Armenia		11.35
2. Azerbaijan		34.92
3. Belarus		60.58
4. Estonia		17.56
5. Georgia		5.27
6. Kazakhstan		184.12
7. Kyrgyzstan		5.66
8. Latvia		8.70
9. Lithuania		15.91
10. Moldova		7.04
11. Russia		1,556.66
12. Tajikistan		6.18
13. Turkmenistan		56.55
14. Ukraine		252.47
15. Uzbekistan		115.18

Middle East		1,687.95
1. Bahrain		31.01
2. Iran		528.60
3. Iraq		108.42
4. Israel		70.06
5. Jordan		19.77
6. Kuwait		83.67
7. Lebanon		15.18
8. Oman		48.89
9. Palestine		3.14
10. Qatar		63.64
11. Saudi Arabia		438.25
12. Syria		58.14
13. U.A.E		193.38
14. Yemen		25.78

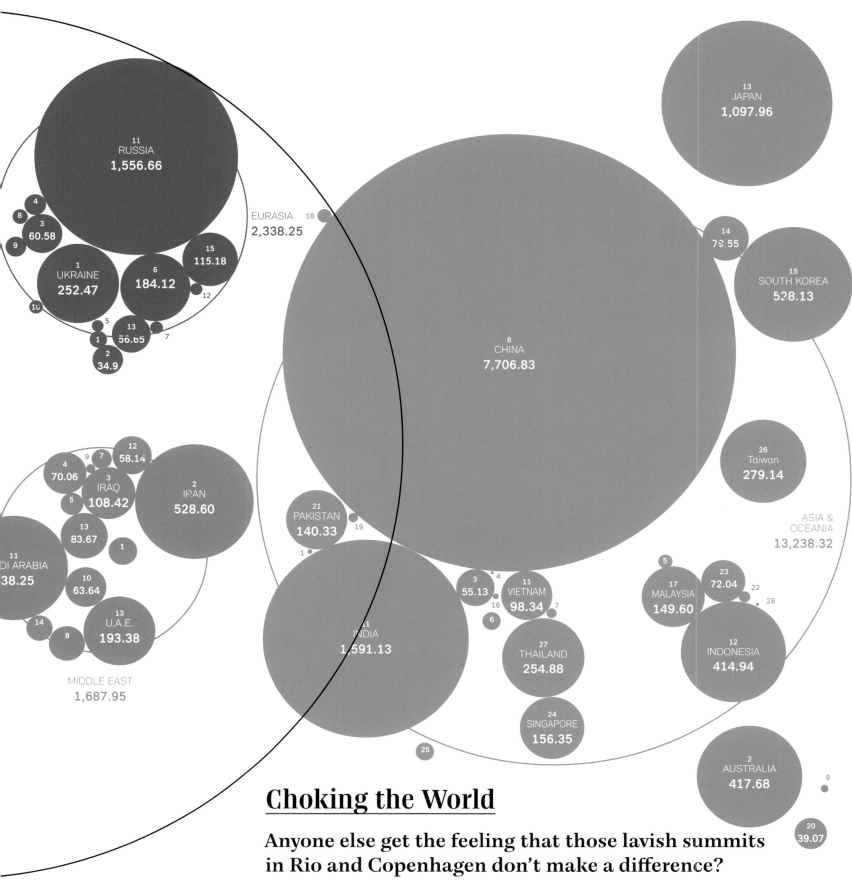

Choking the World

Anyone else get the feeling that those lavish summits in Rio and Copenhagen don't make a difference?

Despite all the negotiations and TV debates, the bottom-line carbon emissions statistics don't seem to change much. It's a depressing fact, but it seems that humanity's urge for growth and expansion is irresistible, no matter how many diplomats argue over emissions trading credits. After decades of painstaking diplomacy, the world's industrial nations just can't agree on a single overarching plan for reducing the emission of greenhouse gases. The United States remains the world's biggest offender compared to size of population, with Russia and the Middle Eastern nations not far behind. Meanwhile, as African and South American countries inevitably expand their consumer classes, they too will begin to choke up the atmosphere in decades to come. And then of course there's China, often blamed for upsetting all the attempts at climate agreements — in its thirst for energy, it seems like the Chinese government just won't rest until it has burned all its vast reserves of coal, and bought as much Middle Eastern oil as it can. In the meantime, the talks go on.

Carbon Emissions

THE WORLD IS NOT ENOUGH   231

Asia & Oceania			13,238.32
1. Afghanistan	0.83	16. Laos	1.24
2. Australia	417.68	17. Malaysia	149.60
3. Bangladesh	55.13	18. Mongolia	7.39
4. Bhutan	0.33	19. Nepal	3.43
5. Brunei	7.58	20. New Zealand	39.07
6. Burma (Myanmar)	12.55	21. Pakistan	140.33
7. Cambodia	3.93	22. Papua New Guinea	4.86
8. China	7,706.83	23. Philippines	72.04
9. Fiji	2.23	24. Singapore	156.35
10. Hong Kong	0	25. Sri Lanka	12.65
11. India	1,591.13	26. Taiwan	279.14
12. Indonesia	414.94	27. Thailand	254.88
13. Japan	1,097.96	28. Timor-Leste	0.38
14. North Korea	79.55	29. Vietnam	98.34
15. South Korea	528.13		

Carbon Equations

It's almost like we're doomed to poison the atmosphere.

You wake up, you answer an email, send a text message, make a cup of tea, and then you go to the local grocery store around the corner, buy an apple, and you go home and eat that apple. Nothing could be more innocent, right? After all, not only are you supporting your local community store, but you're eating something healthy and organic and cholesterol-free that will replenish the nutrients in your body. But no, apparently every little trivial thing we do pumps out more carbon, which slowly traps the heat in the atmosphere, which slowly melts the ice caps and causes more extreme weather conditions, which slowly overwhelm and drown us. We are now, some experts say, officially facing a permanently ice-free world within a hundred years. Enjoy that apple.

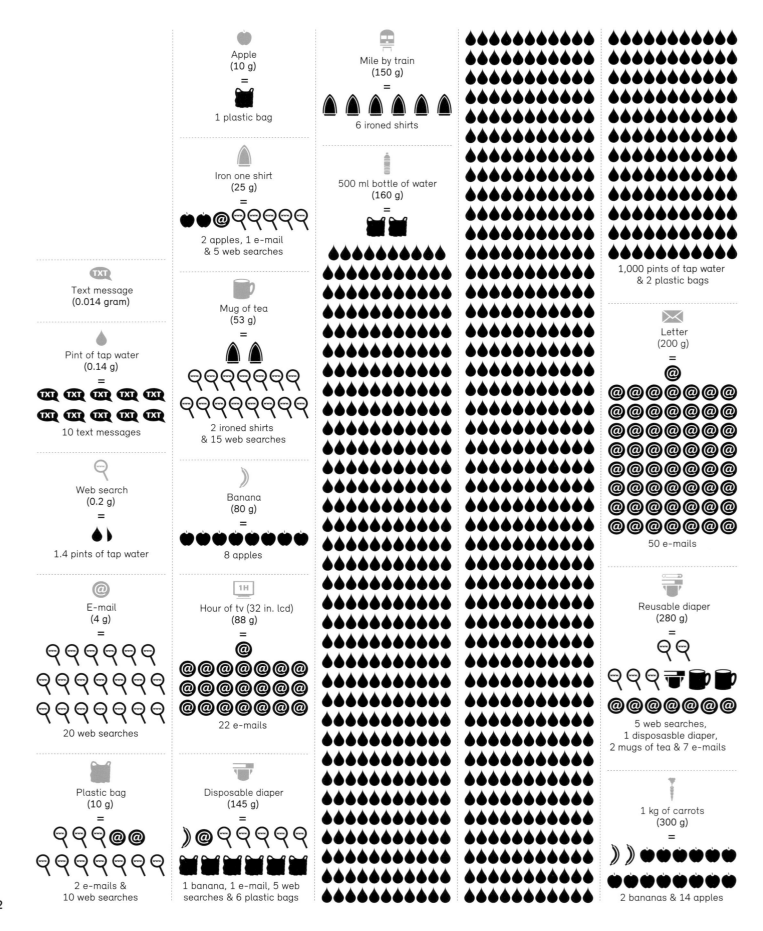

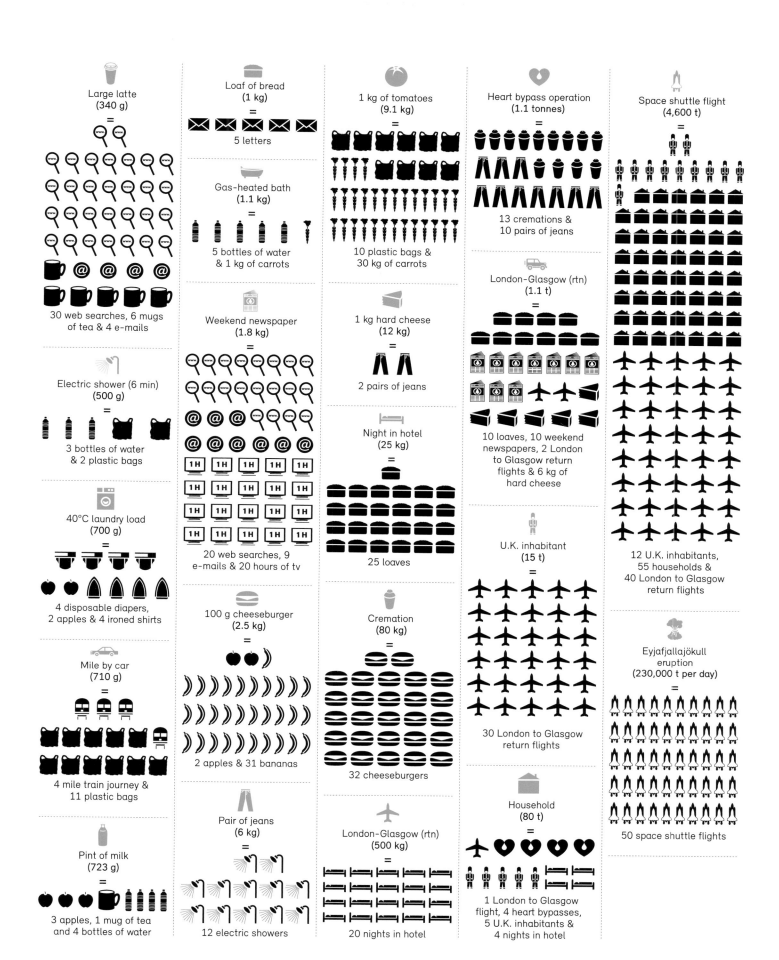

Carbon Footprint

THE WORLD IS NOT ENOUGH

Disaster Costs

How come once-in-a-century floods and hurricanes seem to be happening every other year?

So you thought that the impact of climate change would come sometime in the future, when you're long gone, if at all? In fact, it is already happening. This infographic contains a lot of information, but the bottom line is this: the number of natural disasters that ended up costing the American economy more than $1 billion have doubled compared with the previous 15-year period. The destruction caused by Hurricanes Katrina and Sandy alone — only seven years apart — is estimated to have cost close to $200 billion. That's down to several factors — one of them may be that successive U.S. governments have allowed the country's infrastructure, particularly public transportation, to rot, leaving massive public safety risks. But more than this, the Earth's climate is clearly changing, and the impact is striking the coastal regions of the United States.

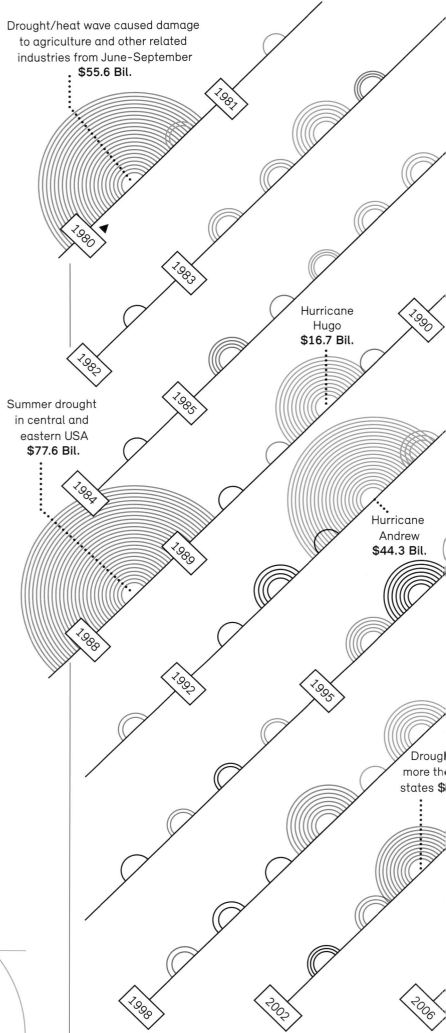

Design: Jennifer Daniel
Published: *Bloomberg Businessweek*

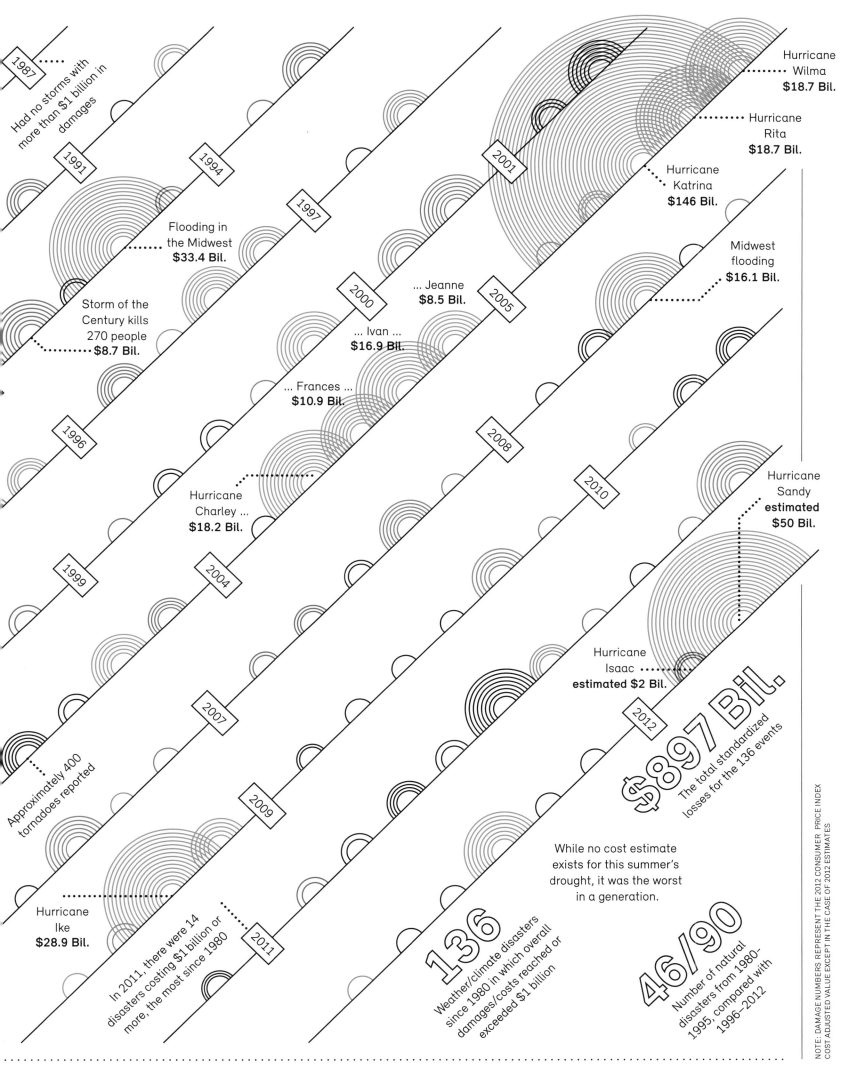

Natural Disasters

THE WORLD IS NOT ENOUGH

Our Growing Plastic Island

There's an unexplored island in the Pacific Ocean six times the size of Texas — but no country wants to claim it.

The problem with plastic is this: while plastic things are often only meant to be used once, the molecules that form plastic are built to last. They're long, strong, and unbreakable — that includes the stuff that goes in the crappy cup you drink that crappy machine coffee from. And the humble plastic shopping bag? A staggering 20,000 of them are made every second, they're used for 15 minutes, but they will most likely survive 20 years. As it turns out, that crappy coffee cup you just threw away could be destined to enter one of the five oceanic "gyres," where the world's plastic garbage swirls around in immense current systems before gathering in "islands," up to 30 meters deep. The largest of these is often called the Great Pacific Garbage Patch, thought to be the size of four Germanys. Once there, your cup will gradually be worn down into smaller and smaller pieces, absorb the sea's toxins, be eaten by fish, and enter our food chain. That's recycling too, but not in a good way.

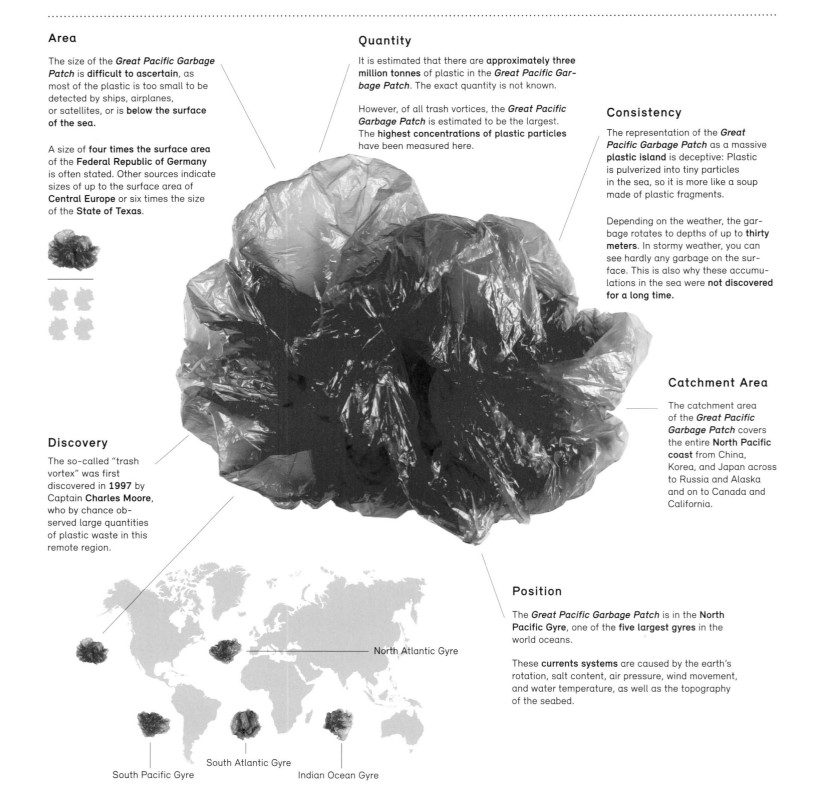

Area

The size of the *Great Pacific Garbage Patch* is **difficult to ascertain**, as most of the plastic is too small to be detected by ships, airplanes, or satellites, or is **below the surface of the sea**.

A size of **four times the surface area** of the **Federal Republic of Germany** is often stated. Other sources indicate sizes of up to the surface area of **Central Europe** or six times the size of the **State of Texas**.

Quantity

It is estimated that there are **approximately three million tonnes** of plastic in the *Great Pacific Garbage Patch*. The exact quantity is not known.

However, of all trash vortices, the *Great Pacific Garbage Patch* is estimated to be the largest. The **highest concentrations of plastic particles** have been measured here.

Consistency

The representation of the *Great Pacific Garbage Patch* as a massive **plastic island** is deceptive: Plastic is pulverized into tiny particles in the sea, so it is more like a soup made of plastic fragments.

Depending on the weather, the garbage rotates to depths of up to **thirty meters**. In stormy weather, you can see hardly any garbage on the surface. This is also why these accumulations in the sea were **not discovered for a long time**.

Discovery

The so-called "trash vortex" was first discovered in **1997** by Captain **Charles Moore**, who by chance observed large quantities of plastic waste in this remote region.

Catchment Area

The catchment area of the *Great Pacific Garbage Patch* covers the entire **North Pacific coast** from China, Korea, and Japan across to Russia and Alaska and on to Canada and California.

Position

The *Great Pacific Garbage Patch* is in the **North Pacific Gyre**, one of the **five largest gyres** in the world oceans.

These **currents systems** are caused by the earth's rotation, salt content, air pressure, wind movement, and water temperature, as well as the topography of the seabed.

North Atlantic Gyre

South Pacific Gyre   South Atlantic Gyre   Indian Ocean Gyre

CAUSES

Worldwide pollution of the oceans has many causes: After plastics, the consumption of tobacco products is the next biggest problem, although these are biologically degraded in a comparatively short period of time.

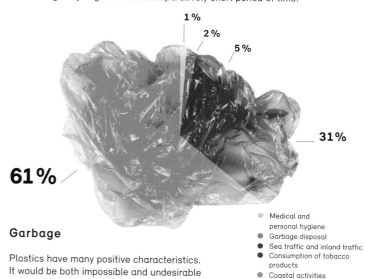

61% Garbage

1%
2%
5%
31%

- Medical and personal hygiene
- Garbage disposal
- Sea traffic and inland traffic
- Consumption of tobacco products
- Coastal activities and leisure activities

Plastics have many positive characteristics. It would be both impossible and undesirable to stop using this material.

However, thoughtless large-scale use has fatal consequences for the environment.

DISTRIBUTION

Light plastics float to the surface, are carried across great distances and reach the coasts. Plastics with a greater density than sea water sink to the seabed. Because of this, what one sees is only the tip of the iceberg.

70%
15%
15%

Seabed

The garbage situation on the seabed has not been researched to any great extent. It is assumed that there are large accumulations of heavy plastic down there, such as PET or HDPE.

It is also feared that the garbage negatively influences the gas exchange between the seabed and the water.

- Arrives at the coasts and beaches
- Moves to the water surface
- Sinks to the seabed

COMPOSITION

Since 1986 the American *Ocean Conservancy* has been organizing the *International Coastal Cleanup*. One day every year in autumn, volunteers all over the world collect garbage from beaches and coasts.

The garbage collected is cataloged in order to get a more exact estimate of actual sea pollution. The following graphs provide an overview of the most common plastic objects and/or garbage in the last 25 years:

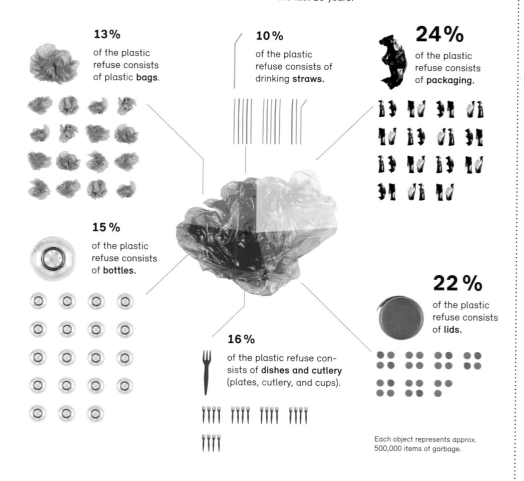

13% of the plastic refuse consists of plastic **bags**.

10% of the plastic refuse consists of drinking **straws**.

24% of the plastic refuse consists of **packaging**.

15% of the plastic refuse consists of **bottles**.

16% of the plastic refuse consists of **dishes and cutlery** (plates, cutlery, and cups).

22% of the plastic refuse consists of **lids**.

Each object represents approx. 500,000 items of garbage.

SERVICE LIFE

Man-made plastic consists of long molecule chains, which are very stable and are not biodegradable. The material, which is often used purely for entertainment purposes, is actually designed to last forever.

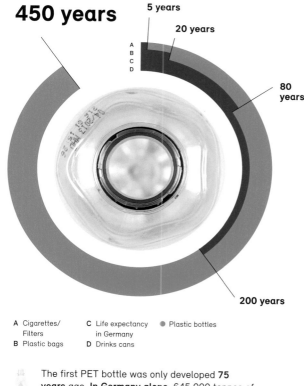

450 years
5 years
20 years
80 years
200 years

A Cigarettes/Filters
B Plastic bags
C Life expectancy in Germany
D Drinks cans
● Plastic bottles

 The first PET bottle was only developed **75 years** ago. **In Germany alone**, 645,000 tonnes of PET were produced. This corresponds to more than **17 billion** 0.5-liter bottles.

Despite its relatively short service life of just **20 years**, the plastic bag is the largest environmental polluter: On average, it is only used for approximately **15 minutes** — and only once.

Plastic

THE WORLD IS NOT ENOUGH

The Heat Is On!

How do humans — and other organisms — cope when the mercury rises? Apart from get out of the kitchen, of course ...

"Mad dogs and Englishmen go out in the midday sun," — as the old Noel Coward song goes. But whatever effect hot temperatures have on canines with mental health issues or western Europe's island people, it's obvious that most people love it when the Sun comes out and they can hang out at the beach all day. But heat is deadly — if your core body temperature rises just four degrees above the normal 37°C, your cells start to die. Two degrees above that, and you start to die. The graphics on this page explore all the ways we have learned to regulate temperature — from technological to evolutionary. As usual, other animals — especially the camel and the gazelle — have developed even more ingenious ways of keeping cool than our messy sweat glands. But of all of them, no one can cope with heat better than single-cell bacteria, some of which can survive temperatures close to 100°C.

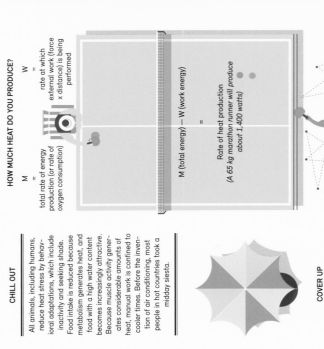

HOW MUCH HEAT DO YOU PRODUCE?

M = total rate of energy production (or rate of oxygen consumption)

W = rate at which external work (force x distance) is being performed

M (total energy) − W (work energy) = Rate of heat production
(A 65 kg marathon runner will produce about 1,400 watts)

Camels can lose up to 25 percent of their body water and remain unaffected. When water is scarce, a camel allows its body temperature to rise by as much as 6°C before it starts sweating, thereby conserving water.

CHILL OUT

All animals, including humans, reduce heat stress by behavioral adaptations, which include inactivity and seeking shade. Food intake is reduced because metabolism generates heat, and food with a high water content becomes increasingly attractive. Because muscle activity generates considerable amounts of heat, manual work is confined to cooler times. Before the invention of air conditioning, most people in hot countries took a midday siesta.

COVER UP

In hot climates, humans regulate clothing, housing, and degree of exposure. Desert people wear several layers of loose-fitting clothes that completely cover the body and act as a heat shield, providing an insulating layer which keeps heat out of the body. Similarly, camels have thick fur, particularly on their backs. Other animals burrow beneath the ground in order to escape the heat, or simply give up the struggle and aestivate (a form of summer hibernation) until temperatures fall.

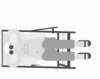

37°C Normal body temperature

41°C Cells begin to be damaged

43°C Lethal body temperature

CORE VALUES

What is commonly meant by body temperature is actually core temperature — that in the tissues of the chest and abdomen. This is kept at about 37°C. Clinically, hypothermia is a temperature below 35°C and hyperthermia above 40°C. Above a core temperature of 42°C, death occurs. Our ability to handle heat relies on thermoregulatory systems that keep our core below 42°C.

LOSING BODY WATER (% OF TOTAL)

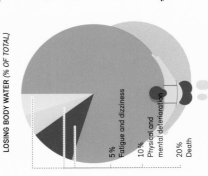

5% Fatigue and dizziness
10% Physical and mental deterioration
20% Death

In 1775 Charles Blagden, the secretary of the Royal Society, and his dog survived for 15 minutes in a room heated to 105°C. The eggs he took in with him were baked hard and a steak was cooked to a crisp. Blagden and his dog were unharmed.

Region	Hottest	Coldest
Antarctica	Vanda Station, Antarctica May 1, 1974 15.0°C	Vostok, Antarctica Jul 21, 1983 −89.2°C
Oceania	Tuguegarao, Philippines Apr 29, 1912 42.2°C	Mauna Kea, Hawaii May 17, 1979 −11.0°C
Europe	Athens, Greece Jul 10, 1977 48.0°C	Ust'-Schugor, Russia Dec 31, 1978 −58.1°C
South America	Rivadavia, Argentina Dec 11, 1905 48.9°C	Sarmiento, Argentina Jun 1, 1907 −32.8°C
Australasia	Oodnadatta, Australia Jan 2, 1960 50.7°C	Charlotte Pass, New South Wales Jun 29, 1994 −23°C
Asia	Tirat Tsvi, Israel Jun 21, 1942 53.9°C	Oimekon, Russia Jun 1, 1907 −32.8°C
North America	Furnace Creek Ranch, US Jul 10, 1913 56.7°C	Snag, Yukon Territory, Canada Feb 3, 1947 −63.0°C
Africa	El Azizia, Libya Jun 21, 1942 53.9°C	Ifrane, Morocco Feb 11, 1935 −23.9°C

CAN'T STAND THE HEAT?

The summer of 2003 was one of the hottest on record in Europe. Britain went barbecue mad. But in France, where temperatures exceeded 40°C for seven days in a row, 14,802 people died from heat-related conditions. Last year, the death rate in Moscow doubled as the worst heat wave in 1,000 years hit the city. The secret of survival in such heat is not to allow your body temperature to rise.

Heat

THE WORLD IS NOT ENOUGH

Desert dwellers build up fat reserves, in case of scarcity of food. As fat is a very good insulator, such reserves are kept in one place. This is why camels have humps, and why the Bushmen of southern Africa have large bottoms.

MINE'S A WATER

The risk of heatstroke is increased if you do not drink enough, as you will then sweat less. Humans can tolerate extreme dry heat, if supplied with sufficient water (and salt) to replace that lost in sweating. It is not so much the temperature but the lack of water that makes the desert a dangerous place.

Honeybees use evaporative cooling to maintain the temperature of their larvae at 35°C, spreading droplets of water over the surface of the honeycomb and fanning their wings to create currents of cooler and drier air.

EXTREME SWEAT LOSS *(millilitres per hour)*

Male rower (32°C) — 2,000
Male rower (10°C) — 1,200
Typical per day — 4

BLOW DRY

Make sure children keep cool, by encouraging them to splash around in paddling pools — the evaporation of the water will help to cool them down. Fans are useful, as they move moisture-laden air away from your sweating skin, so more water can evaporate. If it is very hot and muggy, however, it may be best to seek refuge in an air-conditioned environment such as a shopping mall, library, or museum. Fewer French people might have died in the 2003 heatwave if they had not stayed in their homes but had instead sat out the day in an air-conditioned supermarket.

PASS THE SALT

So if temperatures soar this summer, remember — it's cool to sweat. Make sure you drink plenty of water, of course. Also, if you sweat a great deal, you will lose more salt than usual so you may need to increase (sparingly) your salt intake.

ELECTRICITY USE IN U.S. HOMES *(as a % of total)*

Air-conditioning 15%
Refrigeration 8%

Although no multicellular animals can survive for long above 50°C, single-celled archaea and bacteria can tolerate temperatures close to boiling. They're found in hot springs, in Iceland, Yellowstone Park, and deep-sea volcanic vents.

LEVEL OF UV REFLECTION
Water: <10%
Grass, Soil: 10%
Beach Sand: 15%
Sea foam: 25%
Fresh snow: 80%

STROKE OF BAD LUCK

Heatstroke hits when the body's thermoregulatory system fails and core temperature rises to 41°C or above. Its onset can be rapid. First signs include a flushed face, hot and dry skin, headache and dizziness, followed by confusion and uncoordinated movements. Sweating may cease so that the temperature rises further, up to the fatal 43°C.

DRIEST PLACES ON EACH CONTINENT *(annual precipitation, mm)*

Location	mm
Arica, Chile (South America)	0.76
Wadi Halfa, Sudan (Africa)	2.54
Amundsen-Scott, South Pole (Antarctica)	20.3
Batagues, Mexico (North America)	30.5
Aden, Yemen (Asia)	45.7
Mulka, South Australia (Australasia)	103
Astrakhan, Russia (Europe)	163

SWEAT BUCKETS

Heatstroke often affects old people, as they sweat less. In addition, exercise in a warm climate is a common cause of heatstroke. In sedentary people, heatstroke is usually a result of impaired sweating, which is often preceded by an inflammation of the sweat glands that causes prickly heat, which is characterised by itchy red pimples.

Who Built the Ark?

Why bother saving the climate if we can't save the bees?

There was a time, a few years ago, when conservationists concentrated their campaign efforts on a few beautiful, majestic species: Save the Tiger, Save the Elephant, Save the Whale. "Why should we save them when there's so much else wrong with the world?" a few misanthropes would ask. And the answer would always be, "Erm, because they're beautiful and majestic." Now though, we know there are more urgent — if not exactly better — reasons. We know that the various creatures and plants on the Earth are more interdependent than we realized, and that our ecosystem is more fragile than we imagined. Consequently, environmentalists have been telling us for a while that the threat to biodiversity is at least as much of a danger to the world as climate change. If all the bees die, we really don't know where all the fruit, the coffee, the cotton is going to come from. This graphic delves into the causes and consequences of the threat to global biodiversity — a word that most people hadn't even heard of a couple of years ago.

HOW RICH IS THE BARRIER

In the map, the places where the diversity of living species (marine and terrestrial) is greatest. High levels of biodiversity are recorded mainly in the equatorial forests and along the Great Barrier Reef.

WHAT WILL HAPPEN IF WE PRESERVE...
- protected hotspots
- all hotspots
- all tropical forests

FAMILIES THAT ARE BORN AND DIE
600 million years of evolution
- origin
- extinction

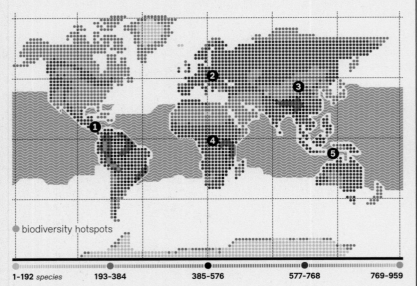

WARNING, VULNERABLE INSECTS

The IUCN (International Union for Conservation of Nature) periodically assesses the species potentially at risk. Here are the percentages of those endangered according to the 2008 Red List.

VERTEBRATES
- **Mammals** 21% — 1,141 out of 5,488 species assessed
- **Birds** 12% — 1,222 out of 9,990 species assessed
- **Reptiles** 31% — 423 out of 1,385 species assessed
- **Amphibians** 30% — 1,905 out of 6,260 species assessed
- **Fish** 37% — 1,275 out of 3,481 species assessed

INVERTEBRATES
- **Insects** 50% — 626 out of 1,259 species assessed
- **Molluscs** 44% — 978 out of 2,212 species assessed
- **Crustaceans** 35% — 606 out of 1,735 species assessed
- **Corals** 27% — 235 out of 856 species assessed

① Americas — Silent epidemic from the pole to the tropics

Ⓐ SEA OTTER — *Enhydra lutris*
The reduction of the populations of sea otters causes the proliferation of urchins, one of their main prey. Too many urchins deplete kelp and macroalgae reserves.

Ⓑ GIANT ARMADILLO — *Priodontes maximus*
A mammal symbol of the Amazonian ecosystem, classified as endangered on the IUCN red list. Indiscriminate hunting and reduction of habitats are threatening it.

Ⓒ GRIZZLY BEAR — *Ursus arctos horribilis*
In sub-polar ecosystems such as Alaska the grizzly bear is at the top of the food chain. An "umbrella species": on its survival depends that of many others.

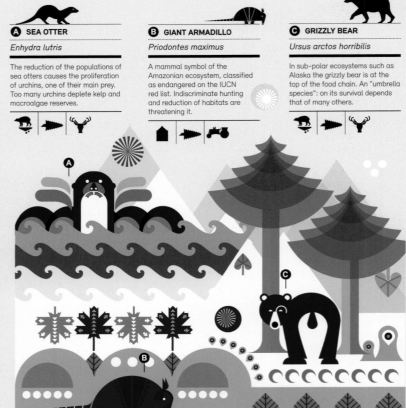

ALL THE CAUSES OF EXTINCTION
Are also indicated in the charts on these pages with reference to the 15 animals analyzed.

URBANIZATION — The growth of urban areas damages and fragments many habitats.

POLLUTION — A serious problem for marine species, amphibians, and insects, sensitive to chemical agents.

ALIEN SPECIES — The artificial introduction of a species in an ecosystem can destroy its equilibrium.

CLIMATE CHANGE — Affects marine species sensitive to temperature and migratory birds.

2 Europe — *The howling returns, the frogs suffer*

A BEE
Apis mellifera

Not yet at risk, but in rapid decline: among the causes are pollution, pesticides, and parasitic epidemics. Their extinction would be a disaster for pollination, on which the life of flowers and fruit depends.

B ITALIAN WOLF
Canis lupus italicus

Prime example of the success of protection strategies, the wolf has returned to populate the Apennines and vast areas of Europe. Poaching was the primary threat to the species.

C PELOBATES FUSCUS
Pelobates fuscus insubricus

The subspecies of pelobates italico, typical of the Po Valley, like many other amphibians suffers the destruction of wetland habitats and is preyed upon by artificially introduced species such as turtles and bullfrogs.

3 Asia — *Where the tigers reigned*

A TIGER
Panthera tigris

It is the symbol species of the fight for biodiversity: there are 3 thousand adults left. Three of the nine subspecies are extinct. A curiosity: 2010, according to the Chinese calendar, is precisely its year.

B GIANT PANDA
Ailuropoda melanoleuca

The total population is somewhere between one thousand and two thousand individuals. It is considered "in danger of extinction." The threats: habitat destruction and hunting for the precious skin.

C TIBETAN ANTELOPE
Pantholops hodgsonii

The population of this species that lives in the Tibetan highlands is 100 thousand heads. It is hunted for its fine wool, the shatoosh. The horns are used in Chinese medicine.

4 Africa — *The destiny of "charismatic" species*

A BLACK RHINOCEROS
Diceros bicornis

Widespread in African savannas up to the mid twentieth century, this mammal has come to the brink of extinction. The horns are used in traditional Chinese medicine.

B LION
Panthera leo

Its extension, once including the whole of Eurasia, has been reduced to sub-Saharan Africa. The IUCN considers it "vulnerable": its reduction threatens the equilibrium of the ecosystem.

C AFRICAN ELEPHANT
Loxodonta africana

Experts estimate that in the wild there are about 400 thousand specimens compared to 1.3 million in the seventies. Poaching for ivory trade is showing a resurgence.

5 Oceania — *If the geography of the atolls changes*

A CORALS
Scleractinia

The order includes 26 families. The degradation of the Great Barrier Reef threatens an entire ecosystem: fish, molluscs, seahorses. Effects also on the geography of the atolls.

B CASSOWARY
Casuarius casuarius

Not only predators but also herbivores may be in danger: the cassowary is crucial because it disperses, through its feces, the seeds of the plants on which it feeds.

C TASMANIAN DEVIL
Sarcophilus harrisii

This marsupial, once common in Australia, today lives only in Tasmania. Endangered due to hunting and the spread of an infectious facial tumor, is threatened with extinction.

ALTERNATIVE MEDICINE — Traditional Chinese medicine still uses parts of animals, especially bones and horns.

LOSS OF HABITAT — Deforestation and habitat reduction strangle biodiversity.

AGRICULTURAL DEVELOPMENT — The expansion of agricultural areas causes the loss of indigenous forests and species.

HUNTING AND FISHING — Poaching and illegal trafficking of animal parts are a global plague.

Animal Biotopes
THE WORLD IS NOT ENOUGH

Water Wheels

The Earth's water turns in vast global cycles of precipitation and evaporation.

One little oxygen atom and two tiny hydrogen atoms: water is a simple, yet deeply mysterious chemical that is essential for life on Earth. It exists in three different states — ice, liquid, and vapor — at least two of which it passes through in slow, immense cycles that gradually replace the vast resources stored in the oceans, the polar ice caps, or in the water tables. Our own water supply usually comes from the ground, and as this graphic shows, increased urbanization has resulted in a depletion of nearby groundwater and the necessity to bring water in from increased distances. Meanwhile, water scarcity remains an acute global problem. There have been improvements in recent decades, but the depressing fact remains that of the seven billion people living on the planet, as many as one billion still don't have access to safe drinking water — and, according to a United Nations "Millennium Development Goals" report from 2008, as many as 2.5 billion people do not have access to hygienic sanitation. It has been estimated that the world needs about 17 percent more water to provide its population with a safe, clean supply.

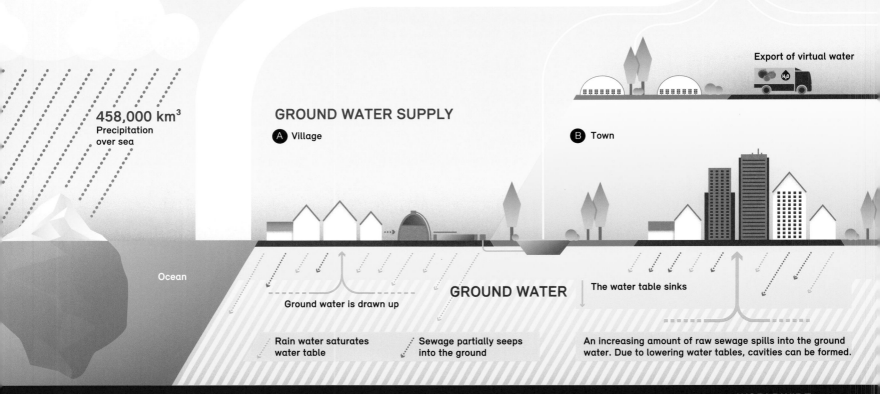

WATER CYCLE

- 50,000 km³ Evaporation on sea
- 74,000 km³ Evaporation on land
- 458,000 km³ Precipitation over sea
- Export of virtual water

GROUND WATER SUPPLY
A Village B Town

GROUND WATER
- Ground water is drawn up
- The water table sinks
- Rain water saturates water table
- Sewage partially seeps into the ground
- An increasing amount of raw sewage spills into the ground water. Due to lowering water tables, cavities can be formed.

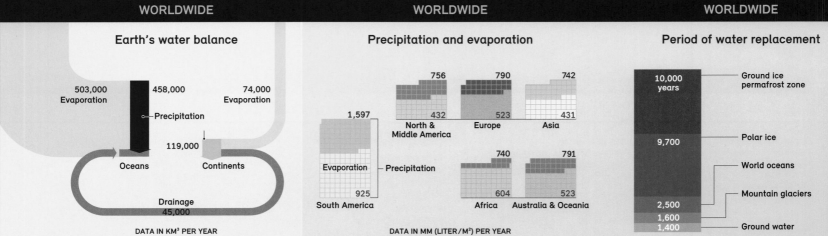

WORLDWIDE — Earth's water balance
- 503,000 Evaporation
- 458,000 Precipitation
- 74,000 Evaporation
- 119,000
- Oceans — Continents
- Drainage 45,000

DATA IN KM³ PER YEAR

WORLDWIDE — Precipitation and evaporation

Region	Evaporation	Precipitation
North & Middle America	432	756
Europe	523	790
Asia	431	742
South America	925	1,597
Africa	604	740
Australia & Oceania	523	791

DATA IN MM (LITER/M²) PER YEAR

WORLDWIDE — Period of water replacement
- Ground ice permafrost zone: 10,000 years
- Polar ice: 9,700
- World oceans: 2,500
- Mountain glaciers: 1,600
- Ground water: 1,400

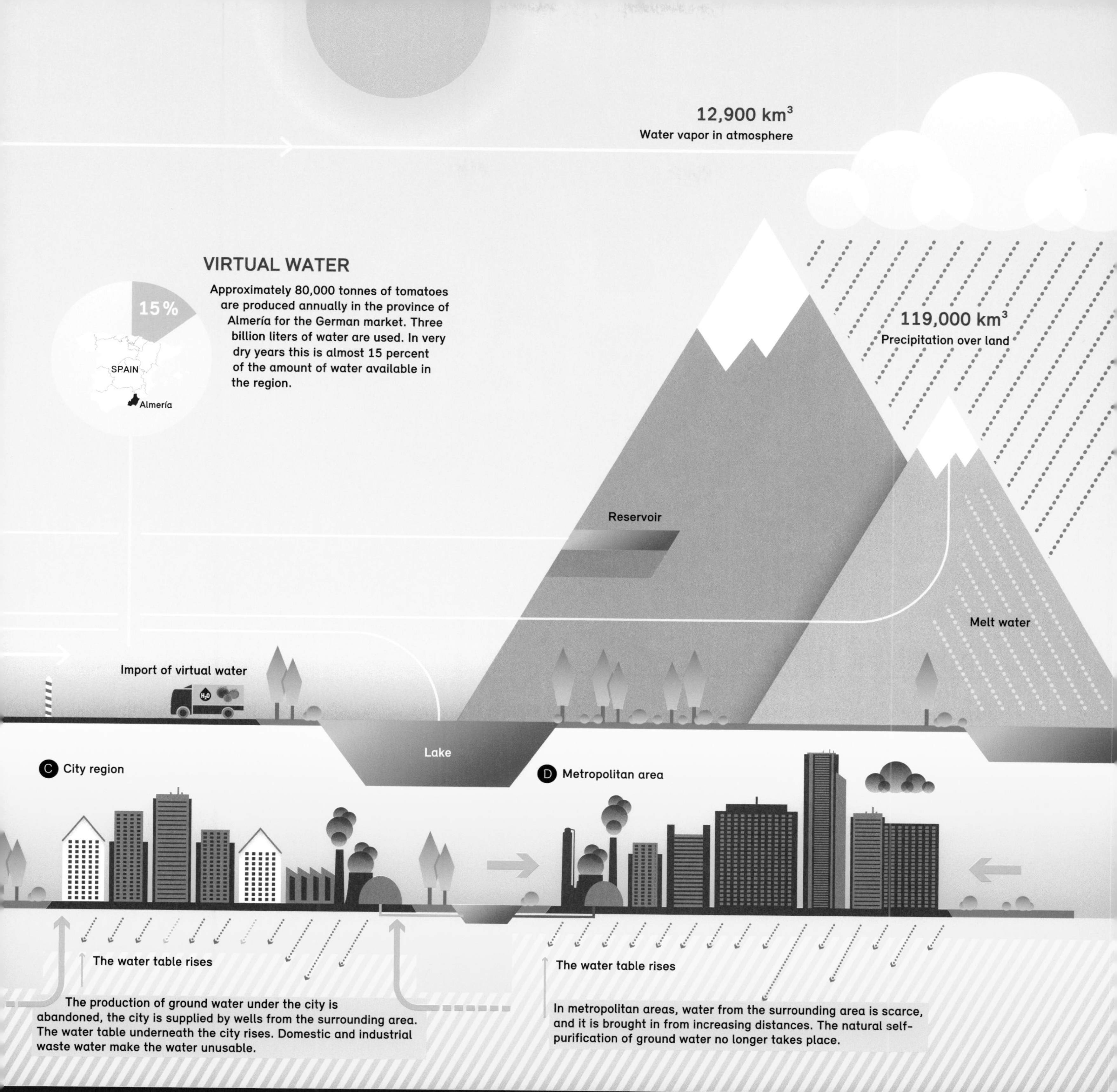

12,900 km³
Water vapor in atmosphere

VIRTUAL WATER

Approximately 80,000 tonnes of tomatoes are produced annually in the province of Almería for the German market. Three billion liters of water are used. In very dry years this is almost 15 percent of the amount of water available in the region.

15%

SPAIN
Almería

119,000 km³
Precipitation over land

Reservoir

Melt water

Import of virtual water

Lake

C City region

D Metropolitan area

The water table rises

The water table rises

The production of ground water under the city is abandoned, the city is supplied by wells from the surrounding area. The water table underneath the city rises. Domestic and industrial waste water make the water unusable.

In metropolitan areas, water from the surrounding area is scarce, and it is brought in from increasing distances. The natural self-purification of ground water no longer takes place.

WORLDWIDE

Access to clean water and sanitation

- More than 95% of the rural population have access
- Good supply, above the global average
- Improved sanitary conditions for 60 to 70%

No access to clean water/sanitation
- More than 50%
- 30 to 50%
- 10 to 30%

- No information

WORLDWIDE

The thirst of the large cities

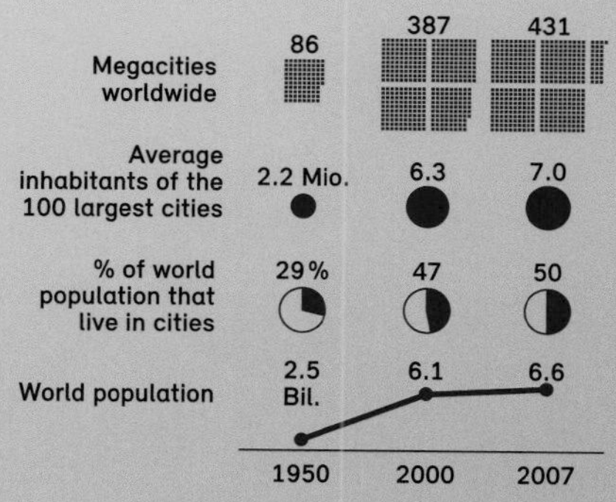

Megacities worldwide	86	387	431
Average inhabitants of the 100 largest cities	2.2 Mio.	6.3	7.0
% of world population that live in cities	29%	47	50
World population	2.5 Bil.	6.1	6.6
	1950	2000	2007

Water

THE WORLD IS NOT ENOUGH

1,338,000,000 km³ | 96.5%
Salt water in the oceans

And Not a Drop to Drink

The world is covered with water — but only three percent of it is drinkable, and clean, safe water is even rarer.

Do you leave the faucet running while you're brushing your teeth? Or maybe you like to add a little more hot water to the bathtub after you've been lying there a while? That's because you can afford to. But one in five people in the world already live in areas where water scarcity is becoming a problem, and managing water resources is likely to become a key sustainability issue in the future. In 2010, the United Nations reacted to the impending problem by making access to water a human right, but that has done nothing to halt the increasing privatization of the supply, raising the prospect that the world's water will one day be controlled by major corporations for profit. The graphics on the next three pages explore some of the problems associated with managing the world's most precious resource.

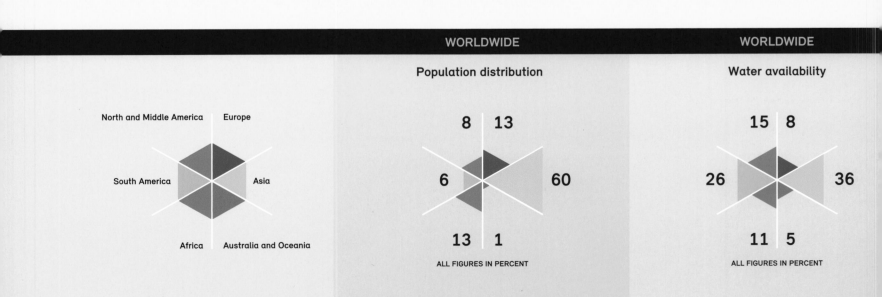

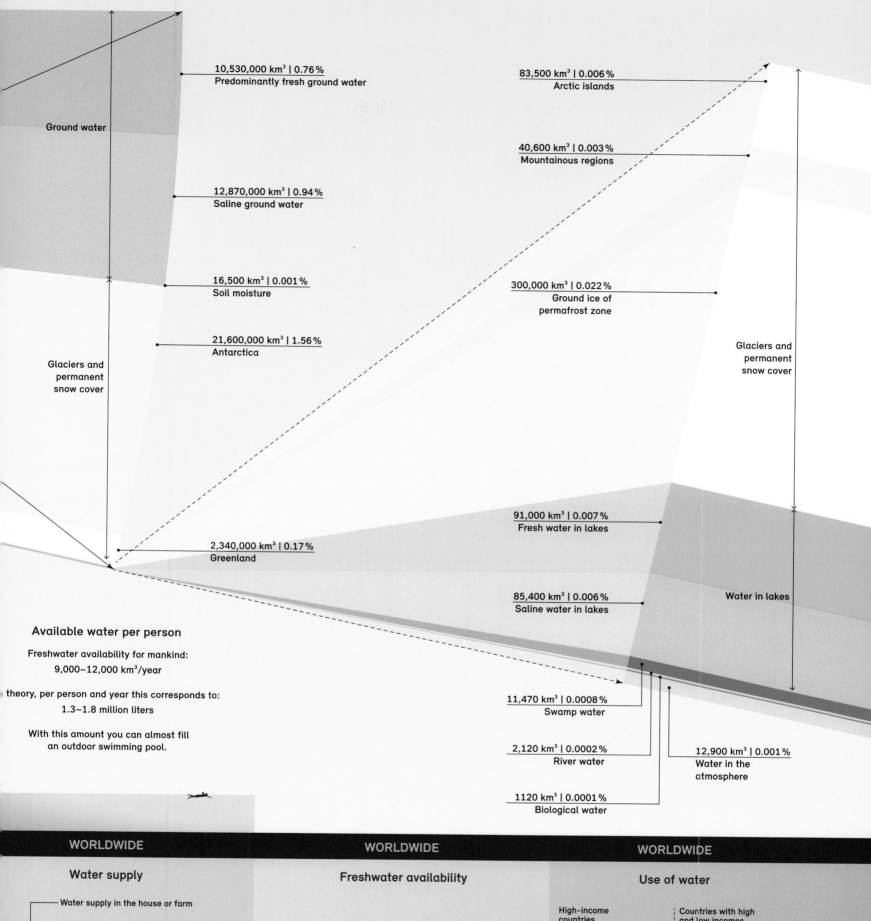

Ground water
- 10,530,000 km³ | 0.76% Predominantly fresh ground water
- 12,870,000 km³ | 0.94% Saline ground water
- 16,500 km³ | 0.001% Soil moisture

Glaciers and permanent snow cover
- 21,600,000 km³ | 1.56% Antarctica
- 2,340,000 km³ | 0.17% Greenland

- 83,500 km³ | 0.006% Arctic islands
- 40,600 km³ | 0.003% Mountainous regions
- 300,000 km³ | 0.022% Ground ice of permafrost zone

Glaciers and permanent snow cover

Water in lakes
- 91,000 km³ | 0.007% Fresh water in lakes
- 85,400 km³ | 0.006% Saline water in lakes

- 11,470 km³ | 0.0008% Swamp water
- 2,120 km³ | 0.0002% River water
- 12,900 km³ | 0.001% Water in the atmosphere
- 1120 km³ | 0.0001% Biological water

Available water per person

Freshwater availability for mankind: 9,000–12,000 km³/year

theory, per person and year this corresponds to: 1.3–1.8 million liters

With this amount you can almost fill an outdoor swimming pool.

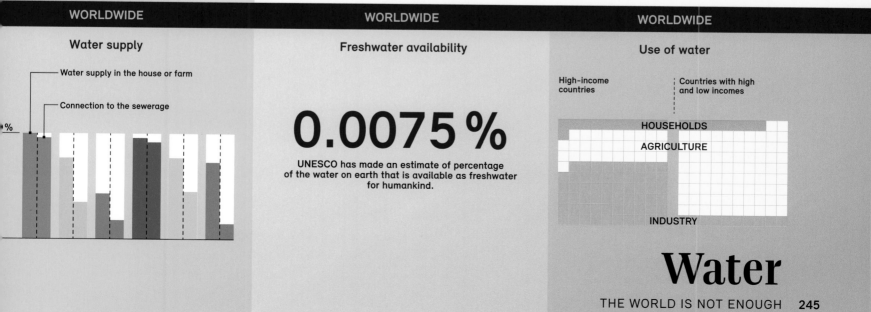

WORLDWIDE — Water supply
- Water supply in the house or farm
- Connection to the sewerage

WORLDWIDE — Freshwater availability

0.0075 %

UNESCO has made an estimate of percentage of the water on earth that is available as freshwater for humankind.

WORLDWIDE — Use of water

High-income countries | Countries with high and low incomes

- HOUSEHOLDS
- AGRICULTURE
- INDUSTRY

Water
THE WORLD IS NOT ENOUGH

Insight into the structure of the Danube basin contracts

In 1862, Austria and Bavaria agreed on the exact boundary line along the river Danube. Since then there have been 17 more agreements to regulate the boundaries and the use of boundary waters for water supply and shipping.

Drawing a Line in the Water

Appeasing the conflicts over our water supply.

If a state downriver suffers flood damage because an upriver state fails to build adequate flood protection — who is liable? Should a downriver state pay for barriers not situated on its own territory? Whether it pools in lakes or courses along rivers, water causes so many legal headaches because it respects no boundaries. The Danube Basin in central and southeastern Europe, for instance, encompasses some 17 countries. The consequences for international diplomacy can also be significant — one reason why Russia and Iran maintain good diplomatic ties is because they both have an interest in keeping former Soviet states like Azerbaijan, Kazakhstan, and Turkmenistan from claiming a share of the Caspian Sea and its oil and gas reserves. Regional peace is so often dependent on complex legal agreements and treaties to balance such interests.

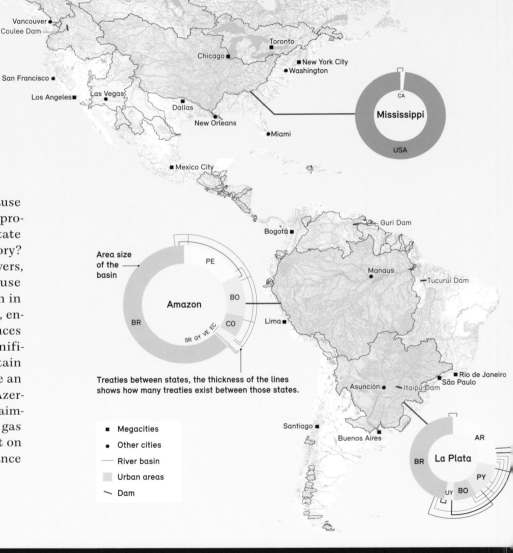

Treaties between states, the thickness of the lines shows how many treaties exist between those states.

- Megacities
- Other cities
- River basin
- Urban areas
- Dam

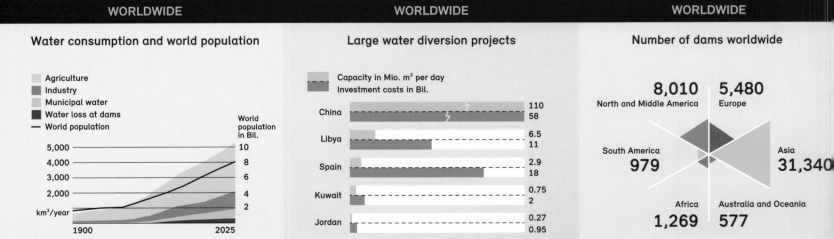

WORLDWIDE
Water consumption and world population

- Agriculture
- Industry
- Municipal water
- Water loss at dams
- World population

WORLDWIDE
Large water diversion projects

Capacity in Mio. m³ per day
Investment costs in Bil.

	Capacity	Investment
China	110	58
Libya	6.5	11
Spain	2.9	18
Kuwait	0.75	2
Jordan	0.27	0.95

WORLDWIDE
Number of dams worldwide

- North and Middle America: 8,010
- Europe: 5,480
- South America: 979
- Asia: 31,340
- Africa: 1,269
- Australia and Oceania: 577

246

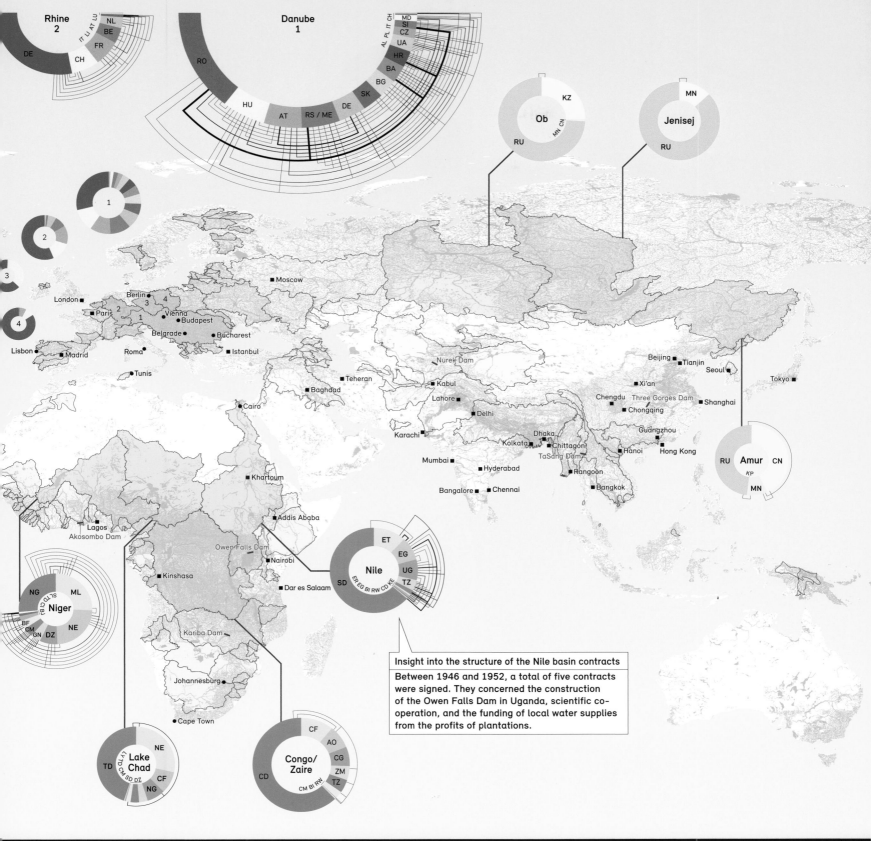

Insight into the structure of the Nile basin contracts

Between 1946 and 1952, a total of five contracts were signed. They concerned the construction of the Owen Falls Dam in Uganda, scientific co-operation, and the funding of local water supplies from the profits of plantations.

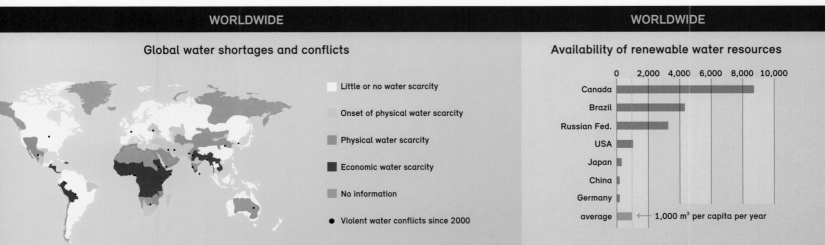

WORLDWIDE

Global water shortages and conflicts

- Little or no water scarcity
- Onset of physical water scarcity
- Physical water scarcity
- Economic water scarcity
- No information
- • Violent water conflicts since 2000

WORLDWIDE

Availability of renewable water resources

| | 0 | 2,000 | 4,000 | 6,000 | 8,000 | 10,000 |

- Canada
- Brazil
- Russian Fed.
- USA
- Japan
- China
- Germany
- average ← 1,000 m³ per capita per year

Water
THE WORLD IS NOT ENOUGH

OUR GREATEST IDEAS

The creators of some of the most influential ideas in history remain anonymous.

We don't know who had the idea for the oil lamp. We don't know who developed the first suspension bridge (though we're pretty sure that it was someone in what is now called China).

Whether or not their names matter is perhaps secondary to the wider message of our selection for this final chapter: that humans have made some pretty incredible technological leaps, which took place thanks to the platforms of knowledge and ideas built by incredible minds before and around those events.

The Large Hadron Collider is a remarkable invention, but it didn't materialize through the dramatic intervention of a God particle — it came about thanks to a confluence of scientific consensus, political will, and a vast number of tiny advances in engineering and theoretical physics.

For what is invention but the next step on the path that was begun long ago, for better or worse? Each step is not inevitable, just as our current state of technology wasn't always going to be where humanity would reach. We choose a direction, and we follow our nose. Sometimes the consequences — such as those that come from living in a fossil-fuel based world — are mixed. Gunpowder enables construction and destruction. Any important invention does the same.

As a society, we celebrate — through our history books and the naming of elements, theories, and equipment — certain people who we have chosen to recognize as having made significant leaps on our behalf. The most famous of these are Nobel Prize winners, who each year are chosen by small anonymous committees in Sweden and Norway, paid for and named after the inventor of dynamite.

Each era has its technological peaks, new heights reached through the science of the age. Right now we are in the midst of the digital wave, and the Internet is the sexiest new creation for the Western world, one that claims to revolutionize lives, at least for those who have easy, reliable access to its wonders. Twenty years ago, and in twenty years time, the achievements we might choose to celebrate will probably be quite different. That's the important thing about brilliant ideas — they only prove to be truly brilliant if they inspire others to have more of them. We hope we have inspired you with these reflections of the world around you. Now get thinking.

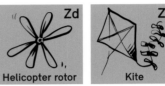 
      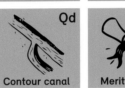 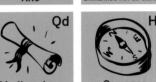

Nc	Neolithic China (before c. 1600 BC)	
Shd	Shang Dynasty (c. 1600–1050 BC)	
Zd	Zhou Dynasty (1034–246 BC)	
Wsp	Warring States Period (403–221 BC)	
Qd	Qin Dynasty (221–206 BC)	
Hd	Han Dynasty (202 BC–220 AD)	
Jd	Jin Dynasty (265–420)	
SNd	Southern and Northern Dynasties (386–581)	
Sud	Sui Dynasty (581–618)	
Td	Tang Dynasty (618–907)	
5d10	Five Dynasties and Ten Kingdoms (907–979)	
Sd	Song Dynasty (960–1279)	
Yd	Yuan Dynasty (1260–1368)	
Md	Ming Dynasty (1368–1644)	
Qgd	Qing Dynasty (1644–1911)	
Pqd	Post-Qing (after 1911)	

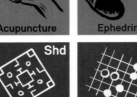

Essential Elements

How China invented our civilization.

Get up, go to the toilet, brush your teeth, post a letter, make a cup of tea, salt your egg, stick a fork in it. You've already used six or seven Chinese inventions and you've not even had your breakfast. The minutiae of our lives would look so different if it hadn't been for that ingenious ancient civilization. The Chinese can also claim responsibility for so many world-changing technologies too — such as paper-making and gunpowder. More recently, the USA usurped the role of the dominant power in the twentieth century, while China slumbered under communist rule. Now capitalist ambition has been added to the potent potion of centralized state control, and China is set to become the world's leading power in this century.

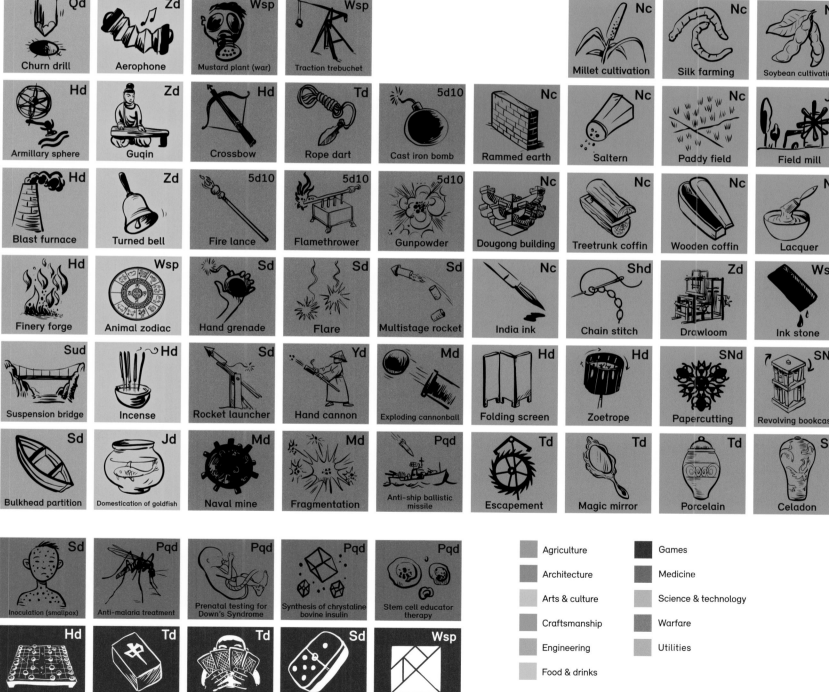

Inventions

OUR GREATEST IDEAS

Noble Winners of the Nobel

The Nobel Prizes analyzed by age and academic achievement.

Guglielmo Marconi remains unique in world history. In 1909, the 35-year-old Italian — inventor of the radio — won the Nobel Prize for Physics without having earned any academic degree beforehand. Other Nobel laureates have made themselves conspicuous on this graphic, which arranges every winner on a scale of age, and by academic degree — for instance, there's Elinor Ostrom, the only woman to be awarded the Nobel Prize in Economics, and Erik Axel Karlfeldt, the only one to be actually dead when he won the Nobel Prize for Literature. Another notable winner is New York City, the hometown of some 51 laureates — more than double its closest rival Paris, which only musters 23. On the right, the world's major universities feed their graduates into the Nobel machine — with the noble exception of self-taught Marconi.

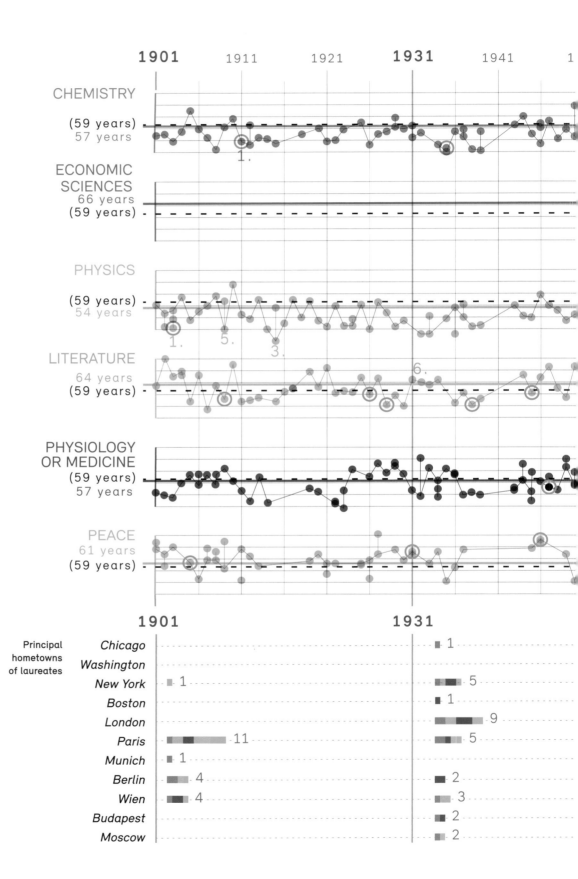

HOW TO READ IT
Each dot represents a Nobel laureate, each recipient is positioned according to the year the prize was awarded (x axis) and age of the person at the time of the award (y axis).

Average age for each for each CATEGORY

Average age of Nobel laureate

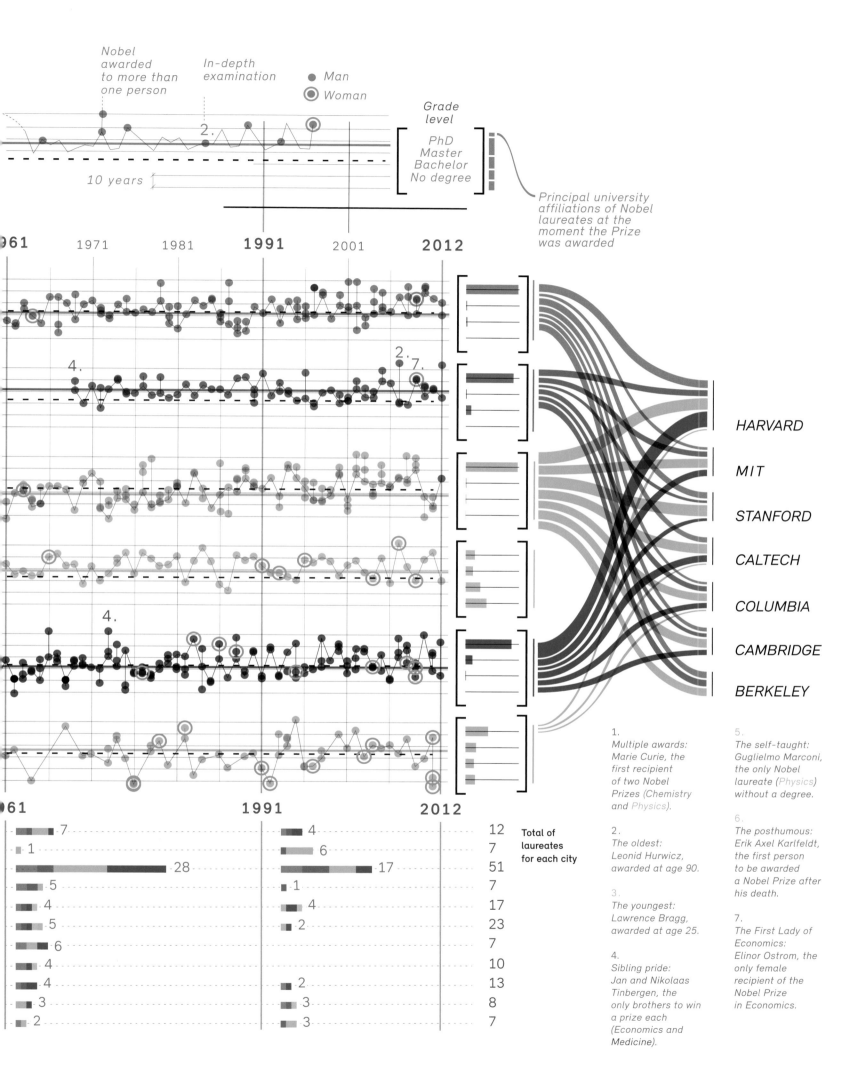

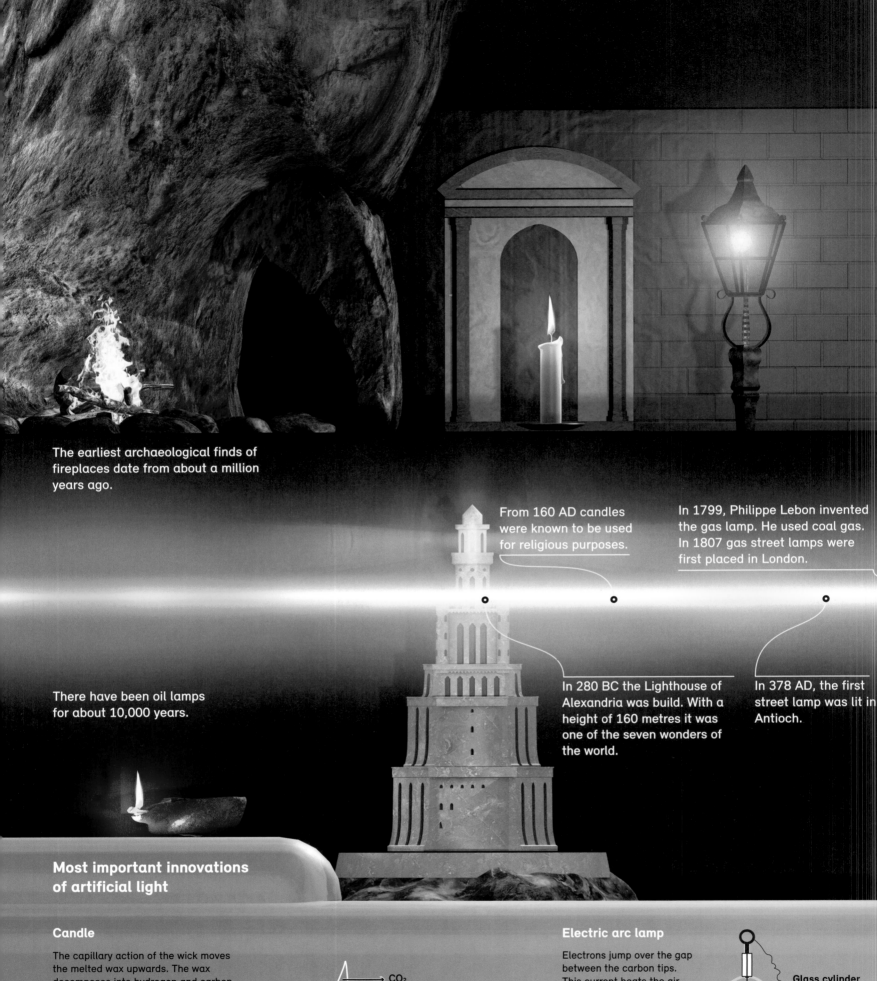

The earliest archaeological finds of fireplaces date from about a million years ago.

From 160 AD candles were known to be used for religious purposes.

In 1799, Philippe Lebon invented the gas lamp. He used coal gas. In 1807 gas street lamps were first placed in London.

There have been oil lamps for about 10,000 years.

In 280 BC the Lighthouse of Alexandria was build. With a height of 160 metres it was one of the seven wonders of the world.

In 378 AD, the first street lamp was lit in Antioch.

Most important innovations of artificial light

Candle

The capillary action of the wick moves the melted wax upwards. The wax decomposes into hydrogen and carbon. This reacts in the flame with oxygen to form water and carbon dioxide and the heat of the flame melts the wax again.

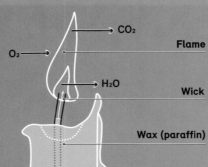

CO_2
O_2
Flame
H_2O
Wick
Wax (paraffin)

Electric arc lamp

Electrons jump over the gap between the carbon tips. This current heats the air inside the glass cylinder to such an extent that it strikes an arc and the carbon tips begin to glow.

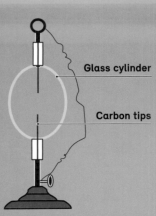

Glass cylinder

Carbon tips

254

From titans to light-bulbs — the eternal struggle to shed light on our lives.

Light couldn't be more basic. That's one of the few things that scientists and the authors of the Bible can agree on. Apart from being an essential element in photosynthesis and the cycle of vegetable life, it is the first thing God decides He needs. He's got to see what he's doing, after all. But of course the same theme emerged in the pagan Greek religions that preceded the Judeo-Christian God's tale, most famously in the story of Prometheus. But maybe even that titan's rebellious efforts and his subsequent tortures at the beak of a liver-loving eagle were all in vain, for scientists tell us that we, that is, the homo sapiens, were not the first creatures to catch and control fire, and so claim some independence from nature. In fact, the earliest evidence for fireplaces dates back some one million years, when other members of the homo genus were still fighting for a spot to call their own. It is thought that homo erectus was the first human species to control fire. After that it took at least another 990,000 years, by which time we humans had established ourselves as the dominant ape species, for the next leap in technology to catch on: the oil lamp. From there, our dark history was forever punctuated with more frequent technological advances in the history of light.

In 1808 Humphry Davy invented the electric arc lamp that was perfected in 1848 by Jean Foucault.

The Polish chemist Ignacy Łukasiewicz is considered to have invented the kerosene lamp in 1853.

In 1878, Thomas Alva Edison invented the carbon filament lamp in the USA.

1850   1900

In 1862, Friedrich Wöhler discovers acetylene for portable light sources like bicycle lamps.

In 1899, David Misell invented the flashlight. The name affirms the extremely short life of the batteries available.

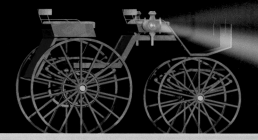

Kerosene lamp

Similar to the candle, kerosene moves through the wick to the flame and reacts. The length of the wick and the oxygen supply control the brightness and heat of the flame.

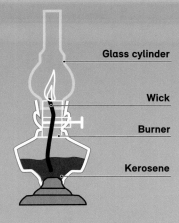

- Glass cylinder
- Wick
- Burner
- Kerosene

Carbon filament lamp

In a glass bulb, an electric current flows through a thin carbon filament that begins to glow; it cannot burn as there is no oxygen. Modern light bulbs use a tungsten filament.

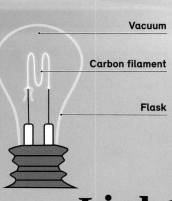

- Vacuum
- Carbon filament
- Flask

Light
OUR GREATEST IDEAS

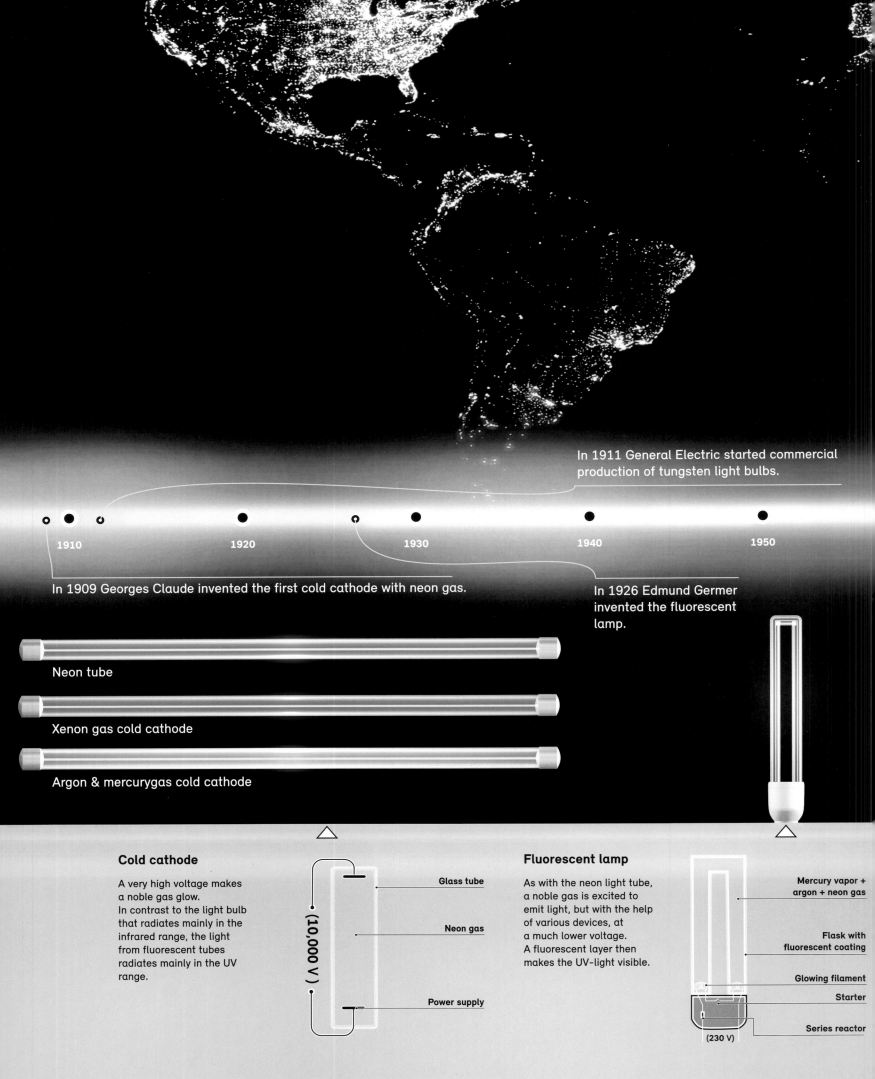

The Light of Science

Light does so much more than help us read in bed.

From space, it looks as if humanity is making a serious attempt to mirror the stars. There seem to be as many lights shining up into the universe as the universe has shining down on us. What is obvious is that the modern world has created many more purposes for light than just to help find the way on a dark night or to signal the position of a harbor to passing ships. As the technology of light has advanced, we have learned more about the complex phenomenon that is light, and scientists and engineers have invented innumerable ways of generating it. At the same time, our efforts to separate light from heat have created new ways to conserve energy. The less heat, the more energy can be invested in creating light. More than all this, light is now the principal means by which we send information — flashing millions of pulses of it every second through fiber-optic cables across continents. By laying those cables across the bottom of the ocean, we have taken that light to the darkest corners of the Earth.

Since 2001 LED spot lights are on cars.

In 2007, LED lamps were sold with a standard fitting for the first time.

1970   1980   1990   2000   2010

In 1959, General Electric patented the halogen lamp.

In 1926, Nick Holonyak developed the first usable light emitting diode (LED). But only in the 1990s was blue and white LED available.

In 1985, Osram launched the first compact fluorescent lamp, or energy-saving lamp.

Halogen bulb

Basically a tungsten light bulb filled with a halogen gas. When tiny amounts of tungsten filament evaporate, the gas prevents deposition on the glass cylinder.

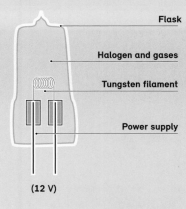

Flask
Halogen and gases
Tungsten filament
Power supply

(12 V)

Light-emitting diode

A semiconductor material, usually gallium is doped, so polluted with tiny amounts of another substance.
When current flows through the diode, this material emits light.

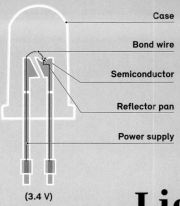

Case
Bond wire
Semiconductor
Reflector pan
Power supply

(3.4 V)

Light
OUR GREATEST IDEAS

Submarine Cable Map 2013

Wiring the World

Information is power — but it also needs a lot of hardware.

There is no escaping the matrix. The matrix is everywhere. Even at the bottom of the ocean. In fact, especially at the bottom of the ocean. For the past century, ships have been sent out to wire the world, reeling out thousands of kilometers of cable, stretched out miles deep across the ocean floor. These fiber-optic cables consist of eight glass cords wound together, and with the help of laser technology, billions of light pulses are flashed along them in a constant stream of data. Altogether,

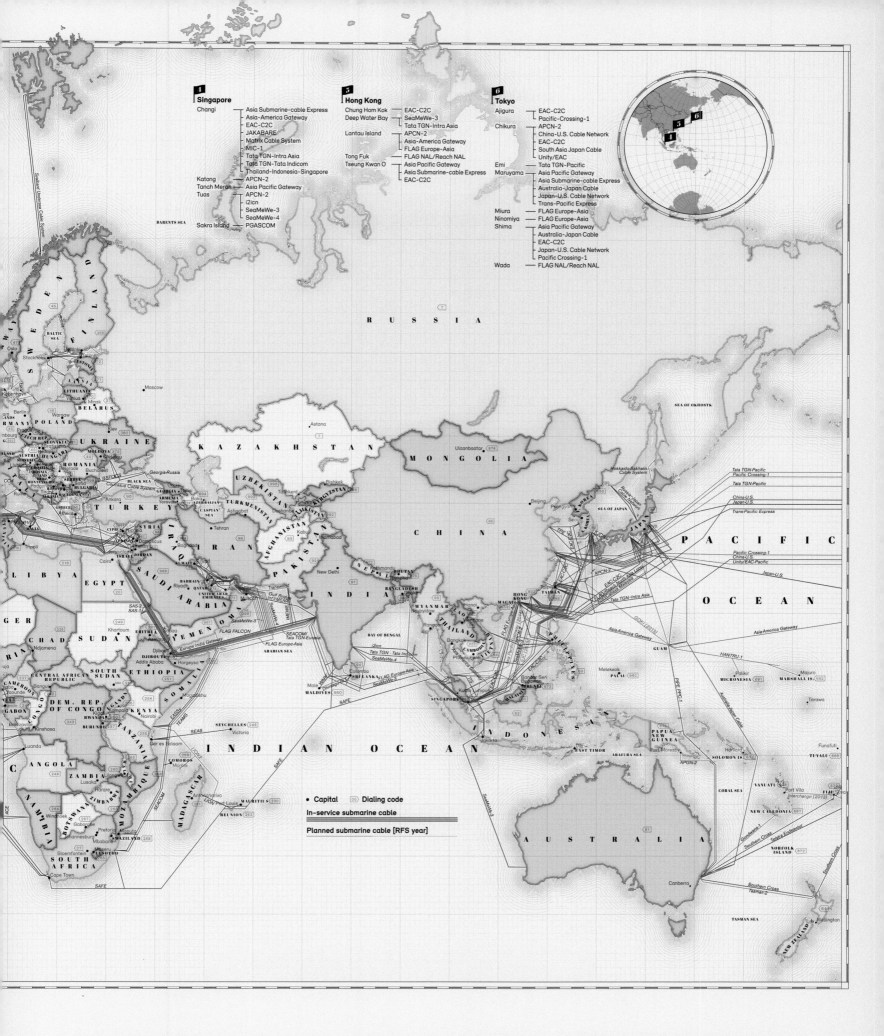

it can amount to five gigabytes of data per cord, per second. Parallel to the fiber optics, copper power cables are also laid out powering signal boosters every 80 kilometers or so to keep the signal strong. These boosters, as it happens, are also weak points in the matrix, because in order to be boosted, the individual fibers must be separated out. This makes them easier to tap for information. And there has been much media speculation that the U.S. government has been using submarines for just this purpose — mining data directly out of the wires and sifting it for information on you. Now do you want the red pill or the blue pill?

Submarine Cables

OUR GREATEST IDEAS

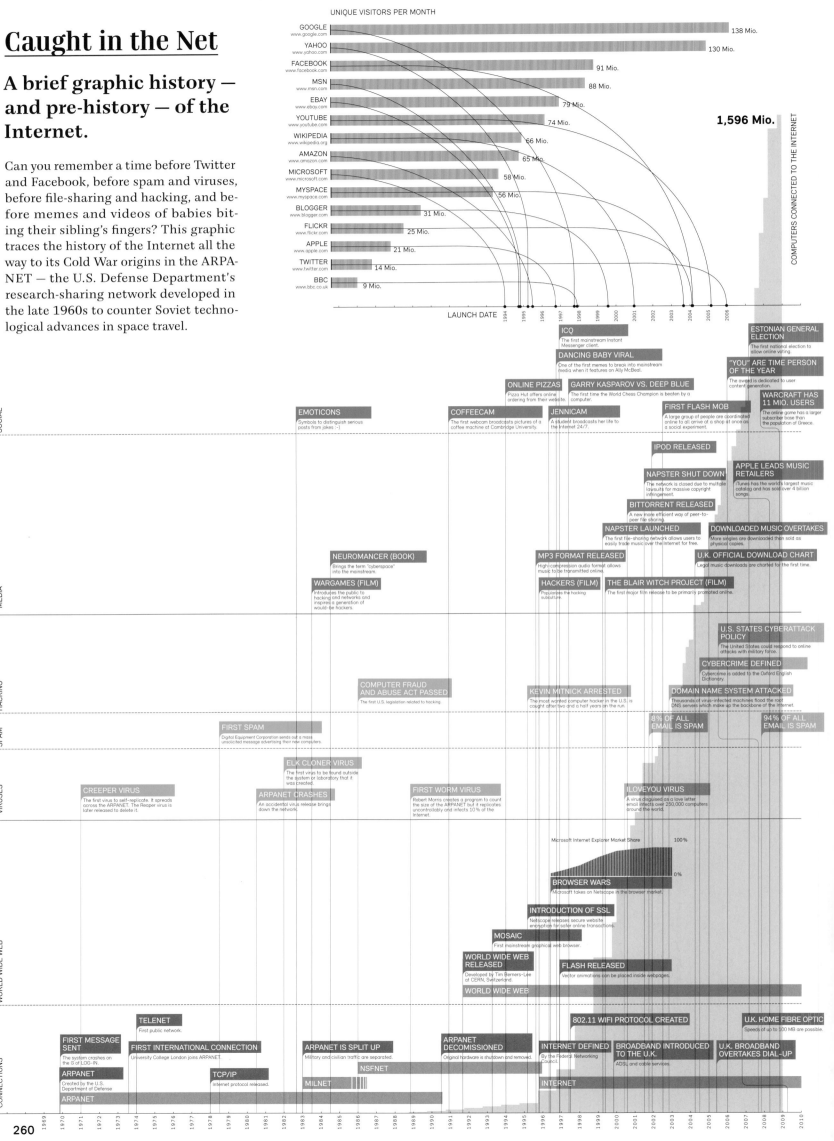

Caught in the Net

A brief graphic history — and pre-history — of the Internet.

Can you remember a time before Twitter and Facebook, before spam and viruses, before file-sharing and hacking, and before memes and videos of babies biting their sibling's fingers? This graphic traces the history of the Internet all the way to its Cold War origins in the ARPANET — the U.S. Defense Department's research-sharing network developed in the late 1960s to counter Soviet technological advances in space travel.

All Plugged In

A few academics linked together sometime in the 1980s, and then look what happened.

The Internet is so big and wide that no one knows when exactly it was switched on. All we know is that it developed out of a U.S. Defense Department research program, and that a British scientist named Tim Berners-Lee created the World Wide Web, (which he almost modestly called TIM — short for The Information Mine) to display information on it. Now the Internet is a phenomenon so ubiquitous that modern life has become virtually synonymous with it. Almost 2.6 billion people on the planet are thought to have used it at least once, and in some liberal European cities, there are people arguing that wireless signals should be provided free of charge by the public authorities. It has become, in some people's eyes, a basic right.

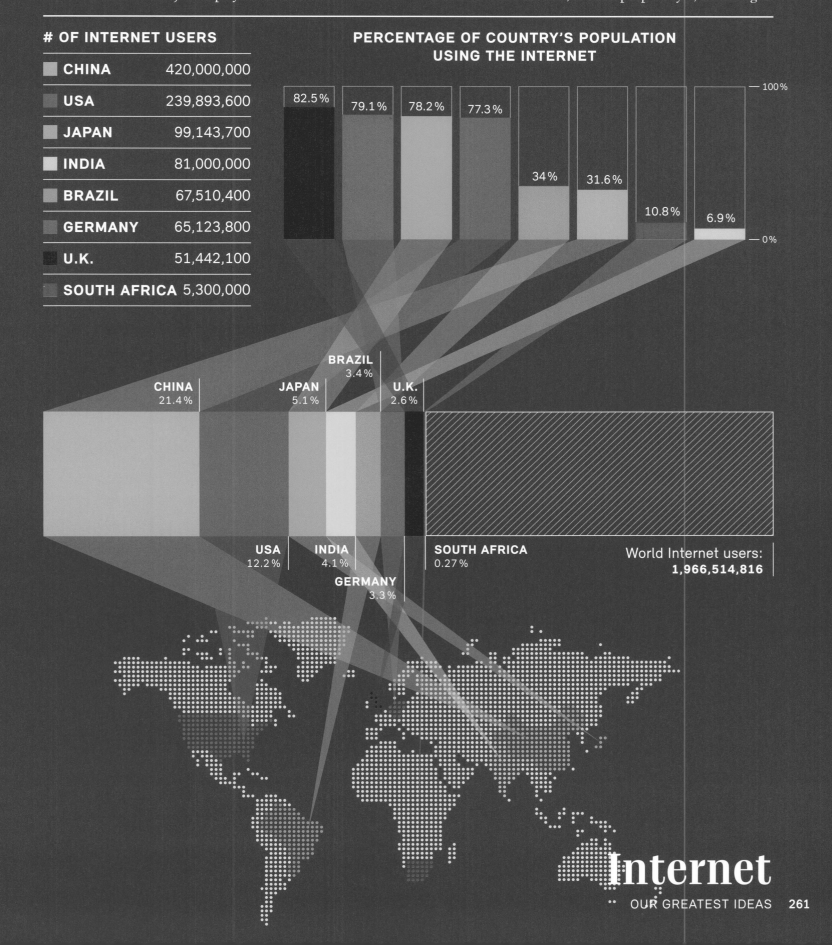

OF INTERNET USERS

- CHINA 420,000,000
- USA 239,893,600
- JAPAN 99,143,700
- INDIA 81,000,000
- BRAZIL 67,510,400
- GERMANY 65,123,800
- U.K. 51,442,100
- SOUTH AFRICA 5,300,000

PERCENTAGE OF COUNTRY'S POPULATION USING THE INTERNET

82.5% · 79.1% · 78.2% · 77.3% · 34% · 31.6% · 10.8% · 6.9%

CHINA 21.4%
JAPAN 5.1%
BRAZIL 3.4%
U.K. 2.6%
USA 12.2%
INDIA 4.1%
GERMANY 3.3%
SOUTH AFRICA 0.27%

World Internet users: 1,966,514,816

Internet
OUR GREATEST IDEAS

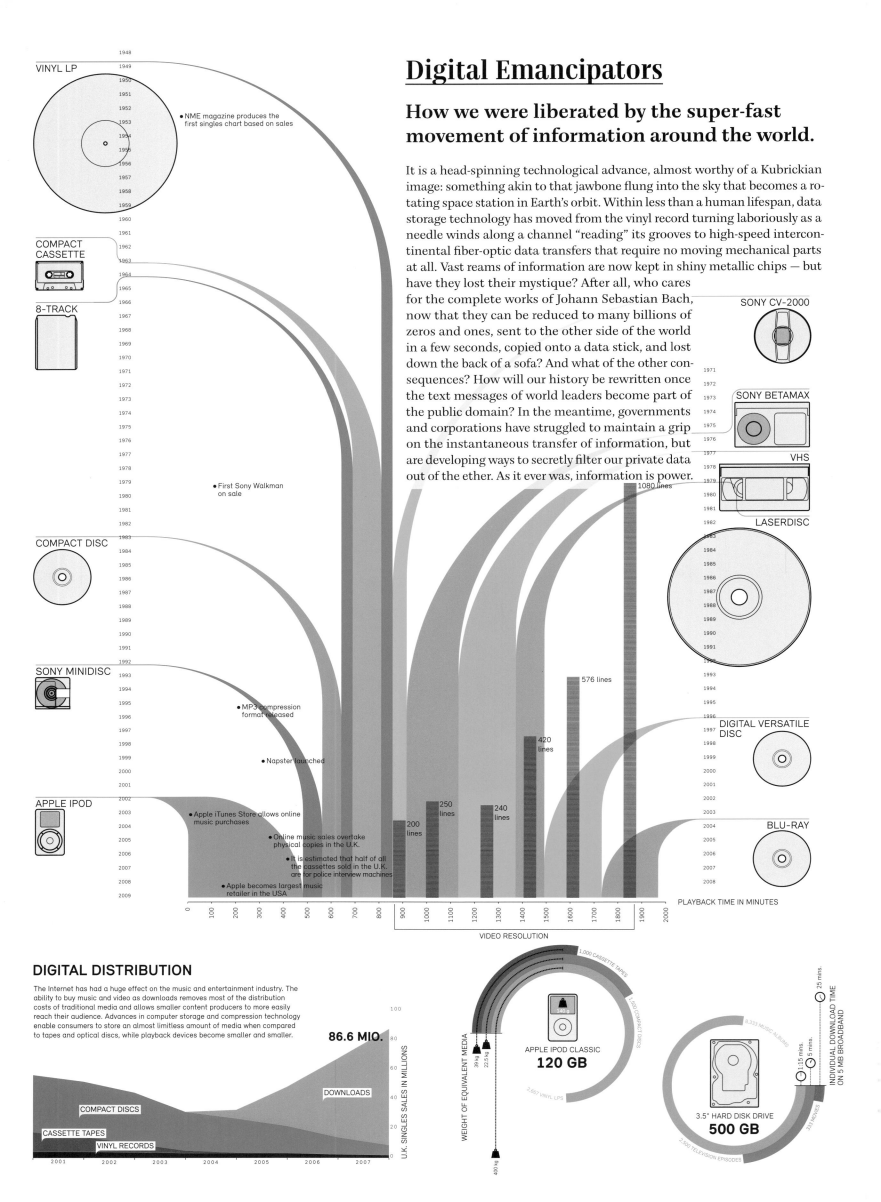

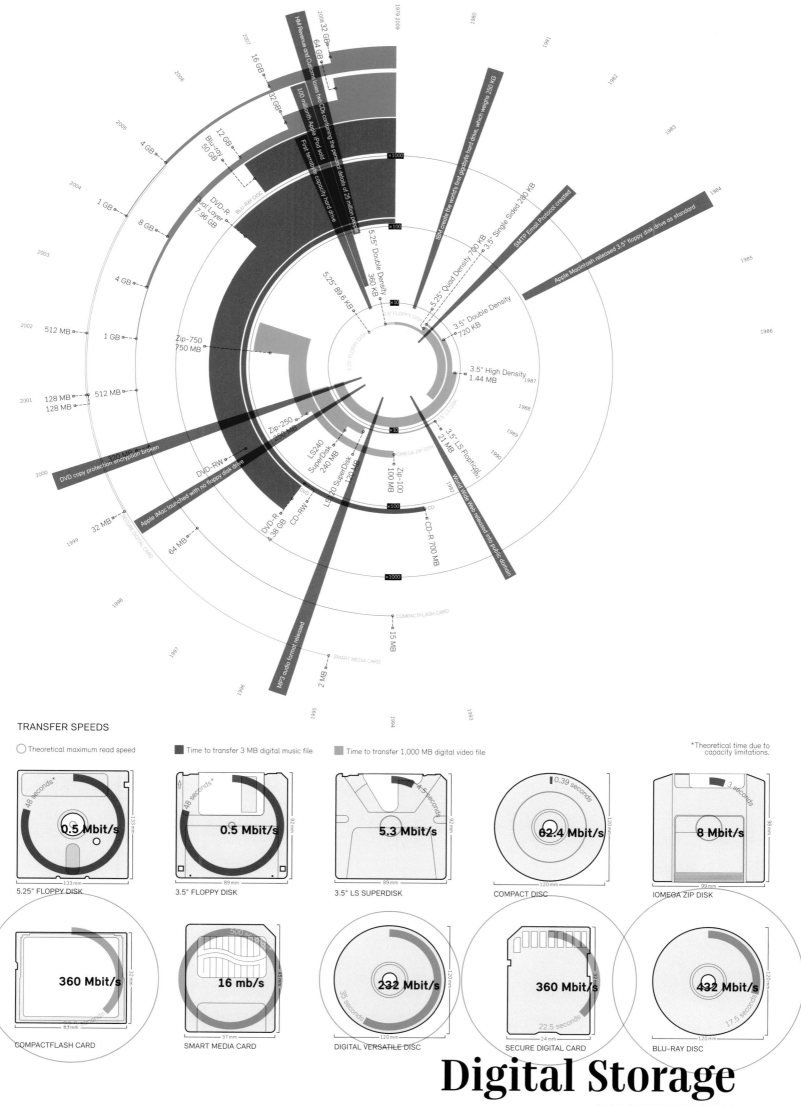

Digital Storage
OUR GREATEST IDEAS

God's Toolbox

The Large Hadron Collider is nothing less than the most complex machine and project ever created. Built by a multinational joint venture of institutions, the CERN project tries to answer some fundamental questions about the creation and composition of our universe. Not bad for people descended from a couple of naked apes.

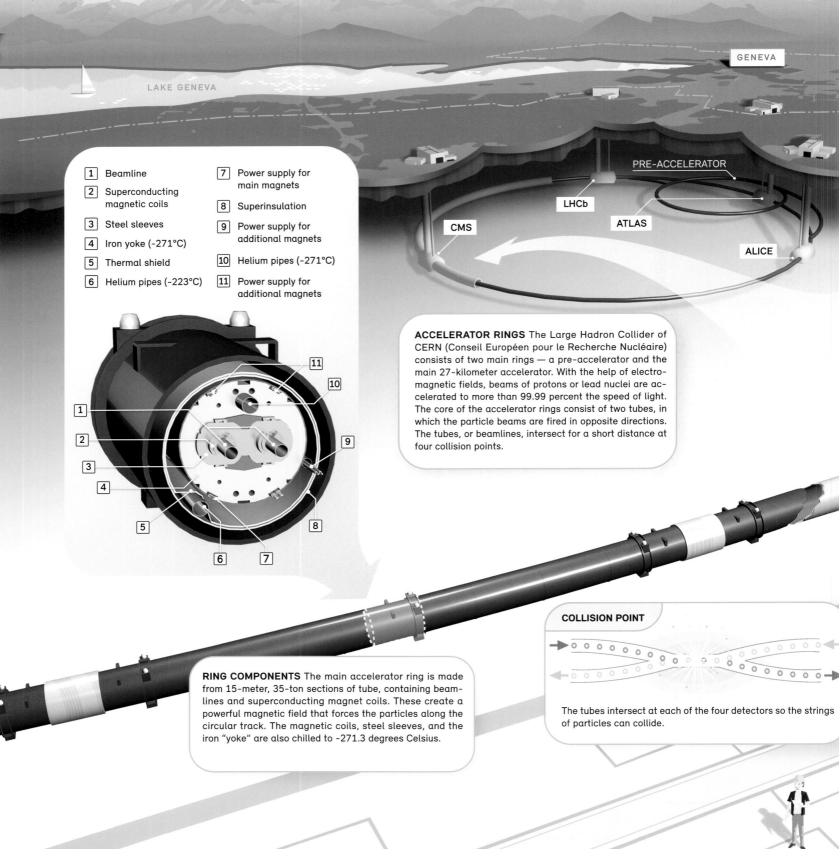

1. Beamline
2. Superconducting magnetic coils
3. Steel sleeves
4. Iron yoke (-271°C)
5. Thermal shield
6. Helium pipes (-223°C)
7. Power supply for main magnets
8. Superinsulation
9. Power supply for additional magnets
10. Helium pipes (-271°C)
11. Power supply for additional magnets

ACCELERATOR RINGS The Large Hadron Collider of CERN (Conseil Européen pour le Recherche Nucléaire) consists of two main rings — a pre-accelerator and the main 27-kilometer accelerator. With the help of electromagnetic fields, beams of protons or lead nuclei are accelerated to more than 99.99 percent the speed of light. The core of the accelerator rings consist of two tubes, in which the particle beams are fired in opposite directions. The tubes, or beamlines, intersect for a short distance at four collision points.

RING COMPONENTS The main accelerator ring is made from 15-meter, 35-ton sections of tube, containing beamlines and superconducting magnet coils. These create a powerful magnetic field that forces the particles along the circular track. The magnetic coils, steel sleeves, and the iron "yoke" are also chilled to -271.3 degrees Celsius.

COLLISION POINT

The tubes intersect at each of the four detectors so the strings of particles can collide.

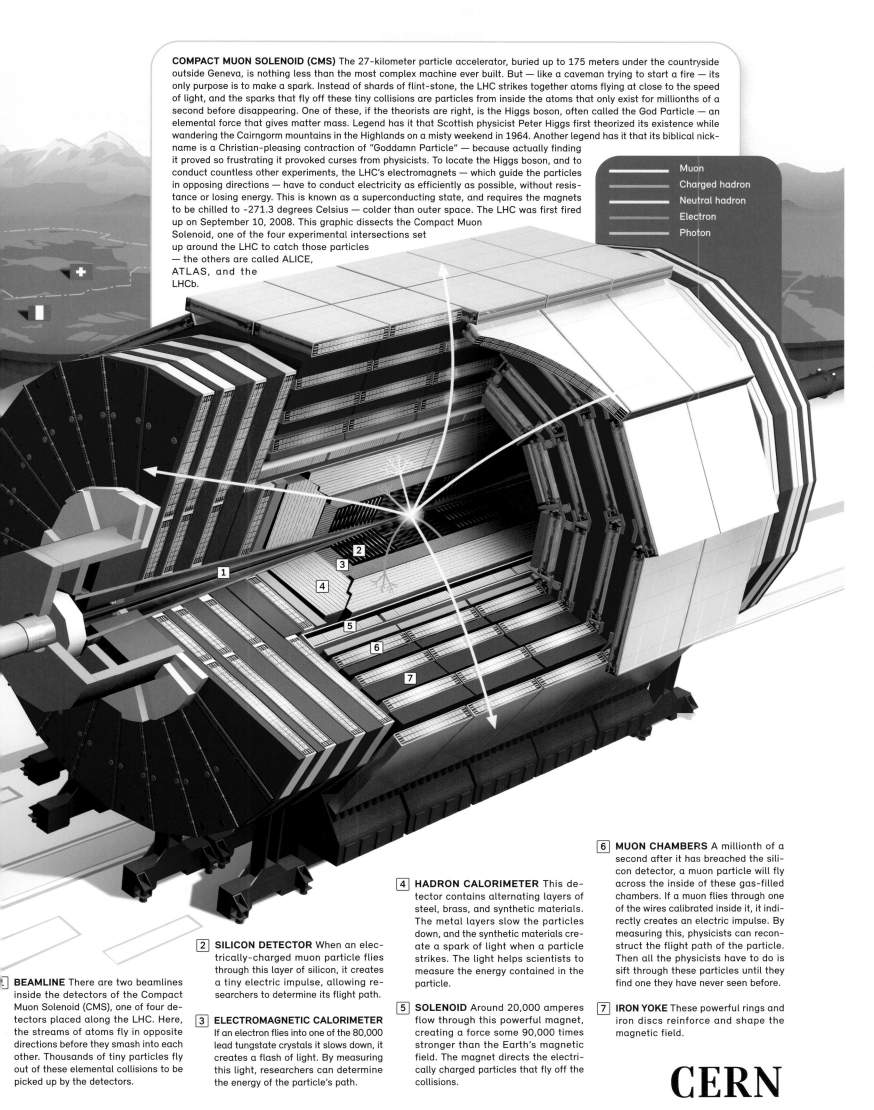

EPILOG

Our planet is tiny. Our world is vast. We have traveled together through space and time, across continents and down the eyepiece of microscopes. We have learned incredible truths through amazing illustrations. Thank you for joining us on this journey.

It may be a small world after all, but Disneyland it's not. There were harsh truths among our pages, unpleasant realities often of humanity's own creation. There were incredible tales of achievement and inspiration too, amazing inventions beyond the imaginations of a few generations ago.

This book is how we have chosen to travel visually across and inside the world. We could have chosen many other routes, other themes, other subjects. We encourage you to look for these in your daily life, in the media you consume and in your home.

This is a complex, fascinating, incredible planet filled with endless stories. Look closer, look further, look up, look down, look inside yourself, look up at the stars — look at everything again with fresh eyes. Take a journey around your existence and explore the narrative of everything you see. Inspiring new adventures await you.

Bon voyage!

INDEX

THE PLACE WE CALL HOME

8 Big Bang
Design: Nathalie Lees
with Zoe Bather (art direction)
www.nathalielees.com
Text: Brian Cox, Andrew Cohen
Sources: ESA, NASA
Published: *Wonders of the Universe* (Brian Cox, 2011)

10 The Universe
Design: Golden Section Graphics
with John Grimwade
www.golden-section-graphics.com
www.johngrimwade.com
Source: NASA
Published: *In Graphics* Volume 5, 2013

12 Planet Earth
Design: National Geographic – 5W Infographics,
Nat Geo Creative
www.nationalgeographic.com
Art research: Patricia Healy
Photos: NASA (moon), Reto Stöckli, NASA Earth Observatory (earth), NASA/ESA/SOHO (sun)
Sources: Stephen Marshak (University of Illinois), W. Sean Chamberlin (Fullerton College), J.T. Kiehl (National Center for Atmospheric Research)

14 Climate
Design: KircherBurkhardt – Jakub Chrobok,
Stefan Fichtel
www.kircher-burkhardt.com
Text: Tom Levin
Research: Tom Levin, Jakub Chrobok,
Stefan Fichtel
Sources: CC, MPI, DKRZ, *Atlas of Climate Change*, *GEO-Kompakt*
Published: *Welt am Sonntag*, 2007

16 Seasons
Design: KircherBurkhardt – Sebastian Müller (designer), Christina Ackermann (art direction)
www.kircher-burkhardt.com
Text / research: Horst Güntheroth
Sources: Uni Münster (hypersoil.uni-muenster.de), mäh-Werk (www.maeh-werk.de), Umwelt Bundesamt für Mensch und Umwelt, www.mtwow.org, www.pilzepilze.de, www.planet-wissen.de, de.academic.ru, www.landtreff.de, Bundesamt für Naturschutz, Institut für Biologie, Lehrstuhl für Entwicklungsbiologie der Friedrich-Alexander-Universität Erlangen-Nürnberg, Plant-Animal Interaction, Springer-Verlag, *European Atlas of Soil Biodiversity*, *Beitrag zur Geschichte des Grünlands Mitteleuropa* (Alois Kapfer, 2010), www.wikipedia.org, Botanik Bochum, Uni Graz, Agro Atlas, Uni Jena, www.dracaena-drachenbaum.de, www.mykoweb.com, www.myheimat.de
Published: *Stern Magazin*, 2012

24 Arctic Ice
Design: National Geographic – Alejandro Tumas (graphics editor), Pablo Loscri and Alejandro Tumas (design), Hernan Cañellas and Pablo Loscri (art), Mollie Bates (production), Robert Stacey, Worldsat International Inc. (satellite image), Theodore A. Sickley (GIS)
www.nationalgeographic.com
Text: Jane Vessels
Research: Kaitlin M. Yarnall
Consultants: Kenneth Jezek, Byrd Polar Research Center, Ohio State University (BPRC, OSU), Clarie L. Parkinson, NASA; Donald K. Perovich, U.S. Army Cold Regions Research and Engineering Lab, Leonid Polyak, BPRCH, OSU, Mark Serreze, Cooperative Institute for Research in Environmental Sciences (CIRES), University of Colorado, Konrad Steffen, CIRES, Muyin Wang, University of Washington
Sources: National Snow and Ice Data Center, University of Colorado (ICE); James Overland, NOAA and Muyin Wang (Temperature); Konrad Steffen and Russell Huff, CIRES (Greenland)
Published: Supplement to *National Geographic*, 2009

26 Great Migrations
Design: National Geographic – Fernando Batista,
Nat Geo Creative
www.nationalgeographic.com

28 Continents
Design: A project by Accurat, directed by Giorgia Lupi, Simone Quadri, Gabriele Rossi with Davide Ciuffi, Matteo Riva
www.accurat.it
Source: *Calendario Atlante De Agostini*
Published: *La Lettura – Corriere della Sera*, 2012

30 The End
Design: Nathalie Lees
with Zoe Bather (art direction)
www.nathalielees.com
Text: Brian Cox, Andrew Cohen
Sources: ESA, NASA
Published: *Wonders of the Universe* (Brian Cox, 2011)

LIVING TOGETHER

34 Greetings
Design: Zsuzsanna Ilijin
www.helloyellowstudio.com
Text / research: Zsuzsanna Ilijin
Self-published, 2007

36 Wedding Traditions
Design: RIA Novosti – Paulina Chemeris, Konstantin Potapov, Alexei Novichkov (author of cliché)
www.rian.ru
Text / research: Alexandra Yarosh
Published: www.ria.ru, 2013

38 Happiness
Design: United States of the Art – Jens Uwe Meyer
www.unitedstatesofheart.com
Text / research: *Stern VIEW*
Source: Gallup World Poll 2006
Published: *Stern VIEW*, 2007

40 Global Population
Design: South China Morning Post –
Simon Scarr/SCMP
www.scmp.com
Source: United Nations
Published: *South China Morning Post*, 2011

42 Migration
Design: MakingUse – Carlo Zapponi
www.makinguse.com
Research: Carlo Zapponi
Sources: The World Bank Open Data (migration data), U.S. Census Bureau, International Data Base (population data)
Published: www.peoplemov.in, 2011

44 Megacities
Design: Golden Section Graphics with Paul Blickle
www.golden-section-graphics.com
Research: Burdayukom, SAP (Five), Harald Sterly (University of Cologne)
Sources: www.megacities-megachallenge.org, www.geographie.uni-koeln.de, www.citypopulation.de, www.geography.about.com
Published: *In Graphics* Volume 01, 2010

46 Skyscrapers
Design: TD – Theo Deutinger, Dario Marino,
Paul Valeanu
www.td-architects.eu
Text: Theo Deutinger
Research: Theo Deutinger, Dario Marino,
Paul Valeanu
Sources: www.wikipedia.org, www.celebritynetworth.com, www.arquitectonica.com, www.e-architect.co.uk, www.huffingtonpost.com, www.eubusiness.com, www.world-most-expensive.com
Published: *MARK* #41, Dec/Jan 2012/13

48 Underworld
Design: National Geographic – Hernan Canellas,
Nat Geo Creative
www.nationalgeographic.com

50 Metro Systems
Design: Francesco Franchi (information designer, art director), Laura Cattaneo (illustrator)
www.francescofranchi.com
Text: Sara Deganello
Research: Sara Deganello, Francesco Franchi
Published: *IL / Il Sole 24 ORE*, 2009

52 Empires
Design: A project by Accurat, directed by Giorgia Lupi, Simone Quadri, Gabriele Rossi with Davide Ciuffi, Stefania Guerra
www.accurat.it
Text / research: Accurat
Sources: www.wikipedia.org, Le Garzantine, Atlante Storico
Published: *La Lettura – Corriere della Sera*, 2012

54 The American Way
Design: Hyperakt – Julia Vakser Zeltser,
Jason Lynch, Wen Ping Huang
www.hyperakt.com
Self-published, 2012

56 The White House
Design / text / research: Cyprian Lothringer
www.cyplot.de
Sources: www.whitehouse.gov, Google Maps
Published: *MAIN-POST*, 2012

58 The Kanzleramt
Design / text / research: Cyprian Lothringer
www.cyplot.de
Sources: www.bundeskanzlerin.de, www.bundes-regierung.de, *Der Spiegel*, *Deutschland verstehen* (Ralf Grauel, Jan Schwochow, 2012)
Published: *MAIN-POST*, 2013

60 The Left Wing
Design: Álvaro Valiño
www.alvarovalino.com
Research: Carlos Prieto
Source: *The Red Flag: A History of Communism* (Bargain Price, 2009)
Published: *Público*, 2010

62 Nuclear Weapons
Design: Richard Johnson with Andrew Barr
www.newsillustrator.com
Text: Adam McDowell
Research: Andrew Barr, Adam McDowell
Sources: *Bulletin of the Atomic Scientists*, *Jane's Weapons Systems Strategic,* www.missilethreat.com, www.globalsecurity.org, www.brahmos.com
Published: *National Post*, 2012

64 Casualties
Design: Marion Kotlarski
www.100yearsofworldcuisine.com
Text / research: Clara Kayser-Bril, Nicolas Kayser-Bril
Sources: Various academic sources
Published: Online, stand-alone project, 2011

66 Religion
Design: Nora Coenenberg
www.ncoenenberg.de
Text / research: Claudia Füßler
Sources: www.islam.de, www.kirchenaustritt.de, www.buddha-infos.de, www.international-daoismus.de, www.sikh-religion.de, www.bahai.de, www.bahai.org, Encyclopedia of Shinto, remid, World Evangelization Research Center, *Die Weltreligionen* (Arnulf Zitelmann, 2002)
Published: *DIE ZEIT*, 2012

68 Religion
Design / text / research: Richard Johnson
www.newsillustrator.com
Sources: Pew Forum on Religion and Public Life, www.adherents.com
Published: *National Post*, 2011

70 Religion
Design / text / research: Similar Diversity – Andreas Koller, Philipp Steinweber
www.similardiversity.net
Source: Internet
Published: First exhibition "Is it possible to touch someone's heart with design?" at Hangar 7 in Salzburg, a number of book publications and other exhibitions followed 2007-2013

72 Jerusalem
Design: National Geographic – Fernando G. Baptista, Mollie Bates (production), Jerome Cookson (timeline maps), Patricia Healy (art)
www.nationalgeographic.com
Text: Don Belt, Jane Vessels
Research: Kathy Maher
Consultants: Dan Bahat, retired Chief Archaeologist of Jerusalem; Eric H. Cline, George Washington University; Steven Fine, Yeshiva University; Oleg Grabar, Institue for Advanced Study, Princeton University; Jodi Magness, University of North Carolina, Chapel Hill; Jerome Murphy-O'Connor, École Biblique, Jerusalem; Leen Ritmeyer, Ritmeyer Archaeological Design
Photo: "Sacrifice of Isaac" by Caravaggio, Uffizi Gallery, Scala/Art Resource, New York
Published: Supplement to *National Geographic*, 2008

74 Picking the Pope
Design / text / research: Antonio Farach
www.linkedin.com/in/antoniofarach
Sources: Universi Dominici Gregis - Pope John Paul II
Published: *Facts*, 2005

76 The Hajj
Design: Golden Section Graphics – Jan Schwochow, Julian Weiß, Juliana Köneke, Mesut Capkin, Alice Rzezonka, Nina Steingrobe
www.golden-section-graphics.com
Research: Jan Schwochow, Sebastian Pittelkow (*Stern VIEW*), Max Rauner (*Die Zeit*)
Sources: *Stern View*, *DIE ZEIT*, Hadsch-Ministerium (www.hajjinformation.com), www.trafficforum.org, www.zikr.co.uk, www.slamonline.net, www.enfal.de, www.islammedia.free.fr, www.ummah.net, www.bible.ca, www.wikipedia.org, www.railway-technology.com, www.saudirailexpansion.com, www.kabahinfo.net, www.eslam.de
Published: *In Graphics* Volume 1, 2010

78 Holidays
Design / research: Golden Section Graphics – Katja Günther
www.golden-section-graphics.com
Sources: www.feiertage-weltweit.com, www.wikipedia.org, www.kalender-365.de, www.karl-may-stiftung.de/kalender
Published: *In Graphics* Volume 3, 2011

80 Family Planning
Design: DensityDesign Research Lab
www.densitydesign.org
Academic faculty: Paolo Ciuccarelli, Stefano Mandato, Donato Ricci, Tommaso Venturini, Salvatore Zingale
Teaching assistants: Michele Mauri, Azzurra Pini, Matteo Azzi
Authors / students: Alberto Barone, Maria Luisa Bertazzoni, Martina Elisa Cecchi, Elisabetta Ghezzi, Alberto Grammatico
Sources: Unmet need for family planning: United Nations, Department of Economic and Social Affairs, Population Division; life expectancy index: list by the CIA: The World Factbook; democracy index: Economist Intelligence Unit, religious regulations and distribution by country: Department of Obstetrics and Gynecology (Queen's University); literacy index: United Nations Development Programme; gender equality index: Social Watch; press freedom index: Reporters Without Borders; development index: United Nations Development Programme, Usage of modern methods contraception: United Nations, Department of Economic and Social Affairs; population division, number of women by country: CIA: The World Factbook; human rights: Human Rights Watch, Italian system focus: Sì alla vita nr 7,8,9,11,12 (2011) and 1,7,8 (2012), Pianificazione Familiare Naturale di Maurizio Guida 1994, Relazione del Ministero della Saluta sulla attuazione della legge contenente norme per la tutela sociale della maternità e per l'interruzione volontaria di gravidanza (2010),
Chinese system focus
Published: Graphics realized during the "Final Synthesis Studio" at Politecnico di Milano, 2013

82 It's a Woman's World
Design: DensityDesign Research Lab.
www.densitydesign.org
Academic faculty: Paolo Ciuccarelli, Stefano Mandato, Donato Ricci, Tommaso Venturini, Salvatore Zingale
Teaching assistants: Michele Mauri, Azzurra Pini, Matteo Azzi
Authors / students: Viviana Ferro, Ilaria Pagin, Sara Pandini, Federica Sciuto, Elisa Zamarian
Sources: World Data Bank: data.worldbank.org/topic/health, UN publications: www.data.un.org, UNFPA financial resources flows and reports: www.unfpa.org/public/global/pubs_rh, www.unfpa.org
Published: Graphics realized during the "Final Synthesis Studio" at Politecnico di Milano, 2013

THE DAYS THE EARTH STOOD STILL

86-88 Titanic
Design: Golden Section Graphics – Jim Dick, Thomas Richter, Jan Schwochow, Dirk Aschoff
www.golden-section-graphics.com
Research: Jim Dick, Thomas Richter
Sources: www.turbosqiud.com, www.encyclopedia-titanica.org, www.worldwideschool.org, www.kenmarshall.com, www.copperas.com, www.titanic100yearsinfo.com, www.wikipedia.org, *Titanic in Pictures* (Henry First, 2011), *Inside The Titanic* (Ken Marshall, 1997)
Published: *In Graphics* Volume 3, 2011

89 Titanic
Design / research: Golden Section Graphics – Dirk Aschoff
www.golden-section-graphics.com
Sources: www.wikipedia.org, www.encyclopedia-titanica.org
Published: *In Graphics* Volume 3, 2011

90 Pearl Harbor
Design: South China Morning Post – Adolfo Arranz/SCMP
www.scmp.com
Sources: *Pearl Harbour 1941: The Day of Infamy* (Carl Smith, 2001), *National Geographic*, *Graphic News*
Published: *South China Morning Post*, 2011

92 D-Day
Design / research: Stern – Andrew Timmins
www.stern.de
Text: Teja Fiedler
Published: *Stern Magazin*, 2004

94 Berlin Airlift
Design: Golden Section Graphics – Jan Schwochow, Mesut Capkin, Felix Waldow, Katharina Erfurth
www.golden-section-graphics.com
Research: Jan Schwochow, Felix Waldow, Katharina Erfurth
Sources: Bewag, *Die Rosinenbomber: Die Berliner Luftbrücke 1948/49, ihre technischen Voraussetzungen und deren erfolgreiche Umsetzung* (Wolfgang J. Huschke, 2008)*; Rosinenbomber über Berlin* (Gerhard Keiderling, 1998); *Landing on Tempelhof. 75 Jahre Zentralflughafen. 50 Jahre Luftbrücke. Ausstellungskatalog* (Bezirksamt Tempelhof, Berlin, 1998); *Auftrag Luftbrücke, Alliierte in Berlin 1945 - 1994: Ein Handbuch zur Geschichte der militärischen Präsenz der Westmächte* (Friedrich Jeschonnek, Dieter Riedel, William Durie, 2004)*;* www.history.army.mil, History House Foundation of the Federal Republic of Germany, www.wikipedia.org, German Weather Service
Published: *Berliner Morgenpost, Welt am Sonntag*, 2008

98 Mount Everest
Design: South China Morning Post – Adolfo Arranz/SCMP
www.scmp.com
Sources: Royal Geographic Society, National Geographic Society, www.8000ers.com, *El Mundo*
Published: *South China Morning Post*, 2013

100 Cuban Missile Crisis
Design / text / research: dpa-infografik – Andreas Brühl
www.dpa-infografik.com
Sources: JFK Presidential Library and Museum, National Security Archive (University of Washington), Institute of Historical Research (University of London), Global Security Org., CIA, US Navy, US Air Force, *The Cuban Missile Crisis* (Don Munton, David A. Welch, 2011), *Kuba-Krise* (Bernd Greiner, 1988)
Published: *Hellweger Anzeiger*, 2012

102 JFK
Design: John Grimwade
www.johngrimwade.com
Text: Clive Irving
Research: Gerald Posner, Joyce Pendola
Published: *Case Closed*, 1993

106 Space Exploration
Design: National Geographic – Sean McNaughton, Nat Geo Creative
www.nationalgeographic.com

108 Space Flight
Design: South China Morning Post – Simon Scarr/SCMP
www.scmp.com
Source: NASA
Published: *South China Morning Post*, 2011

110-113 9/11
Design: Golden Section Graphics – Jan Schwochow, Thomas Richter, Katharina Stipp
www.golden-section-graphics.com
Research: Jan Schwochow
Sources: The 9/11 Commission Report, www.ara.com, National Transportation Safety Board, FAA, www.911research.wtc7.net, www.wikispaces.com, www.honorflight93.org
Published: *In Graphics* Volume 2, 2011

114 9/11
Design / research: Golden Section Graphics – Jan Schwochow (Graphics editor), Ludwig Maertins (Illustration)
www.golden-section-graphics.com
Sources: The 9/11 Commission Report, 9/11 and Terrorist Travel, StaffReport of the National Commission on Terrorist Attacks Upon the United States, Washington Post (note: due to the large amounts of data, we could only visualize a part of it.)
Published: *In Graphics* Volume 2, 2011

116 9/11
Design: South China Morning Post – Simon Scarr/SCMP
www.scmp.com
Research: Alex Nicoll/SCMP
Sources: Watson Institute for International Studies of Brown University, US Congressional Research Service, *Encyclopaedia Britannica*
Published: *South China Morning Post*, 2011

118 Fukushima
Design / text / research: Cyprian Lothringer
www.cyplot.de
Sources: WNN, WNA, CNN, ARD, CNIC, IAEA, Statfor, USGS, BBC, TEPCO, UN, RIU, BMU
Self-published, 2011

120 Killing Osama bin Laden
Design: Jennifer Daniel
www.whatwouldjenniferdo.com
Text / research: Jennifer Daniel, Barrett Sheridan
Sources: Twitter, Bloomberg News
Published: *Bloomberg Businessweek*, 2011

THE GOOD LIFE

124 Family Time
Design: Infografika (*Infografika magazine*) – Dmitry Gorelyshev, Artem Koleganov
www.infographicsmag.ru
Text / research: Nikolay Romanov
Source: www.wciom.com
Published: *Infografika magazine* issue 6, 2011

126 Family Spendings
Design / research: Samuel Granados with Gianluca Seta
www.sgranados.es
www.nascuto.com
Text: Isidoro Trovato
Source: Istat
Published: *La Lettura – Corriere della Sera*, 2012

128 Breakfast
Design: Golden Section Graphics – Katharina Stipp, Philipp Dettmer, Julian Kontor
www.golden-section-graphics.com
Research: Katharina Stipp
Sources: www.bestcuicine.nfo.ph, www.thefoodieshandbook.co.uk, www.tastebuddeluxe.blogspot.com, www.blog.travelpod.com, www.shanghaistuff.com, www.justinaltman.com, www.costaricapages.com, www.wellsphere.com, www.123rf.com, www.en.petitchef.com, www.gocentralamerica.about.com, www.dotsconnected.net, www.buzzle.com, www.wikipedia.org, www.nekobento.com, www.morningbreakfasts.blogspot.com, www.womansday.com, www.zottarella.de, www.stockholm-tourism.com, www.dinnerdiary.org, www.kitchenlove.blogg.se, www.spiegel.de, www.megcasey.com, www.simonfoodfavourites.blogspot.com, www.heconnoisseurs.com
Published: *In Graphics* Volume 3, 2011

130 Food
Design: Ole Häntzschel
www.olehaentzschel.com
Text: Bettina Stiekel
Research: Ansbert Kneip, Klaus Falkenberg
Published: *Dein SPIEGEL*, 2013

132 Tuna
Design / text / research: Paul Blickle with Bettina Achinger, Max Birkl, Katrin Eberhard, Bernd Riedel
www.colorful-data.net
Sources: *Welt am Sonntag*, www.fair-fish.ch, www.fischinfo.de, www.iss-foundation.org, www.welt.de
Published: *In Graphics* Volume 1, 2010

134 Fruit
Design / research: Golden Section Graphics – Katharina Schwochow
www.golden-section-graphics.com
Sources: Statistisches Bundesamt, Eurostat, Bundesministerium für Ernährung, Landwirtschaft und Verbraucherschutz, Deutscher Fruchthandelsverband e.V., Europäische Kommission
Published: *In Graphics* Volume 5, 2013

136 Organic Farming
Design / text / research: Francesco Franchi
www.francescofranchi.com
Sources: *The World of Organic Agriculture, Statistics & Emerging Trends 2009* (Helga Willer, Lukas Kilcher, 2009)
Published: *IL / Il Sole 24 ORE*, 2009

138 Chocolate
Design: Carl DeTorres
www.cdgd.com
Text / research: Jennica Peterson
Sources: International Cocoa Organization, International Confectionary Association and Association of Chocolate, Biscuit and Confectionary Industries of the European Union (CAOBISCO)
Published: *Afar*, 2009

140 Beer
Design: Francesco Franchi (information designer, art director), Laura Cattaneo (illustrator)
www.francescofranchi.com
Text / research: Alessandro Giberti
Source: World Health Organization (Global Status Report: Alcohol Policy)
Published: *IL / Il Sole 24 ORE*, 2010

142 Wine
Design: South China Morning Post – Adolfo Arranz/SCMP
www.scmp.com
Sources: Riedel Glas Austria, *Wine Tasting. Professional Handbook* (Ronald S. Jackson, 2009); *Wine. A Scientific Explanation* (Merton Sandler, Roger Pinder, 2002)
Published: *South China Morning Post*, 2012

144 Liquids
Design: Ole Häntzschel
www.olehaentzschel.com
Text: Bettina Stiekel
Research: Ansbert Kneip, Klaus Falkenberg
Published: *Dein SPIEGEL*, 2013

146 Travel
Design: Condé Nast Traveler – John Grimwade, Section Design, Haisam Hussein
www.johngrimwade.com
Text: Ted Moncrieff
Research: Rena Marie Pacella
Published: *Condé Nast Traveler*, 2012

148 Flight Paths
Design: Condé Nast Traveler – John Grimwade
www.johngrimwade.com
Text: Graham Boynton
Sources: Gander, Shanwick Oceanic Control Centres
Published: *Condé Nast Traveler*, 1996

150 Highways
Design / research: Golden Section Graphics – Jan Schwochow, David Weinberg, Katharina Stipp
www.golden-section-graphics.com
Sources: www.adac.de, www.avd.de, www.motorradreiseforum.de, www.calsky.com, www.autogenau.de, www.meine-auto.info, www.radarforum.de, www.autobahn.wikia.com, www.karawane.de, www.etsc.eu, www.rhinocarhire.com, www.europe.org, www.wordiq.com, www.goruma.de, www.urlaubinfoportal.de, www.jnto.de, www.oesterreich.orf.at, www.australien-info.de, www.camperhireaustralia.au
Published: *In Graphics* Volume 1, 2010

152 Christmas
Design: Kiss Me I'm Polish – Agnieszka Gasparska, Joshua Covarrubias, Louise Ma (illustration)
www.kissmeimpolish.com
Text / research: Morgan Clendaniel, GOOD
Sources: ABC, CardTrak, ComScore, The Federal Reserve, National Candle Association, National Turkey Federation, National Retail Federation, *The New York Times*, NPD Group, U.S. Census, U.S. Postal Service, University of Illinois, University of Wisconsin-Madison
Published: *GOOD Sheet*, print, available at Starbucks from November 20-26th, 2008

154 Fashion
Design: Peter Grundy with Tilly Northedge
www.grundini.com
Text / research: Andrew Hicks
Source: *The Guardian*
Published: *G2 Newspaper*, 2007

156 Pets
Design: Peter Grundy with Tilly Northedge
www.grundini.com
Text / research: Andrew Hicks
Source: *The Guardian*
Published: *G2 Newspaper*, 2007

158 Music
Design: DensityDesign Research Lab.
www.densitydesign.org
Text: Stefania Ulivi
Research: Michele Mauri, Giorgio Uboldi
Source: www.whosampled.com
Published: *La Lettura – Corriere della Sera*, 2012

159 Music
Design: South China Morning Post – Simon Scarr/SCMP
www.scmp.com
Sources: Official Charts Company, RIAA, *Billboard*
Published: *South China Morning Post*, 2012

160 TV Shows
Design: HistoryShots – Larry Gormley
www.historyshots.com
Text / research: Larry Gormley
Sources: Various books, periodicals
and internet resources
Self-published, 2013

162 Twentieth-Century Painters
Design: A project by Accurat, designed and directed by Giorgia Lupi and Michela Buttignol, with Simone Quadri, Gabriele Rossi, Davide Ciuffi, Pietro Guinea Montalvo
www.accurat.it
Text / research: Accurat
Sources: *Le Garzantine – Arte*, Google images
Self-published, 2013

164 Twenty-first-Century Art Sales
Design: Thorsten Lange
www.special-empire.com
Source: www.handelsblatt.com

166 Olympia
Design: South China Morning Post –
Adolfo Arranz/SCMP
www.scmp.com
Sources: *A Brief History of the Olympic Games* (David C. Young, 2004), *The Olympic Games in Antiquity* (Olympic Museum, 2007), University of Texas
Published: *South China Morning Post*, 2012

167 Olympia
Design: Cyprian Lothringer
www.cyplot.de
Text: Christoph Drösser
Research: Alexandra Aschbacher, Christoph Drösser
Sources: *Buch der Sportrekorde*, www.isst-sport.org; sportrecords.co.uk, www.wikipedia.org
Published: *DIE ZEIT*, 2012

168 Football
Design: Golden Section Graphics – Jan Schwochow, Tatiana Lysenko, Niko Wilkesmann
www.golden-section-graphics.com
Research: Cemano Communication GmbH
Sources: Allianz SE, FC Bayern
Published: sponsoring.allianz.com,
In Graphics Vol. 1, 2010

FEAR AND LOATHING

172 Tarantino
Design / research: Golden Section Graphics –
Kamila Olszewska
www.golden-section-graphics.com
Source: www.imfdb.org
Published: *In Graphics* Volume 5, 2013

174 Guns
Design / text / research: Richard Johnson
www.newsillustrator.com
Sources: Individual National Census Reports, Individual National Police Reports, National Ministries of Justice, World Health Organization, www.gunpolicy.org
Published: *National Post*, 2012

176 Phobias
Design / text: i Infografia – Carlos Monteiro
iinfografia.blogspot.com
Research: Ana Kotowicz, Carlos Monteiro
Source: www.wikipedia.org
Published: *i*, 2012

178 Plane Crashes
Design: South China Morning Post – Jane Pong and Adolfo Arranz/SCMP
www.scmp.com
Sources: www.aviation-safety.net, Boeing, Greenwich University, www.aerointernational.de
Published: *South China Morning Post*, 2013

180 Pandemics
Design: Haisam Hussein
www.haisam.com
Text: Simon Apter, Timothy Don
Research: Lucy Medrich, Marie D'Origny, Sarah Stodola
Sources: *Encyclopedia Britannica*,
various reputable newspapers
Published: *Lapham's Quarterly*, 2009

182 Pandemics
Design: Column Five – Van Run
www.columnfivemedia.com
Text / research: Madeleine Nguyen
Sources: Mayo Clinic, Centers for Disease Control and Prevention, World Health Organization, *The New York Times*, National Center for Biotechnology Information
Published: *GOOD Magazine*, 2011

184 Smoking
Design: Column Five – Andrew Effendy
www.columnfivemedia.com
Text / research: Madeleine Nguyen
Sources: American Heart Association, Nation Health Interview Survey, National Center for Health Statistics, World Health Organization, Ohio University
Published: *GOOD Magazine*, 2011

186 Drugs
Design: Kiss Me I'm Polish – Agnieszka Gasparska (art direction), Joshua Covarrubias
www.kissmeimpolish.com
Text / research: Morgan Clendaniel, GOOD
Source: Department of Justice
Published: GOOD.is, 2009

188 New Year's Resolutions
Design: Ernesto Olivares Visual Information
www.ernestoolivares.com
Text: Ernesto Olivares, Visual.ly
Research: Visual.ly
Sources: CDC, Substance Abuse and Mental Health Services Administration (SAMHSA), National Institutes of Health (NIH), U.S. Bureau of Labor Statistics, U.S. Environmental Protection Agency (EPA), Journal of Clinical Psychology, Opinion Research Corporation
Published: Visual.ly, 2012

MONEY MAKES THE WORLD GO ROUND

192 Pyramid of Wealth
Design: KircherBurkardt – Maximilian Nertinger, Stefan Merker, Christina Ackermann
www.kircher-burkhardt.com
Text / research: Michael O'Sullivan
Source: *Credit Suisse Global Wealth Report 2011*
Published: *Bulletin, The Credit Suisse Magazine*, 2011

194 B.R.I.C. States
Design: Peter Grundy
www.grundini.com
Text: Christie Ferdinando
Source: *Sunday*
Published: *Modus magazine*, 2011

196 Development Aid
Design: Hyperakt – Julia Vakser Zeltser, Eric Fensterheim, Edwin Carter
www.hyperakt.com
Self-published, 2012

198 Market Bubbles
Design: A project by Accurat, directed by Giorgia Lupi, Simone Quadri, Gabriele Rossi with Davide Ciuffi, Anna Bassi, Mario Poropora
www.accurat.it
Text / research: Accurat
Sources: www.sharelynx.com/index2.php, http://inflationdata.com/inflation/inflation_rate/historical_oil_prices_table.asp
Published: *La Lettura – Corriere della Sera*, 2012

200 Globalization
Design: Makeshift with Rifle
www.mkshft.org
www.wearerifle.com
Text / research: Steve Daniels
Sources: US Department of Transportation, Source-map, Leopold Center for Sustainable Agriculture, Robert Neuwirth
Published: *Makeshift Magazine*, 2012

202 Drug Trafficking
Design: Nora Coenenberg
www.ncoenenberg.de
Text / research: Camilo Jiménez
Sources: Drug Enforcement Administration, European Monitoring Centre for Drugs and Drug Addiction, *Pschyrembel Klinisches Wörterbuch* 2012, Stockholm International Peace Research Institute, United Nations Office on Drugs and Crime, *World Drug Report 2011*
Published: *DIE ZEIT*, 2012

204 Food
Design: DensityDesign Research Lab.
www.densitydesign.org
Academic faculty: Paolo Ciuccarelli, Stefano Mandato, Donato Ricci, Tommaso Venturini, Salvatore Zingale
Teaching assistants: Michele Mauri, Azzurra Pini, Matteo Azzi
Authors / students: Irene Cantoni, Claudio Cardamone, Sara De Donno, Fabio Matteo Dozio, Arianna Pirola
Sources: *Slaughtering the Amazon* (Greenpeace International, 2009), *Eating up the Amazon* (Greenpeace International, 2006), World Bank, World Development Indicators, United States Department of Agriculture, Food and Agriculture Organization
Published: Graphics realized during the "Final Synthesis Studio" at Politecnico di Milano, 2013

206 Food
Design: United States of the Art – Carsten Raffel
www.unitedstatesoftheart.com
Text / research: Wolfgang Hassenstein (*Greenpeace Magazin*)
Sources: UNEP, World Food Programme, Faostat, Welthungerhilfe, Weltagrarbericht, Statistisches Bundesamt, Bundesanstalt für Landwirtschaft und Ernährung, TGRDEU
Published: *Greenpeace Magazin*, 2010

208 Brands
Design / research: Mario Mensch
www.mariomensch.de
Text: Christoph Drösser
Source: *Global Powers of Consumer Products* (Deloitte, 2012)
Published: *DIE ZEIT*, 2013

210 Brands
Design / text / research: History Shots – Larry Gormley
www.historyshots.com
Sources: Various books, periodicals
and internet resources
Self-published, 2012

212 Nuclear Power
Design: United States of the Art – Carsten Raffel
www.unitedstatesoftheart.com
Text: Wolfgang Hassenstein
(*Greenpeace Magazin*)
Research: *Greenpeace Magazin*
Sources: IAEA, World Nuclear Industry Handbook 2009, World Nuclear Association
Published: *Greenpeace Magazin*, 2010

214 Solar Power
Design / text: Francesco Franchi
www.francescofranchi.com
Research: Francesco Franchi,
Maria Cristina Origlia
Sources: Rapporto 2008 dell'Energy & Strategy Group, Osservatorio MIP-Politecnico di Milano
Published: *IL / Il Sole 24 ORE*, 2009

216 Energy Market
Design: Fraser Lyness
www.fraserlyness.com
Text / research: David Reay
Sources: Parliamentary Office of Science, National Grid, Networks
Published: *Eureka Magazine — The Times Newspaper*, 2011

218 Oil
Design: National Geographic – 5W Infographics, Nat Geo Creative
www.nationalgeographic.com
Source: Global Trade Information Services, Inc.

220 Energy Flow
Design: DensityDesign Research Lab.
www.densitydesign.org
Academic faculty: Paolo Ciuccarelli, Stefano Mandato, Donato Ricci, Tommaso Venturini, Salvatore Zingale
Teaching assistants: Michele Mauri, Azzurra Pini, Matteo Azzi
Authors / students: Giulio Bertolotti, Elia Bozzato, Gabriele Calvi, Stefano Lari, Gianluca Rossi
Sources: UN Intergovernmental Panel on Climate Change (IPCC), UN Framework Convention on Climate Change (UNFCCC), International Energy Agency (IEA), Climate Change Adaptation Database, US Energy Information Administration (EIA)
Published: Graphics realized during the "Final Synthesis Studio" at Politecnico di Milano, 2013

THE WORLD IS NOT ENOUGH

224 Waste of Energy
Design / text: Nigel Holmes
www.nigelholmes.com
Research: Nigel Holmes, *GOOD Magazine*
Source: United States Department of Energy
Published: *GOOD Magazine*, January 2008

226 Resources
Design: Piero Zagami with Duncan Swain (creative director)
www.pierozagami.com
Text: Duncan Swain
Research: Miriam Quick
Sources: UN TEEB, US Geological Survey, BP, Worm et al (2006), London Metal Exchange. Figures are worldwide. Living natural resources dates are worst-case based on published estimates. Minerals and fossil fuel data based on known reserves currently economical to extract, assuming fixed % increase in usage per year. No provision made for changes in demand caused by new technologies, discoveries of new reserves or market forces. Agricultural land means land suitable for rainfed cultivation net of other land usage. Thirty year historic agricultural expansion rates are applied.
Published: Information is Beautiful Studio for BBC Future, 2012

228 Resources
Design: Miriam Baña García
Source: www.footprintnetwork.org
Published: *In Graphics* Volume 5, 2013

230 Carbon Emissions
Design / text / research: Richard Johnson
www.newsillustrator.com
Source: United States Energy Administration
Published: NationalPost.com, 2012

232 Carbon Footprint
Design: Fraser Lyness
www.fraserlyness.com
Text: Mike Berners-Lee
Research: David Reay
Source: *How bad are bananas?* (Mike Berners-Lee, 2010)
Published: *Eureka Magazine - The Times Newspaper*, 2011

234 Natural Disasters
Design / text / research: Jennifer Daniel
www.whatwouldjenniferdo.com
Sources: NOAA, IHS Global Insight
Published: *Bloomberg Businessweek*, 2012

236 Plastic
Design / text / research: Stefan Zimmermann
www.deszign.de
Sources: Greenpeace, IK Industrievereinigung Kunststoffverpackungen e. V., Museum für Gestaltung Zürich, Naturschutzbund Deutschland (NABU) e. V., NOAA Marine Debris Programm, Ocean Conservancy's International Costal Cleanup, PlasticsEurope Deutschland e. V., 2013
Self-published, 2013

238 Heat
Design: Nathalie Lees with Fraser Lyness (art direction)
www.nathalielees.com
Text: Frances Ashcroft
Source: *The Life of Extremes* (Frances Ashcroft, 2001)
Published: *Eureka Magazine - The Times Newspaper*, 2011

240 Animal Biotopes
Design: Francesco Franchi (information designer, art director), Laura Cattaneo (illustrator)
www.francescofranchi.com
Text: Daniele Lorenzetti
Research: Francesco Franchi, Daniele Lorenzetti
Sources: Iucn Red List 2008, Global Biodiversity Outlook, WWF, Nature, *Biodiversity: Extinction by numbers* (Stuart L. Pimm, Peter Raven, 2000)
Published: *IL / Il Sole 24 ORE*, 2009

242 Water
Design / research: Golden Section Graphics – Katharina Stipp
www.golden-section-graphics.com
Sources: *Le Monde diplomatique – Atlas der Globalisierung* (2006 & 2009), Technische Universität Dresden, I.A., *World Water Resources at the Beginning of the Twenty-First Century* (Shiklomanov and John C. Rodda, 2003), *The Urban Agglomerations* (UN Population Division Department of Economic and Social Affairs, 2007), SAM
Published: *In Graphics* Volume 5, 2013

244 Water
Design / research: Golden Section Graphics – Jan Schwochow
www.golden-section-graphics.com
Sources: UNESCO, *World Water Resources at the Beginning of the Twenty-First Century* (I.A. Shiklomanov, John C. Rodda, 2003)
Published: *In Graphics* Volume 5, 2013

246 Water
Design / research: Golden Section Graphics – Katharina Stipp
www.golden-section-graphics.com
Sources: *Atlas of International Freshwater Agreements* (2002), www.unep.org, www.wwf.de, www.transboundarywaters.orst.edu, www.worldwater.org, *GWI: Global Water Market* (2011 and 2010), *Water – a shared responsibility* (UNESCO), *The United Nations World Water Development Report 2* (2006), www.unesco.org
Published: *In Graphics* Volume 5, 2013

OUR GREATEST IDEAS

250 Inventions
Design: South China Morning Post – Adolfo Arranz/SCMP
www.scmp.com
Source: www.wikipedia.org
Published: *South China Morning Post*, 2013

252 Nobel Prize
Design: A project by Accurat, directed by Giorgia Lupi, Simone Quadri, Gabriele Rossi with Davide Ciuffi, Federica Fragapane, Francesco Majno
www.accurat.it
Text / research: Accurat
Sources: www.nobelprize.org, www.treccani.it, www.wikipedia.org
Published: *La Lettura – Corriere della Sera*, 2012

254 Light
Design: Golden Section Graphics – Thomas Richter, Tatjana Lysenko, Dirk Aschoff
www.golden-section-graphics.com
Research: Michael Hopp (for Redaktion 4), Golden Section Graphics for EVONIK
Sources: www.basics-de.de, www.ledshift.com, www.alte-technologie.de, www.artikel32.com, www.swisseduc.ch, www.vip-bremen-nord.de, www.apod.nasa.gov, www.siemens.com, www.wikipedia.org
Published: *In Graphics* Volume 3, 2011

258 Submarine Cables
Design: TeleGeography – Markus Krisetya, Larry Lairson
www.telegeography.com
Text: TeleGeography
Research: The maps are based on research conducted by TeleGeography
Self-published on www.telegeography.com, interactive version on submarine-cable-map-2013.telegeography.com, 2013

260 Internet
Design / text / research: Section Design – Paul Butt
www.sectiondesign.co.uk
Sources: Various internet sources
Self-published, 2009

261 Internet
Design: Kiss Me I'm Polish – Agnieszka Gasparska
www.kissmeimpolish.com
Text / research: Anthony Giddens, Mitchell Duneier, Richard P. Appelbaum, Deborah Carr, Karl Bakeman (editor), Rebecca Charney (W.W. Norton assistant editor), Kate Feighery (W.W. Norton p roject editor), W.W. Norton
Source: Internet World Stats 2010b
Published: *Introduction to Sociology*, W.W. Norton & Company, 2012

262 Digital Storage
Design / text / research: Section Design – Paul Butt
sectiondesign.co.uk
Sources: Various internet sources
Self-published, 2009

264 CERN
Design: Mario Mensch
www.mariomensch.de
Text / research: Klaus Bachmann
Sources: CERN
Published: *GEO*, 2007

AROUND THE WORLD

THE ATLAS FOR TODAY

This book was conceived, edited, and designed by Gestalten.

Edited by Andrew Losowsky, Sven Ehmann, and Robert Klanten

Preface, epilog, and chapter introductions by Andrew Losowsky
Texts by Ben Knight

Layout and cover by Matthias Hübner and Jonas Herfurth
Layout assistance by Lisa Charlotte Rost

Chapter illustrations by Dermot Flynn
Typefaces: Relevant by Mika Mischler and Nik Thoenen, foundry: www.gestaltenfonts.com;
Fayon by Peter Mohr
Proofreading by Rachel Sampson

Printed by Nino Druck GmbH, Neustadt/Weinstraße
Made in Germany
Published by Gestalten, Berlin 2013 ISBN 978-3-89955-497-7
© Die Gestalten Verlag GmbH & Co. KG, Berlin 2013

All rights reserved. No part of this publication may be reproduced or transmitted in any form or by any means, electronic or mechanical, including photocopy or any storage and retrieval system, without permission in writing from the publisher.

Respect copyrights, encourage creativity!

For more information, please visit www.gestalten.com.

Bibliographic information published by the Deutsche Nationalbibliothek.
The Deutsche Nationalbibliothek lists this publication in the Deutsche Nationalbibliografie; detailed bibliographic data are available online at http://dnb.d-nb.de.

None of the content in this book was published in exchange for payment by commercial parties or designers; Gestalten selected all included work based solely on its artistic merit.

This book was printed on paper certified by the FSC®.

Gestalten is a climate-neutral company. We collaborate with the non-profit carbon offset provider myclimate (www.myclimate.org) to neutralize the company's carbon footprint produced through our worldwide business activities by investing in projects that reduce CO_2 emissions (www.gestalten.com/myclimate).